WITHDRAWN

SPACE:
From Gemini to the Moon & Beyond

SPACE:
From Gemini to the Moon & Beyond

Edited by Robert W. Peterson

FACTS ON FILE, INC.   NEW YORK, N. Y.

SPACE:
FROM GEMINI TO THE MOON
& BEYOND

©Copyright, 1972, by Facts on File, Inc.

All rights reserved. No part of this book may be reproduced in any form without the permission of the publisher except for reasonably brief extracts used in reviews or scholarly works. Books in the Interim History series are published by Facts on File, Inc., 119 West 57th Street, New York, N.Y. 10019.

Library of Congress Catalog Card Number: 78-190545

ISBN 0-87196-157-1

9 8 7 6 5 4 3 2 1

PRINTED IN THE UNITED STATES OF AMERICA

CONTENTS

	Page
INTRODUCTION	1

1965 — 3

MAN IN SPACE 4
  First 'Space Walks' (4); Gemini 5 Sets New Records (4); U.S. Spacecraft Rendezvous (9); Air Force Space Lab Authorized (14); Scientists Named for Space Flight (16); Moon Plans Discussed (16)

INTERPLANETARY & LUNAR MISSIONS 17
  U.S. Probe Photographs Mars (17); Soviets Probe Moon (21); Soviet Probes Head for Venus (23)

OTHER U.S. DEVELOPMENTS 24
  Scientific & Military Launchings (24); Budget Cut, Programs Reduced (29); 1965 Called Most Successful Year (30)

INTERNATIONAL DEVELOPMENTS 31
  Soviet Launchings (31); Soviets Display 'Orbital Missiles' (34); Activity by Other Nations (35)

1966 — 39

MAN IN SPACE 40
  Gemini 8 'Docks' in Space (40); Gemini 9 Fails in Docking Attempt (41); Gemini 10's Complex Mission (45); Gemini 11 Sets Records (47); Gemini 12 Mission Ends Series (49); Budget Cut Threatens Moon Landing (52)

LUNAR & INTERPLANETARY PROBES 53
  Soviet & U.S. Craft 'Soft Land' on Moon (53); U.S. Surveyor 1 Soft Lands on Moon (56); USSR's 2d Soft Landing on Moon (59); Luna 10 Orbits Moon (60); First U.S. & 2d Soviet Probes in Moon Orbit (61); Soviet Probe Hits Venus (64); Mariner 4 Reports (65)

U.S. PROGRAM 65
  NASA & Military Launchings (65)

	Page
INTERNATIONAL DEVELOPMENTS	72

U.S. Urges Moon Treaty (72); USSR Orbits 2 Dogs (73); Eldo Fires First Rocket (74); International Congress (76); Soviets Launch Satellites (77); Non-Soviet Reds View Launchings (78); 3d Soviet Space Base Reported (79); Cosmos Launchings (80); Space Shots by Other Nations (81)

1967 — 85

DEATHS IN U.S. & USSR PROGRAMS ... 86
3 Astronauts Killed in 'Rehearsal' (86); Equipment Failure Blamed (87); Congressional Committee Hearings (89); Shake-Up in Top Echelon (92); Cosmonaut Killed in Crash (93)

LUNAR PROBES ... 95
3d U.S. Probe Orbits Moon (95); Surveyor 3 Digs Into Moon (96); 4th Lunar Orbiter Circles Moon (97); 5 More U.S. Lunar Probes (98); Lunar Findings (101); USSR & U.S. Send Probes to Venus (102)

INTERNATIONAL DEVELOPMENTS ... 105
Arms-Ban Treaty Signed (105); World Weather Watch Established (106); USSR Hints Eventual Collaboration (106); Consortium Launches 3 Satellites (107); 2 International Launchings Fail (108)

ORBITAL & OTHER MISSIONS ... 109
Unmanned Soviet Spacecraft 'Dock' (109); USSR Lofts 3 Communications Satellites (110); 58 Cosmos Satellites Orbited (111); U.S. Scientific & Military Launchings (114); Errant Satellites (118); Launchings by Other Nations (118)

OTHER DEVELOPMENTS ... 121
More Astronauts Selected (121); Radiation Hazard Seen on Long Flights (122); Civilian Use of Navy Satellites Approved (122); Communications Policy Review Set (123); U.S. Lag in Rocket Power Reported (124); Red Chinese Progress Reported (125); USSR Nuclear Missile Development Hinted (125); MIRV Development Revealed (127); Canadian Space Agency Urged (127); French Developments (128); Soviet Developments (128); European Rocket Cooperation (129)

	Page
1968	131

MAN IN SPACE 132
   U.S. Moon Module Orbited (132); Apollo Test Fails (133); Apollo Spaceship Tested in Earth Orbit (133); Soviets Disclaim Race to Moon (136); U.S. Astronauts Fly Around Moon (137); Cosmonaut Orbits Earth 64 Times (142)

UNMANNED LUNAR ORBITAL MISSIONS 144
   Soviet Probes Return After Moon Orbits (144); Luna 14 Orbits Moon (147); Last Surveyor Lands on Moon (148)

OTHER U.S. DEVELOPMENTS............................ 149
   U.S. Space Program 'Gearing Down' (149); NASA Sets Up New Units (150); U.S. Tests Hybrid Rocket (151); UFO Study Proposed (151); Unmanned Interplanetary Flights Urged (151); 2 Biosatellites Canceled, X-15 Program Ends (152); U.S. Launchings (153)

INTERNATIONAL DEVELOPMENTS 157
   USSR Orbits Satellites (157); Canadian Plans (162); Astronaut Rescue Pact (163); European Disagreements & Cooperation (163); Red Network Planned (165); Intelsat Activities (165); Life Forms Reported in Meteorites (166); Gagarin Dies in Plane Crash (166)

1969	169

REHEARSING FOR MOON LANDING 170
   Moon Lander Flown in Test (170); Apollo 10 Orbits Moon (174)

MAN ON THE MOON 179
   2 Americans Walk on the Moon (179); Scientists Report on Moon Rocks (196); 2d U.S. Crew Walks on Moon (198); Moon Rocks Differ in Age (204)

OTHER MEN IN SPACE 204
   Soviets Dock Manned Ships (204); 7 Cosmonauts Fly 'Cosmic Troika' (207); USSR Delays Moon Landing (210)

	Page
LUNAR & INTERPLANETARY PROBES	211

2 Soviet Probes Reach Venus (211); Luna 15's Purpose Not Explained (212); 2 U.S. Probes Pass Mars (215); Soviet Zond Circles Moon and Returns (218); Soviet Moon Rocket Reported Destroyed (218)

U.S. PROGRAM 219

Space Budget Cut (219); 11 Moon Landings & 'Space Shuttle' Requested (220); Slowdown Urged (222); Nixon Accepts Mars-Landing Goal (224); Fund Authorization Drops, Mol Canceled (226); Solar Probe Fails (226); U.S. Satellites (227); Borman Visits USSR (231); NASA Personnel Changes (232); Dearth of Scientist-Astronauts Charged (233); Court OKs Prayers in Space (234); Air Force Ends UFO Study (234); Boeing Gets Moon-Car Contract (235)

INTERNATIONAL DEVELOPMENTS 235

Soviet Launchings (235); Soviets Orbit International Satellites (236); Space Debris Hits Japanese Ship (237); Quebec-France Link Dispute (237); U.S. Launches Canadian Satellite (238); No Intelsat Agreement (238); U.S.-Soviet Communications Merger Urged (240); German & British Satellites Launched by U.S. (241); 4th Japanese Failure (241); Europa Shot Fails (242)

1970   243

MAN IN SPACE 244

Crippled Apollo 13 Fails to Land on Moon (244); NASA & 2 Contractors Blamed (250); Cosmonauts Set Endurance Record (250)

SOVIET LUNAR & VENUS PROBES 253

Unmanned Mission Gets Moon Rocks (253); Soviets Put Robot Car on Moon (256); Zond 8 Circles Moon and Returns (258); Soviet Probe Reaches Venus (258)

U.S. PROGRAM 260

Moon Landings Reduced, Budget Cut (260); Nixon Sets Goals for '70s (261); Germs Survived on Moon (262); U.S. Orbital Launchings (263)

	Page
INTERNATIONAL DEVELOPMENTS	265

China Orbits First Satellite (265); Soviet Orbital Launchings (266); Japan's First Success (268); Franco-German Project (269); British Skynet & Orba Fail (269); Italy Orbits U.S. Satellite (270); U.S. Firm Gets Canadian Contract (270); Intelsat Orbits 2 Satellites (271); U.S. & Soviet Experts Confer (272)

1971 — 275

EXPLORING THE MOON 276
3d U.S. Landing on Moon (276); Astronauts Drive Car on Moon (282); 2 Soviet Probes Orbit Moon (289); Moon Car Stops Working (289); Lunar Findings (290)

OTHER MANNED FLIGHT DEVELOPMENTS 291
Cosmonauts Dock With Space Station (291); 3 Cosmonauts Die as Flight Ends (293); Shuttle's Military Role (296); Cosmic Ray Threat (296)

INTERPLANETARY PROBES 296
U.S. Satellite in Mars Orbit (296); Soviets Land Capsules on Mars (298)

EARTH SATELLITES 300
International Communications Satellites (300); China's 2d Satellite (301); Japan Scores 2 Successes (302); U.S. Launchings (303); Soviet Launchings (303); Other Satellites (304)

OTHER DEVELOPMENTS 305
U.S. & Soviet Scientists Cooperate (305); Treaty Drafted (306); NASA Budget Declines (306); Soviet Space Expert Defects (307)

INDEX 309

INTRODUCTION

M AN'S EVOLVING CONQUEST OF OUTER SPACE has been one of the most absorbing dramas of science and technology. Probably no single scientific feat has so captured the imagination of the world as did the first landing of men on the moon in July 1969. The civilized world was intrigued by the televised sight of 2 Americans, Neil A. Armstrong and Edwin E. Aldrin Jr., walking in space suits on the lunar surface.

Before that historic day, and in the months and years since, there have been scores of new thrills as American and Soviet spacemen ventured away from earth and unmanned probes were sent vast distances to explore earth's neighbors in the solar system.

The space age had begun Oct. 4, 1957, when the Soviet Union launched *Sputnik 1*, the first man-made earth satellite. The launching put new urgency into the U.S.' rather modest commitment to rocket and missile technology and led ultimately to a determination to become first in space. By 1958 the U.S. had approved plans for Project Mercury, the nation's first attempts to send astronauts into space. Americans achieved earth orbital flight in 1962, but the Soviets were there first, orbiting a cosmonaut in Apr. 1961.

The Mercury project was followed by Gemini, a series of 2-man flights conducted as a necessary prelude to manned voyages to the moon. The Gemini series, comprising 12 flights of 2-man spaceships (10 with astronauts aboard), ended in Sept. 1966. Project Apollo, the moon-landing program, then began.

The story of the space age from 1957 through mid-1965 is told in an earlier INTERIM HISTORY book: *Space: From Sputnik to Gemini*. This volume picks up the story midway in the Gemini series, in June 1965, and carries it through 1971. It is the story of 8 intrepid U.S. astronauts who walked on the moon, of tragedy in the deaths of 3 Americans who died in a fire on the launching pad and 4 Soviet cosmonauts who died in space flight, of scientific probes of the moon, Venus, Mars and the sun, and of hundreds of man-made satellites circling the earth in orbital flight. As the book ends, U.S. scientists were planning a "space shuttle" to transport astronauts to and from orbiting laboratories. They were preparing to send probes into the outer reaches of the solar system, and the USSR was monitoring TV signals from an instrument package landed on Mars.

This book is a journalistic narration of these events. An attempt has been made to cover in detail all the reported launchings of space vehicles and other important developments in the continuing story of space exploration. Virtually all of the material herein is from the record compiled by FACTS ON FILE.

1965

Men first "walked in space" during 1965, leaving the security of their spacecraft for periods of up to 20 minutes. The first space walk, by a Soviet cosmonaut in March, lasted 10 minutes. Within 3 months a U.S. astronaut left his 2-man Gemini space capsule for a 20-minute space walk. This was during the 2d manned flight of the Gemini series. U.S. astronauts set a record of 8 days in space in the 3d manned Gemini flight, but before 1965 was over, U.S. astronauts in another Gemini had set a record of more than 330 hours in space; and, on this mission, 2 Gemini capsules rendezvoused in orbit, approaching within a few feet of each other. The USSR made no further manned flights during 1965, concentrating instead on trying to "soft-land" an unmanned spacecraft on the moon; but 4 apparent attempts failed. The Soviets, however, did succeed in taking the first photos of the hidden side of the moon from an unmanned probe. The U.S. took the first "closeup" photos of Mars with a Mariner probe that flew within 6,000 miles of the red planet.

MAN IN SPACE

First 'Space Walks'

Soviet cosmonaut Aleksei Arkhipovich Leonov, 30, became the first man to "walk in space" when he stepped out of his 2-man *Voshkod 2* spaceship Mar. 18 for 10 minutes. Leonov maneuvered and performed various experiments while tethered to the spacecraft at a maximum distance of 16 feet. The *Voshkod 2* flight, the 8th manned Soviet flight, ended Mar. 19 when Leonov and his fellow cosmonaut brought their craft down safely in the area of Perm, an industrial city in the Ural Mountains.

The U.S. launched its first manned Gemini flight Mar. 23 with Maj. Virgil I. Grissom, 38, as command pilot and Lt. Cmndr. John W. Young, 34, as co-pilot. 2 unmanned flights of Gemini capsules had preceded this one, which took place in the spaceship *Gemini 3*, a conical capsule $18\frac{1}{2}$ feet high and 10 feet in diameter at its base. Grissom and Young maneuvered their craft in orbit—the first time this had been done in manned flight. After 3 revolutions around the earth, they brought *Gemini 3* down safely in the Atlantic.

Maj. Edward H. White 3d, 34, made the 2d "space walk" in history during the 62-orbit flight of *Gemini 4* June 3-7. During the 3d orbit, White left the capsule June 3 in his spacesuit at the end of a 25-foot lifeline and maneuvered in space by using a small oxygen-jet propulsion gun. He was outside the capsule for 20 minutes. White and the command pilot, Maj. James A. McDivitt, 35, set a new endurance record for 2-man flights of 97 hours, 57 minutes.

The flight of *Voshkod 2* and the first 2 manned flights in the Gemini series were the space highlights of the first half of 1965. Full details on these and developments in space exploration from 1957 through mid-1965 will be found in the INTERIM HISTORY book *Space: From Sputnik to Gemini*, edited by Lester A. Sobel (FACTS ON FILE, 1965).

Gemini 5 Sets New Records

Gemini 4's endurance record lasted only $2\frac{1}{2}$ months. A new mark of 190 hours 56 minutes was set by Lt. Col. Leroy Gordon Cooper Jr., 38, and Lt. Cmndr. Charles (Pete) Conrad Jr., 35, who were launched into orbit from Cape Kennedy, Fla. Aug. 21 aboard the U.S. space capsule *Gemini 5*. Despite initial trouble with their electrical power system, the 2 astronauts remained in orbit for a record 8 days and completed a record 120 revolutions around the earth—just one revolution short of the total originally planned.

Gemini 5, with the astronauts inside, came down safely in the Atlantic Ocean Aug. 29. The astronauts, flown by recovery helicopter to the waiting aircraft carrier *Lake Champlain*, were found to be in excellent physical condition. Dr. Charles A. Berry, chief physician for the mission, said the flight had shown that man was "qualified" to stay in space long enough to go to the moon and back without harm.

A number of space flight records were established by the *Gemini 5* mission. It was the longest manned space flight in distance covered—about 3 million ground miles—as well as in duration. It increased total U.S. manhours in space to a record 639 hours 48 minutes (compared with the Soviet total then of 507 hours 16 minutes). Cooper, *Gemini 5*'s command pilot, became the first person to go into orbit on more than one flight; he had completed 22 revolutions around the earth in the *Faith 7* (Project Mercury) capsule May 15-16, 1963. Cooper also increased his total time in space to a record 225 hours 16 minutes and the number of his revolutions in orbit to a record 142.

As in all U.S. manned space flights, plans for the mission had been publicized in detail for months before the flight, and the flight was covered in exhaustive detail by press and broadcasting media. (This full publicity policy continued for all manned U.S. flights recorded in this book.)

The 7,950-pound *Gemini 5*, with Cooper and Conrad strapped inside in form-fitting couches, had been launched from Cape Kennedy's Launch Complex 19 at 10 a.m. EDT Aug. 21 in the nose of a 2-stage Titan-2 booster rocket. Speeded up to a velocity of 25,817 feet a second on separation from the booster rocket's 2d stage, *Gemini 5* went into an initial orbit with an apogee of 217.4 miles, a perigee of 100.6 miles, a period of 90 minutes and a 33° angle of inclination to the equatorial plane. At 10:56 a.m. Cooper fired maneuvering rockets to increase the perigee to 105 miles.

An innovation in *Gemini 5* was the use of fuel cells to provide electrical power for all in-flight needs. The cells, which replaced the previously used storage batteries for all purposes except re-entry, produced electricity through the reaction of oxygen and hydrogen. Water was produced as a waste by-product (plans called for astronauts to drink the water on future flights). Cooper reported during the 2d orbit that the power system appeared to be in danger because of a decrease in pressure in one of the tanks from which the fuel cells received oxygen. The pressure drop was caused by the failure of a heater whose function was to transform the oxygen from liquid to gas. Pressure in the tank dropped from 800 pounds a square inch to 60, a point at which it remained stable for a while, and the 2 astronauts began to conserve power by shutting off all possible electricity-using equipment. Gemini flight director Christopher Columbus Kraft Jr., 41, analyzed the situation at the Manned Spacecraft Center in Houston, Tex. and then authorized Cooper and Conrad at 5:39 p.m. Aug. 21 to remain in orbit for at least 24 more hours (18 revolutions). By Aug. 22 pressure had risen to about 80 pounds, and the fuel cells surprisingly produced electricity at normal levels. It was therefore decided to continue the *Gemini 5* flight for the full 8 days.

New fuel-cell trouble developed by Aug. 26, when it was reported that the cells were producing too much water. To make sure that excess water did not drown the cells, the astronauts were ordered to economize on electricity for the rest of the flight. The fuel cell troubles and the failure Aug. 26 of 2 of the 8 thruster rockets made it necessary to abandon several experiments. The failure of the 2 thrusters, while not dangerous, made it difficult (and costly in fuel) to execute maneuvers that required stabilization of the capsule for fine sighting and aiming. To conserve rocket fuel and electricity, the astronauts allowed *Gemini 5* to drift, tumble and roll. They reported that the motion made vision of outside objects difficult and that unfastened equipment would "whip over to the side" instead of floating inside the capsule. A corkscrew effect was added to the capsule's motion by a leak of hydrogen from a fuel-cell tank, but this venting ended Aug. 28, and the astronauts reported a lessening of the undesirable motion.

The fuel cell problem resulted in the abandonment of one of the mission's first major experiments—an attempt to rendezvous with another satellite. Shortly after noon Aug. 21, at 12:07 p.m. of the first day of the flight, Cooper pulled a switch and thereby ejected from *Gemini 5*'s adapter section a Rendezvous Evaluation Pod (REP), a 76-pound package of blinking lights, radar and electronic equipment. The REP pulled away from *Gemini 5* in its own slightly lower orbit. The original mission plans had called for the astronauts, using radar and an on-board computer, to calculate the necessary orbit changes and to pilot *Gemini 5* to within 20 feet of the REP. Because of the fuel-cell trouble, the attempt at rendezvous with the REP was "scrubbed," and by the time it was found that the fuel cells were producing adequately, the REP was too far away for the experiment to be conducted. In an effort to get some experimental data on space rendezvous, however, the *Gemini 5* crew was ordered Aug. 23 to conduct maneuvers for rendezvous with an imaginary Agena rocket. The experiment began at noon, during *Gemini 5*'s 31st revolution, when the astronauts received data from a ground computer indicating that their hypothetical target was in orbit 480 miles away. The astronauts used about 90 pounds of rocket fuel by 3 p.m. as they altered both the altitude and plane of *Gemini 5*'s orbit to make it "co-elliptical" with that of the imaginary Agena and to bring *Gemini 5* to a distance of $17\frac{1}{4}$ miles from the non-existent target rocket. The experiment, which lowered the apogee of *Gemini 5*'s orbit by 13 miles and raised the perigee by 10 miles, was completed during the 33d revolution.

2 Minuteman rockets launched by Air Force crews at Vandenberg Air Force Base, Calif. Aug. 24 and 25 were seen and photographed by the astronauts from distances of more than 200 miles. They also took measurements with their infra-red heat sensor, a device that might serve as a missile launching detector. A test rocket sled was fired at Holloman Air Force Base, N.M. Aug. 25 in another successful test of the heat sensor, but efforts to detect the sled's heat later in the flight failed because the capsule was tumbling too much.

Conrad reported as *Gemini 5* passed over Florida Aug. 24 that "I can see Jacksonville and all the streets in it, and the Cape [Cape Kennedy] and all the way down to Miami." But Conrad was unable to recognize the patterns of huge rectangular panels placed on the ground near Laredo, Tex. as a sort of giant "eye chart" designed to show just how well an orbiting astronaut could see and recognize objects on the ground.

The astronauts sighted and reported on storms they saw developing in the Pacific and Atlantic oceans. They also took a number of ground, weather and astronomical photographs.

During the flight neither astronaut ate as much or slept as much as had been expected. Both complained occasionally that they had been given too many tasks to do in the limited time available. Mission physicians on the ground nagged them several times to do the isometric and bungee-cord exercises that the doctors hoped would keep them from becoming weak during prolonged weightlessness in space. Cooper lost 7 pounds during the flight, and Conrad lost $8\frac{1}{2}$.

Cooper began the computer-aided landing procedure by firing the capsule's 4 retro-rockets at 8:27 a.m. EDT Aug. 29 while over the Pacific 700 miles north of Hawaii. The astronauts had been ordered to come down after 120 revolutions—instead of the 121 originally planned —because a hurricane was moving toward the Atlantic recovery area. The computer, however was given data later described as incorrect. As a result, *Gemini 5*, eased down by parachute, splashed into the Atlantic about 103 miles from the intended landing area.

The capsule hit the water at 8:56 a.m. EDT Aug. 29 some 760 miles east of Cape Kennedy. Navy frogmen, brought to the floating spaceship by helicopter, attached a flotation collar around it to keep it from sinking, and the 2 astronauts were flown by helicopter to the carrier *Lake Champlain* for preliminary medical checks. They were then transferred Aug. 30 to Cape Kennedy to begin 11 days of medical tests and "debriefings" at the cape and at the Manned Spacecraft Center in Houston, Tex.

The first stage of *Gemini 5*'s Titan-2 launching rocket had been spotted floating in the Atlantic shortly after the launching Aug. 21, and it was picked up the same day by a U.S. destroyer about 640 miles southwest of Bermuda. This was the first time a U.S. rocket had come down intact after boosting a manned capsule into space. (The Titan-2's 2d stage went into an independent orbit after separation from the capsule. It fell out of orbit Aug. 24 after its 50th revolution and burned on re-entering the atmosphere.)

Shortly after the astronauts arrived aboard the *Lake Champlain*, Pres. Lyndon B. Johnson phoned to congratulate them. "You have certainly proved once and for all that man has a place in the exploration of the great frontier of space," the President told them. At a press conference at his Texas ranch Aug. 29, Johnson said the *Gemini 5* flight showed "that man is in space to stay." "Only 7 years ago we [the U.S.] were neither first nor 2d in space," Johnson continued. "We just weren't in space at all.... This afternoon the capacity of this country for leadership in this realm is no longer in valid question or dispute.... [I] renew ... America's invitation to all nations to join together to make ... [space exploration] a joint adventure.... No national sovereignty rules in outer space. Those who venture there go as envoys of the entire human race. Their quest, therefore, must be for all mankind, and what they find should belong to all mankind."

Johnson held that the "successful conclusion" of *Gemini 5*'s "journey of peace" provided "a very fitting opportunity ... to renew our pledge" to search for peace. "To demonstrate the earnestness of that pledge," he said, "and to express our commitment to the peaceful uses of space exploration, I intend to ask as many of our astronauts as possible, when ... their schedule and program will permit, to visit the various capitals of the world. Some, I hope, will be able to journey abroad very soon."

The Soviet press, however, had a few harsh words for *Gemini 5*. The Soviet press agency Tass charged Aug. 22, in a dispatch from its New York correspondent, Leonid Ponomarev, that there were "haste and definite risk" in the *Gemini 5* mission but that "the flight program directors" went ahead with it anyway because they "have been given the task to 'beat the Soviet Union' at any price with regard to duration of the orbital flight." *Gemini 5* mission control flight officer John Hodge in Houston Aug. 22 denied that there were orders to beat the USSR at any price. "That's not true," he said. "... We weren't told to.") *Krasnaya Zvezda (Red Star)*, the Soviet Defense Ministry newspaper, charged in an Aug. 25 article that *Gemini 5*'s principal objective was military and was not, as U.S. officials had claimed, chiefly directed at perfecting techniques to be used for manned flight to the moon. "The main purpose is testing the capability of intercepting artificial satellites and conducting reconnaisance from space," *Krasnaya Zvezda* asserted. It pointed out that "the spaceship will pass 11 times over Cuba, 16 times over ... [North] Vietnam and 40 times over [Communist China]." 6 of the spacecraft's 17 experiments "have been planned by the Department of Defense and are being kept secret," the newspaper declared. (*Gemini 5* back-up pilot Elliott M. See said in Houston Aug. 25 that he knew of "only 4"—not 6—experiments designed by the Defense Department for the mission. The 4 experiments were photography of stars, measurement of the Rendezvous Evaluation Pod, radiation and recognition of objects on the ground. "We've discussed them quite freely," he declared. "There's no hesitation to describe or talk about all of them, none of which is classified in any way.")

U.S. Spacecraft Rendezvous

The next U.S. manned space flight succeeded where *Gemini 5* had failed when 2 spaceships rendezvoused in orbit. The feat was achieved by the 6th and 7th missions of the Gemini series.

The rendezvous was made Dec. 15 by *Gemini 6* and *Gemini 7*. The former mission was flown by Navy Capt. Walter Marty Schirra Jr. and Air Force Maj. Thomas Patten Stafford; the latter mission had Air Force Lt. Col. Brank Borman and Navy Cmndr. James A. Lovell Jr. aboard. Schirra piloted his spaceship to within a few feet of *Gemini 7*, and the 2 capsules then flew in formation—an exercise called "station keeping"—for 6 hours. *Gemini 7* had been in orbit for 11 days when rendezvous was achieved, while *Gemini 6* had been launched only that day—Dec. 15.

Schirra brought *Gemini 6* out of orbit Dec. 16 to a virtually perfect splashdown in the Atlantic about 630 miles southwest of Bermuda. *Gemini 7* continued on in orbit, setting new records for space flight, until Dec. 18, when command pilot Borman brought it down in the same Atlantic target area. All 4 astronauts were plucked speedily out of the ocean and put aboard the waiting aircraft carrier *Wasp*, where preliminary medical checks showed them to be in excellent physical shape.

The *Gemini 7–Gemini 6* mission produced a number of records for manned space flight: (a) Borman and Lovell aboard *Gemini 7* set an endurance record of 330 hours 35 minutes 26 seconds for manned space flight. (b) They circled the earth a record 206 times in orbit. (c) They traveled a record 5,129,400 miles in space. (d) Borman and Lovell, with their 330 hours 35 minutes in space, became the individuals with the most time in space flight. (The previous record-holder was L. Gordon Cooper Jr.) (e) The *Gemini 7/6* flights increased the total time spent by U.S. astronauts in space to 1,352 hours 42 minutes. (f) The flights raised the number of U.S. manned space missions to 11. (g) As a result of the *Gemini 7/6* mission, the U.S. had 13 astronauts who had gone into space (3 of them twice), compared with 11 for the USSR (none more than once). (h) A record 4 men were in orbit simultaneously. (i) The 2 spaceships made a record close approach to each other. (According to the initial NASA announcement, they came within 6 to 10 feet of each other. But Borman said Dec. 19, after seeing movies taken by *Gemini 6*'s astronauts, that at one point they seemed to be no more than 3 feet apart.) (j) The mission raised the number of manned flights by a single nation in one year to 5 and the number of men sent into orbit by one nation in a single year to 10. (k) The rendezvous and formation flying achievements were also firsts.

The 8,069-pound *Gemini 7* was the heaviest manned U.S. spacecraft launched so far. It was sent up by means of a Titan-2 booster rocket launched from Cape Kennedy's Launch Complex 19 at 2:30:-03 p.m. Dec. 4. 6 minutes after launching *Gemini 7* separated from the upper (2d) stage of the booster rocket and hurtled into an orbit with an initial apogee given as 203 miles, a perigee of 100 miles and a period of one hour 35 minutes.

Plans for the launching—and subsequent alterations of the plans—had been announced in detail months in advance. The preparations and events preceding, surrounding and following the launching, as well as the launching itself, were covered exhaustively by TV, radio and the press. Pres. Johnson, who saw the launching on TV at his Texas ranch, said: "Their [Borman's and Lovell's] voyage will be a continuous reminder that the peaceful conquest of space is the only form of conquest in which modern man can proudly and profitably engage. In this struggle, all men are allies, and the only enemy is a hostile environment. The victory ... will belong not just to Americans but to the world."

In preparation for the attempt to rendezvous with the still-earthbound *Gemini 6*, Borman altered *Gemini 7*'s orbit Dec. 7 during the spacecraft's 44th revolution around the earth. Borman fired the thruster rockets for $16\frac{1}{2}$ seconds. The thrusters delivered 12.4 feet a second of thrust and raised the perigee to 146.4 miles while the apogee remained at the adjusted altitude of 196 miles. Borman altered the orbit again by similar maneuvers Dec. 9 and put *Gemini 7* into a more circular orbit with an apogee of 188.3 miles and perigee of 185.3 miles. By Dec. 10 the spacecraft was in a virtually circular orbit 185.1 miles high.

The *Gemini 7* flight marked the first trip in space for either Borman or Lovell. Lovell became the first American to fly in space without protective space clothes. In a planned experiment, he removed his bulky $17,000 space suit Dec. 6 and flew more comfortably in long underwear until Dec. 10. Lovell donned his space suit Dec. 10 while Borman removed his protective gear to take his turn flying in his underwear. Lovell was authorized to remove his space suit again Dec. 12.

Lovell failed twice Dec. 5, during 2 passes over Texas, to identify correctly the patterns in which large panels had been placed on the ground near Laredo. The panels formed a sort of giant eye chart used in a test of an astronaut's ability to see objects on earth from orbit.

During their 31st revolution around the earth Dec. 6, the 2 astronauts saw and photographed the launching of a Polaris A-3 missile by the submerged nuclear submarine *U.S.S. Benjamin Franklin*. Chris Kraft, directing Gemini operations from the NASA (National Aeronautics & Space Administration) Manned Spacecraft Center in Houston, Tex., estimated that the astronauts were 135 miles from the 31-foot missile when they caught sight of it. Lovell, who took the photos, also measured the infra-red "signature" of the Polaris. The astronauts successfully observed an even more difficult target when they photographed and took infra-red readings of an 80-foot Minuteman missile fired in darkness from Vandenberg Air Force Base, Calif. early Dec. 14 (23 minutes after midnight Dec. 13). As the Minuteman headed west, the eastward-bound *Gemini 7* passed it at a combined velocity of some 29,000 mph., but the 2 speeding projectiles never came closer than perhaps 140 miles to each other.

Originally, astronauts Schirra and Stafford had been scheduled to pilot *Gemini 6* into rendezvous and to "dock" with a modified Agena rocket Oct. 25. The 7,090-pound Agena, however, failed to go into orbit, and the *Gemini 6* launching therefore was scrubbed less than an hour before the spaceship was due to start in pursuit of the Agena.

The plan for the *Gemini 7/6* mission called for the launching of *Gemini 6* Dec. 12. But that launch was scrubbed after the rocket engines had actually started. The *Gemini 7* astronauts reported seeing the "light up" and "shut down" during the unsuccessful launching attempt Dec. 12. Preparations to send *Gemini 6* up from Cape Kennedy's

Launch Complex 19 had been started Dec. 4 almost as soon as *Gemini 7* had left the ground. Technicians reported that the launch pad had suffered no significant damage from the *Gemini 7* launching and therefore should be fit for a launching Dec. 12. *Gemini 6*'s computer suffered electrical damage Dec. 7, but a new one was installed in time for a Dec. 12 launching.

With everything in apparent readiness and the 2 astronauts strapped to their form-fitting seats in the capsule, the Titan-2 engines were started at 9:54 a.m. Dec. 12 to hurl *Gemini 6* into orbit. But the engines halted abruptly seconds later because a $108 electric plug at the base of the rocket was shaken loose 2.2 seconds before it would have been pulled from its socket by the launching. The disconnecting of the plug automatically shut off the engines. Schirra, as command pilot, analyzed the available facts and quickly decided that it would not be necessary to activate the escape mechanism—a device installed to safeguard astronauts from such dangers as an explosion during launching. Had he used the escape mechanism, he and Stafford would have been ejected and dropped by parachute hundreds of feet from the rocket, but the effect on the capsule would have ended the chances of *Gemini 6*'s being used in the current mission. His decision made *Gemini 6*'s Dec. 15 launching possible.

In rechecking the Titan-2, NASA technicians Dec. 12 discovered a tiny 2¢ plastic dust cover left in the rocket's fuel-pump system when the rocket engines were installed in the rocket at the Martin Co. plant in Baltimore. NASA officials said Dec. 13 that the dust cover would have prevented a launching in the abandoned Oct. 25 *Gemini 6* mission as well as in the Dec. 12 shot. The NASA officials said that neither the dust cover nor the faulty electric plug had endangered the astronauts; their effect could be only the stopping of the engines.

Gemini 6 was finally sent aloft from Launch Pad 19 at 26 seconds after 8:37 a.m. Dec. 15 as *Gemini 7* sped overhead. The 2-stage booster hurled *Gemini 6* into an initial orbit with an apogee of 161 miles and perigee of 100 miles. *Gemini 6* began to chase *Gemini 7* from a distance some 1,200 miles behind its target and in an orbit with an average altitude of about 50 miles below *Gemini 7*'s. Command pilot Schirra closed in on his target by making 7 orbit-changing maneuvers. Basing his movements on calculations provided by computers on the ground and in the capsule, Schirra fired his on-board rockets to raise the altitude of *Gemini 6*'s orbit and to move the spaceship slightly to the south.

By the middle of *Gemini 6*'s 4th trip around the earth, at about 2:27 p.m. EST, Schirra had brought *Gemini 6* to within 10 feet of *Gemini 7*, close enough so that the astronauts could see each other through their portholes. The 2 spaceships were then in an orbit with an altitude of 182.8 miles to 187.4 miles. "There seems to be a lot of traffic up here," Schirra observed during the flippant radio conversation that took place throughout the maneuvers. "Call a policeman," Borman retorted. Schirra remarked on the beards Borman and Stafford had grown during their 11 days in space.

After achieving close rendezvous, the 2 spaceships flew in formation for 6 hours at the orbital speed of about 17,500 mph. and at distances of 20 to 100 feet. They maneuvered over, under and around each other, and the astronauts took still and motion pictures. All of the maneuvering to achieve rendezvous and most of the maneuvering while "station keeping" was done by *Gemini 6*, which was fresh in space and had more fuel than *Gemini 7*. After 6 hours of formation flying, *Gemini 6* was ordered into a new orbit 20–30 miles lower than *Gemini 7*'s.

Gemini 6 was brought down in the Atlantic by Schirra after 16 revolutions around the earth. Its descent eased by parachute, the capsule hit the water 13.8 miles from the waiting carrier *Wasp* at 10:29:09 a.m. Dec. 16 after 25 hours 51 minutes in space. As the *Wasp* sped to the spaceship, a recovery helicopter dropped 3 frogmen into the water, and the frogmen attached a flotation collar to *Gemini 6* to make sure it did not sink. The capsule with its 2 astronauts inside was then hauled aboard the carrier.

Gemini 7 ended its record-breaking flight with a parachute-eased splashdown at 9:06 a.m. Dec. 18 about 17 miles from the *Wasp*. Borman and Lovell were taken to the carrier by helicopter, and *Gemini 7* was hauled aboard the *Wasp* later. The 2 astronauts, stiff, bearded and dirty after 2 weeks in their cramped capsule, were given quick, preliminary medical exams. Dr. Charles A. Berry, mission medical director, announced in Houston that the 2 astronauts were "in very good shape . . . beyond what we could have hoped for." He said that "there's no evidence whatsoever their physical conditions deteriorated during the flight." Borman had lost 9.6 pounds and Lovell 5.9 pounds while in space.

(The *Gemini 7* flight was harassed by occasional malfunctions of a power-providing fuel cell, but the trouble was not serious enough to cut short the flight. 17 unsuccessful attempts were made by *Gemini 7*'s crew members during the flight to communicate with ground stations by means of an experimental laser device.)

Pres. Johnson, who had watched the launching of *Gemini 6* on television and had kept track of the rendezvous developments, sent NASA Administrator James E. Webb a telegram congratulating the astronauts and others involved. "You have all moved us one step higher on the highway to the moon," he declared. The President Dec. 18 ordered one-grade promotions for Borman, Lovell and Stafford under a policy (started in August) of giving such a promotion (but not beyond the grade of colonel in the Air Force or captain in the Navy) to each astronaut on the completion of his first successful space flight. (Schirra had received his promotion after a Mercury flight in the capsule *Sigma 7*.)

Soviet Pres. Nikolai V. Podgorny sent Johnson a congratulatory message Dec. 18.

Johnson had announced Aug. 25 that the Soviet Academy of Sciences would be invited to send observers to watch the launching of *Gemini 6*, which was then scheduled for October. The invitation was sent Aug. 27, in a letter from NASA administrator James E. Webb, and was declined Sept. 8, in a letter from Prof. Mstislav V. Keldysh, president of the Soviet academy.

U.S. officials said the Russians had been informed that their observers would be welcome at launchings that had preceded *Gemini 6* but had never sent observers, presumably because they did not want to have to reciprocate by inviting Americans to Soviet launchings.

At a Soviet embassy reception in Athens Sept. 17 during the sessions of the 16th International Astronautical Congress, a Soviet scientific spokesman was asked whether the USSR would let foreign newsmen watch Soviet launchings. The scientist replied: Soviet launching rockets were closely connected with the USSR's defense; "for the time being, due to the tension in international relations, we don't consider it appropriate to invite foreigners to watch our rocket launchings"; foreign observation "will be feasible when the whole international situation changes."

Air Force Space Lab Authorized

Pres. Johnson ordered the Air Force Aug. 25 to start work on its projected 2-man Mol (manned orbiting laboratory). He authorized the AF to build and launch 5 Mols.

The 1\frac{1}{2}$ billion Mol project had been delayed for months because of controversy over the roles to be played by the military and civilian agencies in the exploration of space. The project would link a modified Gemini capsule made by McDonnell Aircraft Corp. with a Douglas Aircraft Co.-made cylinder, 42 feet long and 10 feet in diameter, that would provide both laboratory facilities and living quarters for the 2-man Mol crews. The solid-fueled Titan-3C, manufactured by Martin Marietta Co., would be used as the launching rocket. As then planned, the Air Force crew was to ride into orbit in the Gemini capsule and then transfer to the cylinder, where the crewmen would work and live in "shirt sleeves environment" during assignments of perhaps 30 days' duration. On completing their assignments, the crewmen would return to earth by Gemini capsule and probably leave the Mol cylinder in orbit. It was expected that the Mols would operate in orbits at altitudes of up to 350 miles, and it was contemplated that a 2d crew might be sent up in a Gemini capsule for a new Mol assignment after the first crew had left the Mol. The first manned Mol mission was scheduled for 1968.

Pres. Johnson, announcing his order at his press conference Aug. 25, said that "we intend to live up to our agreement not to orbit weapons of mass destruction." A Defense Department statement pointed out, however, that one of the Mol's basic purposes was to "learn more about what man is able to do in space and how that ability can be used for military purposes."

Col. Gen. Vladimir Tolubko, deputy commander of USSR strategic rocket troops, charged in the Soviet magazine *Za Rubezhom* (*Life Abroad*) Sept. 9 that "the Pentagon," in planning the Mol, "wants to use space laboratories not only for espionage but also to accomplish direct combat tasks." Accepting a view that the Mol was to have a bomb-dropping mission, Tolubko derided any suggestion "that the Pentagon generals intend to drop conventional aerial bombs from outer space. Of course not. Surely, nothing but nuclear bombs are implied."

Authority for the Air Force to go ahead with the Mol project "without further delay" had been urged June 3 in a report of the U.S. House Government Operations Committee. The committee asserted that the USSR was "substantially ahead" of the U.S. in military space development, and it held that "the compelling need of the moment" was to overcome this alleged disadvantage. The report was based on a study made by the committee's Military Operations Subcommittee, headed by Rep. Chet Holifield (D., Calif.).

Air Force Secy. Eugene M. Zuckert Aug. 31 appointed Gen. Bernard A. Schriever, commander of the Air Force Systems Command, to the additional job of director of the Mol project.

The Air Force Nov. 12 named the first 8 of 20 military pilots chosen for training as astronauts for the Mol. The pilots—6 from the Air Force and 2 from the Navy—all had physical-science or engineering degrees. They were to receive their aerospace research training at Edwards Air Force Base, Calif. The Air Force pilots were Maj. Michael J. Adams, 35, of Sacramento, Calif.; Maj. Robert M. Crews, 36, of Alexandria, La., who had previously been chosen for the since-canceled DynaSoar program; Capt. Richard E. Lawyer, 33, of Inglewood, Calif.;

Capt. Lachlan Macleay, 34, of Redlands, Calif.; Capt. F. Gregory Neubeck, 33, of Washington, D.C., and Capt. James M. Taylor, 34, of Lewisville, Ark. The Navy pilots were Lt. John L. Finley, 29, of Memphis, and Lt. Richard H. Truly, 28, of Meridian, Miss.

Scientists Named for Space Flight

A geologist, 2 physicians and 3 physicists were named by NASA June 28 to be trained as scientist-astronauts for the Apollo program. One of the physicians, Dr. Duane E. Graveline, 34, a former flight surgeon at the NASA Manned Spacecraft Center in Houston, Tex., resigned from the program Aug. 18 after his wife filed a divorce suit (which she later withdrew). Those who remained in the program were Owen K. Garriott, 34, associate physics professor at Stanford University; Edward K. Gibson, 29, senior research scientist at the Applied Research Laboratories of Philco Corp.'s Aeronutronic Division in Newport Beach, Calif.; Lt. Cmndr. Joseph P. Kerwin, 33, a Navy flight surgeon; Frank Curtis Michel, 31, assistant space sciences professor at Rice University, and Harrison H. Schmitt, 29, U.S. Geological Survey astrogeologist.

The 6 were selected from a group of 1,492 volunteers whom NASA had first reduced to 422 and the National Academy of Sciences had further narrowed down to 16. Kerwin and Michel were qualified jet pilots. The others were assigned to take a year of Air Force pilot training before continuing with their training at the Manned Spacecraft Center.

Moon Plans Discussed

U.S. and Soviet delegates to the 16th International Astronautical Congress, held in Athens Sept. 14-18, met Sept. 14 to discuss U.S.-Soviet cooperation in establishing an international laboratory on the moon. The congress Sept. 16 discussed projects to be assigned to such a laboratory. Such projects included meteorological studies of the earth and possible correlations of terrestrial weather and solar activity.

German-born Dr. Wernher von Braun, chief of the U.S.' Saturn rocket project, predicted at the congress Sept. 14 that the U.S.' Apollo program would meet its schedule of placing "men on the moon before the end of this decade," and he reported that more than 50% of the work required for Apollo had been done. Von Braun said there was evidence that the lunar surface was hard enough to support the U.S.' lunar vehicle. Soft spots had been identified and would be avoided, he said. He estimated that men could be landed on Mars 15 years after the manned lunar landing. (Mstislav V. Keldysh, president of the Soviet

Academy of Sciences, had said at a press conference in Moscow Aug. 23 that he was "sure... so far, that no one in the world can set a time limit [fix a specific date]" for a manned landing on the moon or on the planets. He called the problem too "stupendous, requiring enormous funds and an enormous scientific and engineering effort.")

Soviet cosmonaut Aleksei Leonov reported at the congress Sept. 16 that the USSR planned to orbit "many [permanent satellite] space laboratories, with crews being periodically exchanged," before attempting a manned landing on the moon. He indicated that Soviet space planners contemplated using a larger vehicle for the lunar landing than the U.S.' 3-man Apollo lunar spacecraft.

INTERPLANETARY & LUNAR MISSIONS

U.S. Probe Photographs Mars

The 575-pound *Mariner 4*, an unmanned probe launched from Cape Kennedy Nov. 28, 1964, sped past Mars July 14 and took a series of close-up photos of the planet as it passed.

Transmission of the photos back to earth began July 15, and Jet Propulsion Laboratory (JPL) scientists at Pasadena, Calif. were then able to confirm that the mission was a success. Since it took 8 hours 34 minutes to transmit each picture, the JPL personnel were unable to tell at first whether *Mariner 4* had taken all 21 (and a fraction) photos planned for the experiment.

The first 3 *Mariner 4* photos covered an area in which terrestrial telescopic photos had indicated the presence of 3 of the puzzling Martian "canals" (Hades 1, Hades 2 and Orcus). But Dr. Robert B. Leighton, 45, California Institute of Technology (Caltech) physics professor directing *Mariner 4* photo interpretation, said July 17 that "I see no evidence on any of the 3 frames that leads me to believe that I was looking at a canal." (JPL was operated by Caltech for the National Aeronautics & Space Administration. JPL managed the Mariner program.)

Mariner 4 had begun photographing Mars July 14 at 8:20 a.m. EDT (11 a.m. Martian time) after traveling 325 million miles in 228 days. *Mariner 4* intercepted and photographed Mars when the planet was 134 million miles from earth. The first photo was taken from a distance of 10,500 miles, and the photography continued until 8:45 p.m., when *Mariner 4* was perhaps 7,000 miles from the planet's surface. *Mariner 4*'s closest approach to the planet, 6,118 miles, was made at about 9:03 p.m. The probe disappeared behind Mars at 10:19 p.m. and reappeared on the planet's other side 52 minutes later. After passing Mars at a speed of some 11,500 mph., *Mariner 4* continued on in independent orbit around the sun as a tiny man-made planet. *Mariner 4*'s trajectory had been plotted to make sure that the probe did not crash into the planet.

The *Mariner 4* pictures of Mars were squares consisting of 40,000 dots—200 rows with 200 dots in each row. A series of numbers transmitted to earth indicated the shade (black through gray to white) of each dot. The first photo showed about 192 miles of the northern edge of Mars and what appeared to be haze or dust clouds some 50 miles above the horizon. One corner of the photo extended far above the planet's edge, and the opposite corner extended perhaps 600 miles below the edge. The 2d photo, overlapping the first, took in an area perhaps 186 miles wide and 550 miles long. The 3d photo, taken from an altitude of some 9,500 miles, covered an area of about 175 miles by 310. Each new photo was more distinct than the previous one because the distance from camera was progressively less, because the camera angle was more direct and because the angle of the sun's rays provided more shadow to outline surface features. The first photo showed part of the area noted on Martian maps as Phlegra. The other 2 showed part of the eastern edge of Trivium Charontis from Elysium to Amazonis. The 3d, the best photo of the 3, seemed to show evidence of shallow craters and a possible valley.

Mariner 4 finished transmitting back to earth the 21-plus photos July 24. In addition to the photos, the $126 million *Mariner 4* experiment provided much scientific information about solar wind, cosmic dust and other phenomena as well as about Mars. *Mariner 4* disclosed that Mars has an ionosphere. It showed that Mars has little or no magnetic field, probably because it has no liquid core. The planet's magnetic moment was estimated at 1/1,000 to 1/10,000 of the earth's, an estimate indicating "essentially no [magnetic] field at all." Since it has no magnetic field, Mars has no Van Allen radiation belts such as those around the earth, which trap radiation from the sun and space and thus act as a partial radiation shield for much of the earth.

Mariner 4 showed also that Mars' atmosphere is extremely thin. Atmospheric pressure on Mars decreases with altitude much more sharply than had been anticipated. Atmospheric pressure at the surface of Mars, according to Dr. A. J. Kliore, JPL scientist in charge of atmospheric observations, amounts to only 19 to 25 millibars, or about 1%-$2\frac{1}{2}\%$ of that on earth at sea level. This thinness of atmosphere—coupled with the lack of Van Allen belts—contributes to the intensity of the radiation dosage received on the planet's surface. Although this dosage was estimated at perhaps 50 times more intense than the dosage received on the earth's surface, JPL and other scientists said the radiation intensity did not preclude the possibility of life on Mars.

NASA scientists reported their early evaluations of the data from the Mariner probe July 29 at the White House and at a press conference later the same day. They said the photos indicated that Mars was a dry planet with little atmosphere and a crater-pocked surface on which there was little likelihood that life had developed. The scientists conceded, however, that the photos, which had "sampled only about 1% of the Martian surface," had not answered the question of whether there was life on Mars. The initial statement of the photo evaluation team, headed by Prof. Leighton of Caltech, said: "Mariner photos neither demonstrate nor preclude the possible existence of life on Mars."

At the White House July 29, Pres. Johnson received copies of the Mars photos, and he presented NASA medals to these 3 space officials: Dr. William H. Pickering, director of JPL; Oran Nicks, director of NASA lunar and planetary programs, and Jack N. James, director of the Mariner project at the JPL. Commenting on *Mariner 4*'s findings, the President said: "I believe it is very clear that in this day, when we are reaching out among the stars, the earth's billions will not set their compass by dogmas and doctrines which reject peace and embrace force and rely upon aggression and terror for fulfillment. . . . It may just be that life, as we know it, with its humanity, is more unique than many have thought, and we must remember this."

The initial evaluation report said:

> Man's first close-up look at Mars has revealed the scientifically startling fact that at least part of its surface is covered with large craters. This is a profound fact which leads to far-reaching fundamental inferences concerning the evolutionary history of Mars and further enhances the uniqueness of earth within the solar system. . . . The 70 craters clearly distinguishable on Mariner photos Nos. 5 through 15 range in diameter from 3 to 75 miles. . . . [They] have rims rising a few hundred feet above the surrounding surface and depths of a few thousand feet below the rims. . . . The number of large craters per unit area on the Martian surface is closely comparable to the densely cratered upland areas of the moon. If the Mariner sample is representative of the Martian surface, the total number of craters of the sizes so far observed is more than 10,000, compared to a handful on earth. . . . The Martian craters closely resemble impact craters on earth . . . and the craters of the moon. Craters of widely different degree of preservation and, presumably, age are distinguishable.
>
> A few elongated diffuse markings are present on the Mariner photos, but . . . no conclusions can be offered concerning them. . . . In southern sub-polar latitudes, where the season is late mid-winter, some craters appear to be rimmed with frost. . . .
>
> . . . Although the flight line crossed several 'canals' sketched from time to

time on maps of Mars, no trace of these features was discernible. ... No earth-like features, such as mountain chains, great valleys, ocean basins or continental masses were recognized. Clouds were not identified. ...

Following are some of the more fundamental inferences to be drawn from the *Mariner 4* photos:

(1) In terms of its evolutionary history, Mars is more moon-like than earth-like. Nonetheless, because it has an atmosphere, Mars may shed much light on early phases of earth's history.

(2) ... Much of the heavily cratered surface of Mars must be very ancient—perhaps 2 to 5 billion years old.

(3) The remarkable state of preservation of such an ancient surface leads us to the inference that no atmosphere significantly denser than the present very thin one has characterized the planet since that surface was formed. Similarly, it is difficult to believe that free water in quantities sufficient to form streams or to fill oceans could have existed anywhere on Mars since that time. ...

(4) The principal topographic features of Mars photographed by Mariner have not been produced by stress and deformation originating within the planet, in distinction to the case of the earth. ... Evidently Mars has long been inactive [internally]. ...

(5) As we had anticipated, Mariner photos neither demonstrate nor preclude the possible existence of life on Mars. ... If the Martian surface is truly in its primitive form, that surface may prove to be the best—perhaps the only—place in the solar system still preserving clues to original organic development, traces of which have long disappeared from earth. ...

Prof. Bruce C. Murray, Caltech member of the photo evaluation team, said at the press conference July 29 that in examining the pictures, "we can see no man-made features on Mars," but "the same kind of camera photographing the earth from an equivalent altitude would see nothing man-made on earth, with very few exceptions." Leighton pointed out at the press conference that "we have never expected ... to be able to tell in any certain way [as a result of *Mariner 4*'s photos] whether there is vegetation" on Mars. Dr. Pickering said at the press conference, however, that evidence of Martian life produced by ground-based observations was "still there," "and we have to investigate it."

Murray reported at the press conference that tentative measurements showed an apparent rim of what might be a very large crater to be about 13,000 feet high. This was the highest feature noted in the photos. He said the craters appeared to be "eroded, rounded-off ... to some extent," apparently as a result of wind since there is no liquid water to cause water erosion. Leighton added that "in view of the fact that Mars has a known atmosphere which does move material around on the surface and has somehow been eroding the craters away, I think it ... very surprising that in the time that must have elapsed since those craters were formed that there is so much of them left."

Telemetry reports from *Mariner 4* were discontinued Oct. 1 on radioed orders from the Goldstone (Calif.) Space Communications Station. *Mariner 4* sent its last telemetry from a record distance of

191,059,922 miles from the earth. The Goldstone signal switched *Mariner 4*'s transmitter from its high-gain directional antenna to its low-gain all-direction antenna, which could be used only for tracking the probe in its 567-day orbit around the sun. The change was made because *Mariner 4* was moving too far from the earth for data transmission. (But the probe and the earth came close enough in mid-1967 for several months of telemetry transmission; the nearest approach was about 29 million miles Sept. 7, 1967.)

Between the time of its launching and the antenna switch-over, *Mariner 4* traveled 418,749,386 miles in 307 days in its curving orbit and sent back 50 million scientific and engineering measurements. *Mariner 4*'s orbit had an aphelion (maximum distance from the sun) of 146 miles and a perihelion (minimum distance from the sun) of 103 miles.

Soviets Probe Moon

Scientists got a fresh view of the hidden side of the moon July 29 when an unmanned Soviet spaceship began sending back photos after flying behind the moon July 20. (Because the moon rotates as it revolves around the earth, about 40% of its surface can never be seen from earth.)

The space vehicle, named *Zond 3*, had been launched by Soviet scientists July 18 into a "heliocentric orbit" (orbit around the sun). Tass announced Aug. 14 that the spaceship had photographed the hidden side of the moon as it passed within 6,000 miles of the lunar surface July 20 and had successfully transmitted back to earth high-definition TV pictures of parts of the moon never before photographed. 2 of the photos, covering 1,832,200 square miles of previously unknown lunar surface, were released Aug. 16.

The announcement of the launching July 18 had said nothing of the lunar photography mission. It reported that the last stage of a multi-stage carrier rocket had been put into parking orbit around the earth and that "the automatic station *Zond 3*" was then launched from the orbiting stage into its heliocentric orbit. "The aim of this launching," according to the July announcement, "is to check the station's systems in conditions of prolonged space flight and the holding of scientific studies in interplanetary space."

In its Aug. 14 announcement, the USSR finally disclosed that *Zond 3* "carries equipment for taking pictures in outer space and sending them back ... from great distances." "To test the probe's photo and television equipment and radio channels for the transmission of images," the announcement said, "the station's trajectory was so chosen as to pass near the moon, which made it possible to photograph its surface in passing." Photographs were taken for 68 minutes July 20 and transmitted back to earth in 38 communication periods beginning July 29 when the probe was 1,365,000 miles from the earth, Tass reported. Tass said that these photos, combined with those taken by the USSR's *Luna 3* in 1959 at a distance of some 40,000 miles from the moon's surface, gave Soviet scientists a nearly complete photographic record of the side of the moon always turned away from the earth. Soviet astronomer Grigory Leikin said that features 3 miles or less in diameter were visible in the photos.

The photos released Aug. 16, covering $\frac{1}{5}$ of the moon, showed that the back of the moon was more mountainous and had fewer *maria* (dark, depressed, waterless "seas") than the visible side.

At a press conference held in Moscow Aug. 23, Soviet scientists revealed that *Zond 3* had taken "about 25" photos covering 19 million square kilometers (11,799,000 square miles) and that "around $1\frac{1}{2}$ million square kilometers [931,500 square miles] still remain to be photographed."

Soviet Prof. Aleksandr Michailov told delegates to the 16th International Astronautical Congress in Athens Sept. 14 that *Zond 3*'s photos of the hidden side of the moon gave support to the theory that the moon had a volcanic history. He said 10 craters measuring more than 100 kilometers in diameter each had been revealed by the photos.

During the last 3 months of 1965, Soviet scientists made 2 unsuccessful attempts at "soft-landing" an instrument package on the moon. Earlier in the year, 2 other tries had also failed.

The 3d attempt during 1965 was by a 1,506-kilogram (3,321-pound) lunar probe called *Luna 7*. It was launched from the USSR's Baikonur cosmodrome in Kazakhstan Oct. 4 and crashed into the moon at 1:08:24 a.m. Moscow time Oct. 8. The announcement of the launching was made Oct. 4 after the lunar probe had first been put into parking orbit around the earth aboard the last stage of a "multi-stage booster" and then had been transferred successfully into the 225,000-mile trajectory toward the moon. A mid-course correction was carried out Oct. 5 by signal from the earth.

Luna 7 crashed Oct. 8 in the dry Sea of Storms, west of the Kepler Crater. The Russians kept silence for almost 14 hours after the flight ended, and then Tass announced: "Most of the operations necessary for a soft-landing were fulfilled during the approach to the moon. Some operations, however, were not carried out in accordance with the program and need additional development. . . ."

Sir Bernard Lovell, director of Britain's Jodrell Bank radio-telescope observatory, said Oct. 8 that variations in *Luna 7* signals monitored by Jodrell Bank indicated that the probe's retro-rockets had been ignited one hour and 10 minutes before the probe crashed, had burned for 6 minutes, had been reignited 16 minutes later and had then burned until the crash.

Another unmanned instrument package, dubbed *Luna 8*, was sent aloft from an undisclosed Soviet site Dec. 3. The 1,552-kilogram (3,442-pound) payload, launched with no advance announcement, reached the moon early Dec. 7 ($51\frac{1}{2}$ minutes past midnight of Dec. 6 Moscow time) in what appeared to be the 4th successive Soviet failure to achieve a "soft landing" on the lunar surface.

Luna 8's quarter-million-mile flight ended in what appeared to be the target area in the moon's arid Sea of Storms. The impact point was given by Tass as Latitude 9° 8′ N. and Longitude 63° 18′ E. Tass reported that when *Luna 8* "was approaching the moon, an overall check-up ... of the systems ensuring soft landing ... showed that the systems ... functioned normally during all stages of the lunar landing, except the final one." "As a result of this flight," Tass declared, "a further step has been made in carrying out soft landing."

Sir Bernard Lovell speculated Dec. 7 that *Luna 8* had made a "hard" landing but had not crashed into the moon. Judging by *Luna 8* radio signals picked up by Jodrell Bank, Lovell suggested that the probe's failure was due to a flaw that showed up 4 minutes before impact. "The Russians narrowly missed complete success," he said.

Soviet Probes Head for Venus

The Soviet Union also continued studies of the planet Venus, sending at least 2 unmanned vehicles in its direction in late 1965. Soviet space experts launched *Venus 2* and *Venus 3* Nov. 16. Tass reported that the weight of *Venus 3* was virtually identical to the weight of the 2,123-pound *Venus 2* but that the design of *Venus 3* "differs somewhat from the design of *Venus 2* in make-up of its scientific apparatus."

Both probes were put into "parking orbit" around the earth aboard the final stages of their booster rockets before they were sent into 180 million-mile trajectories that brought them to the vicinity of Venus in Mar. 1966. In each case, the launching attempt, from an undesignated site, was not revealed until the probe seemed to be successfully on its way toward Venus.

(The USSR had announced only one previous shot toward Venus, the unsuccessful 1,418-pound *Venus 1* in 1961, but Western sources reported at least 4 other Soviet failures. The only probe to reach the vicinity of Venus successfully was the U.S.' *Mariner 2*, which transmitted back to earth a tremendous amount of scientific data before, during and after its flight past Venus Dec. 14, 1962.)

The possibility that the Soviets had launched a 3d vehicle toward Venus a few days after *Venus 2* and *Venus 3* was raised by the *N.Y. Times* Dec. 9. The newspaper said that "official sources" in Washington thought that a Nov. 23 launching announced by the Soviets as another in its extensive series of Cosmos earth satellites was actually an attempt to send a 3d probe to Venus in 11 days. As with previous Venus probes, these sources speculated, the Nov. 23 projectile was put into parking orbit around the earth preparatory to the maneuver that was designed to put it into a correct trajectory for Venus. The prospective *Venus 4*, however, exploded in parking orbit, according to the Washington analysis, and was publicly given the name *Cosmos 96* to hide the failure. It was pointed out that the inclination of *Cosmos 96*'s orbit was only .03° different from those of *Venus 2* and *3*. U.S. radar stations detected at least 8 orbiting objects related to *Cosmos 96* but only 4 for *Venus 3* and 3 for *Venus 2*.

OTHER U.S. DEVELOPMENTS

Scientific & Military Launchings

While most public attention on the space program was directed at the manned missions, many unmanned satellites and space probes were sent up by U.S. scientists during the last half of 1965. Most of the announced launchings were by NASA; a few were military. The shots, in chronological order:

June 30—A 70-foot antenna was sent more than 1,000 miles into space by means of a 4-stage Journeyman (Argo D-8) rocket launched by NASA experts from Wallops Island, Va. at 1:34 a.m. EDT. The shot was primarily a radio-astronomy experiment. The main object of the 137-pound payload was to measure the intensity of radio frequency energy originating largely from outside the solar system on the frequencies of 750, 1,225 and 2,000 kilocycles. Secondary objectives included investigation of radio noise and measurement of electron density in the top of the ionosphere.

July 2—The 280-pound *Tiros 10* weather satellite was sent into near-polar orbit by means of a 3-stage Thor-Delta rocket launched by NASA from Cape Kennedy at 12:07 a.m. EDT. The 18-sided hatbox-shaped satellite achieved a sun-synchronous orbit with an initial apogee of 517 miles, perigee of 458 miles, period of 100.6 minutes and 81.4° angle of inclination. *Tiros 10* was to serve as an interim operational satellite (until an operational weather satellite system would be started in 1966). Its mission was to obtain photo data on areas where hurricanes and typhoons are born. *Tiros 10*, for which the Weather Bureau had paid 4\frac{1}{2}$ million, was the first satellite funded by the Weather Bureau. In its sun-synchronous orbit, *Tiros 10* drifted westward about 1° daily—and thereby matched the direction and rate of the earth's movement around the sun. The satellite's orbital plane remained at a constant angle to the earth-sun position. *Tiros 10*, therefore, was in maximum sunlight during each orbit and thus in the best position for photography and for charging its batteries.

July 20—The Air Force sent 3 satellites into orbit by means of a single 2-stage Atlas-Agena rocket fired from Cape Kennedy. 2 of the satellites were 524-pound payloads whose mission was to detect any secret nuclear explosions in space in violation of the treaty banning nuclear tests in space. Each of the sentry satellites carried detection equipment, a rocket motor and a gas-electric maneuvering jet. The U.S. already had 4 other satellites in space on a similar mission. The 3d satellite was a 12-pound package of scientific instruments designed to check on natural radiation in space. All 3 satellites went into orbits with initial apogees of 69,570 miles and perigees of 121 miles, but 2 days of maneuvering changed the orbits of the sentry satellites to circular orbits at the higher altitudes.

July 30—The *Pegasus 3* meteoroid-recording satellite was launched from Cape Kennedy at 9 a.m. EDT atop a 2-stage Saturn-1 booster rocket. It achieved an orbit with an apogee of 336 miles, perigee of 323 miles, period of 95.25 minutes and 28.9° angle of inclination. The NASA launching was the 10th in 10 attempts with Saturn-1s, and it brought to an end the program of Saturn-1 launchings. The first stage of the Saturn-1 in the July 30 shot developed 1,504,000 pounds of thrust. The total weight put into orbit was 33,200 pounds in 2 sections. The major section, 23,100 pounds, was made up of the booster rocket's spent 2d (S-IV) stage weighing 14,500 pounds, the Pegasus satellite (3,200 pounds), the Pegasus support structure adapter (2,800 pounds) and the instrument unit (2,600 pounds). The aggregate weight of 2 Apollo modules and associated "hardware" put into a separate orbit was 10,100 pounds. On launching, the Pegasus satellite had been folded inside the Apollo service module. After reaching orbit, the Apollo modules were separated from the Pegasus, and the Pegasus' winglike meteoroid-recording panels were unfolded. The Pegasus, with its panels extended, was 96 feet long by 17 feet high. The attached

S-IV stage was 70 feet long and 18 feet in diameter. 48 of the Pegasus' aluminum panels were detachable (nicknamed "coupons") for possible recovery by a future astronaut.

Aug. 10—A solid-fueled 4-stage Scout booster rocket was launched by NASA from Wallops Island to evaluate improvements made in the Scout vehicle. As a "bonus," the booster put a 45-pound Army mapping satellite of the Secor (sequential collation of range) series into an orbit with an apogee of about 1,725 miles, perigee of about 690 miles and inclination of about 65°.

Aug. 11—A 2,100-pound model of a Surveyor lunar probe was launched from Cape Kennedy by means of a 2-stage Atlas Centaur rocket at 10:31 a.m. EDT in an NASA test of techniques to be used in a later attempt to "soft-land" instruments on the moon. The 94-inch probe, aimed at a "simulated moon" in an empty space some 240,000 miles away, achieved its objective and went into an elliptical earth orbit with a 509,829-mile apogee, 105-mile perigee, 31-day period and 28.55° angle of inclination. The total weight orbited was 6,230 pounds, including the 4,130-pound burned-out Centaur stage of the booster rocket, which was in orbit separately. The test was described as a vehicle development flight whose primary mission was to show whether the hydrogen-fueled Centaur could put a Surveyor into a lunar-transfer trajectory.

Aug. 13—A "secret satellite" was launched from Vandenberg Air Force Base, Calif. by an Air Force-industry team by means of a Thor-Able-Star rocket.

Aug. 24—NASA failed in an attempt to orbit a 620-pound satellite as the 3d in the orbiting solar observatory (Oso) series. The scientific instrument package was launched from Cape Kennedy at 11:17 a.m. by means of a 3-stage Thor-Delta booster rocket. NASA officials attributed the failure to premature firing of the rocket's 3d stage, and they said the prospective satellite had probably fallen into the South Atlantic. Plans had called for putting the Oso into a nearly-circular orbit about 345 miles high so that its instruments could collect data on solar X-rays and solar gamma, ultra-violet and infra-red rays.

Oct. 5—An Atlas-D rocket fired by Air Force personnel at Vandenberg Base was used to launch a satellite into an orbit with a 1,300-mile apogee. The satellite carried experiments of various types, including one on shielding astronauts from radiation in space.

Oct. 14—A 1,150-pound scientific satellite dubbed *Ogo 2* (for orbiting geophysical observatory) was launched into polar orbit by means of a thrust-augmented Thor-Agena-D rocket fired from Vandenberg Base. Because of guidance system troubles, *Ogo 2* went into an orbit with an apogee of 933 miles (instead of the 575 miles scheduled) and a perigee of 257 miles (instead of the 207 miles planned). The orbital period was given as one hour 44 minutes. A major mission of *Ogo 2* was to secure data on radiation that might endanger astronauts. The satellite's 20 experiments, concentrating on near-earth space phenomena, emphasized global mapping of the geomagnetic field and measurements of radiation. The data it sought was to be correlated with solar emissions and other events to provide knowledge on a variety of phenomena, including (a) the connection between the entrapment of energy particles and the occurrence of aurora, (b) the pressure, temperature, density and chemistry of the neutral atmosphere surrounding the earth, (c) the relationship between solar X-ray and ultra-violet emissions and their effect on the earth's ionosphere, atmosphere and airglow. *Ogo 2*, like its predecessor *Ogo 1* (launched Sept. 4, 1964), consisted of a main body about 6 feet long, 3 feet wide and 3 feet deep, from which instrument-laden booms and solar panels were unfolded to a total span of 80 feet after the satellite went into orbit. NASA reported Oct. 26 that *Ogo 2* was "tumbling" uncontrollably in orbit, that its power was almost all gone and that there was little chance of "ever achieving useful operation" of the satellite.

Oct. 15—A Titan-3C rocket was launched by Air Force personnel at Cape Kennedy, but the transstage (3d stage) broke apart in orbit and failed in its program. The mission had called for the transstage to ignite and stop its engine 10 times to shift course 4 times and to orbit 2 "hitch-hiker" satellites—a 44-inch radar sighting sphere and a satellite with instruments to check on radiation hazards to astronauts.

Oct. 28—NASA announced that it had launched 12 meteorological rockets in a 15-day period in October from Point Barrow, Alaska, Churchill Research Range in Canada and Wallops Island, Va.

Nov. 6—A 385-pound mapping satellite named *Geos 1* (*Explorer 29*) was launched from Cape Kennedy at 1:30 p.m. by a 3-stage thrust-augmented improved Delta (TAID) launching rocket. *Geos 1*, the first satellite to be launched by a TAID, achieved an orbit with an apogee of 1,300 miles, perigee of 700 miles and 59° angle of inclination. The apogee was about 400 miles higher than planned because the guidance system failed to turn off the rocket's 2d stage on time. *Geos 1*'s instruments were designed to determine the earth's shape with great accuracy and to provide precise measurements of bulges, depressions and distances between various points on the earth's surface.

Nov. 18—The IQSY (International Quiet Sun Year) solar explorer satellite *Explorer 30* was sent into orbit by means of a 4-stage Scout rocket launched from Wallops Island. The 125-pound scientific instrument package, designed by the Naval Research Laboratory to measure and monitor solar X-ray emissions during the final portion of IQSY, achieved an orbit with an apogee of about 630 miles, a perigee of about 430 miles and a 60° angle of inclination. (NASA ended its IQSY launchings Dec. 15 when it used a Nike-Apache rocket fired from Wallops Island to carry a 51-pound instrument payload to an altitude of 113 miles. All required data were received by radio before the instrument package was lost in the Atlantic.)

Dec. 16—A 140-pound scientific space probe dubbed *Pioneer 6* was sent into orbit around the sun by means of a 3-stage TAID launched from Cape Kennedy at 2:31 a.m. *Pioneer 6* was a cylindrical instrument package 35 inches long and 37 inches in diameter. It was encrusted with 10,368 solar cells to provide power, and it carried 6 experiments whose instruments weighed 35 pounds. Its mission was to radio back data on the "solar wind," the sun's magnetic fields, the boundary region between the sun's atmosphere and interstellar space, the physics of the sun and the basic interactions of high-energy charged particles and magnetic fields. *Pioneer 6* was the first of a series of 4 probes that were to be launched at 6-month intervals to study the sun and other cosmic phenomena. *Pioneer 6* was described as the most "magnetically clean" spacecraft ever built. The purpose of this "cleanliness"—achieved by the use of such non-magnetic materials as plastics and non-ferrous metals (principally aluminum)—was to permit the probe's magnetometers to sense interplanetary fields without interference from the spacecraft. *Pioneer 6* was also the first spacecraft to attempt a radio propagation experiment in interplanetary space. The spacecraft's orbit was positioned between the orbits of the earth and Venus. The period of revolution was given as 311 days, the perihelion (closest approach to the sun) as 75.6 million miles, the probe's speed in orbit as 59,800 mph.

Pioneer 6's launching was the 35th for a Delta rocket booster but the Delta's first interplanetary mission. The booster sent the probe 4,900 miles from Cape Kennedy to a "target" 346 miles above Africa. *Pioneer 6* then separated from the 3d stage of the booster and continued in a trajectory that put it into its independent orbit around the sun. Final orientation adjustments were made by the spacecraft at a distance of some 230,000 miles from the earth at 10:57 p.m. EST Dec. 17, when it acted on 157 separate commands radioed from the earth.

Dec. 21—A Titan-3 rocket was launched by the Air Force from Cape Kennedy Dec. 21 in an unsuccessful attempt to put 4 satellites into circular orbits 20,930 miles above the equator. The rocket's 3d stage—or transtage—went into an orbit with an apogee of 20,900 miles and perigee of 120 miles. But the transtage failed to restart its engines at the maximum altitude in a maneuver designed to transform the elliptical orbit into a circular orbit at the perigee's altitude. 3 of the satellites were ejected from the orbiting transtage 12 to 30 minutes late, and they continued on in orbits similar to that of the transtage. The 4th satellite apparently remained attached to the transtage. The most successful of the 3 ejected satellites was the 35-pound *Oscar 4*, which had been built by "ham" (amateur radio) operators in Los Angeles from $200 worth of parts. *Oscar 4* emitted the signal "hi" in Morse code from its beacon transmitter as it traveled in orbit, and ham operators throughout the world used it to amplify and relay radio messages to each other. The satellite believed attached to the transtage was a 427-pound sophisticated instrument package carrying 15 sun-study experiments. No signals from it were detected. The other 2 orbiting space vehicles were experimental communications satellites; they sent weak signals that were received by ground stations, but neither of them operated properly.

Budget Cut, Programs Reduced

Even while planning the Apollo program of manned flights to the moon, NASA began cutting back on the development of programs beyond that because of budgetary reductions.

A huge, solid-fueled rocket motor that was cancelled because of funding problems was tested Sept. 25 because of pressure from Congress members who supported solid-fueled rockets. The rocket motor— 260 inches in diameter and the largest ever fired—developed a record $3\frac{1}{2}$ million pounds of thrust Sept. 25 in a static test in a 180-foot concrete-lined pit in Homestead, Fla. The 60-foot device, topped by a 20-foot exhaust nozzle, was actually only a "half-length" motor; a full-length one would develop about twice as much thrust. The 1,680,000 pounds of red rubber-like propellant, cast in a single piece, burned at a rate of more than 6 tons a second. 112 seconds of the 124 seconds of burning time was at near-peak thrust.

In an economy move earlier in 1965 NASA had cancelled the program for developing big solid-fueled rockets, but pressure by several Congress members and the enthusiasm of NASA supporters of the idea had resulted in a 2-test revival of the program.

NASA Dec. 15 cited "budgetary considerations" in announcing that it had canceled further development of its 1,250-pound advanced orbiting solar observatory, or Aoso. Aoso development had begun in Nov. 1963, and $39 million had been budgeted for the program through fiscal 1966. But NASA reported that "some of the $24,900,000 appropriated" in fiscal 1966 "will be recoverable."

In another economy move, NASA announced Dec. 22 that the first mission of its unmanned Voyager interplanetary spacecraft had been deferred until 1973. NASA had planned to attempt to put its first Voyagers into orbit around Mars in 1971 and to use a Voyager in 1973 to land instrumented capsules on the planet. In an effort to continue interplanetary exploration at lower cost, NASA decided not to end its Mariner program with the smaller *Mariner 4*. No further Mariner missions had been scheduled after *Mariner 4* executed its successful fly-by of Mars in July. The Dec. 22 announcement said 3 new Mariner flights to Venus and Mars had been scheduled.

Interplanetary missions planned by the U.S. under the revised program announced Dec. 22: (1) A Mariner would be launched in mid-1967 by means of an Atlas-Agena rocket on a 3-to-4-month voyage that, it was hoped, would take it closer to Venus than the 21,600-mile approach made by *Mariner 2* in 1962. (2) 2 "somewhat heavier" Mariners would be launched early in 1969 by means of Atlas-Centaur rockets on 9-month missions to Mars. (3) "NASA is considering the launch of 2 identical Voyager spacecraft to Mars early in [1973] ... with a single Saturn-5 launch vehicle," but "Voyager missions will not be finally committed until a decision is made to proceed to hardware development"; under this plan, both Voyagers would orbit Mars and send down "large capsules ... to search for evidence of life ... and make other scientific studies." The Jet Propulsion Laboratory would continue to manage both the Mariner and the Voyager projects.

1965 Called Most Successful Year

Pres. Johnson reported to Congress (Jan. 31, 1966) that 1965 was the most successful year in the U.S.' aero-space history. He said that in 1965 the U.S. had orbited more spacecraft than in any previous year and had put astronauts into space for a total length of time (10 U.S. astronauts spent a total of 1,297 hours 42 minutes in space during 5 Gemini flights in 1965) greater than the previous total for all manned U.S. space flights. But he pointed out that the USSR was also "far from idle" during 1965.

The National Aeronautics & Space Council said in a separate chapter of the President's report: During 1965 the U.S. "successfully placed into earth orbit 94 spacecraft as compared with 66 by the USSR. This was an increase over 1964 of 36% for the United States and 83% for the USSR. ... [The Soviets], with their 7 escape mission flights, more than doubled the United States activity in lunar and planetary exploration for the year."

INTERNATIONAL DEVELOPMENTS

Soviet Launchings

2 Molniya-1 (Lightning) satellites were launched into orbit Apr. 23 and Oct. 14 in Soviet experiments in the use of active re-transmitter satellites for 2-way long-distance TV and phono-telegraph-radio communications. Tass said Oct. 14 that a goal of the launchings was to verify "the possibility of organizing a communications system with the simultaneous use of several sputniks." In announcing the 2d launching, after the satellite's success had been verified, Tass said the 2d Molniya-1 had already been used for phone calls and an exchange of TV programs between Moscow and Vladivostok. The announcement said the satellite was in an orbit with an apogee of 40,000 kilometers (24,840 miles) in the Northern Hemisphere and a perigee of 500 kilometers ($310\frac{1}{2}$ miles) in the Southern Hemisphere. The period of revolution was given as 11 hours 59 minutes; the orbital inclination as 65°.

An unmanned satellite, *Proton 1*, weighing a record 26,896 pounds (not counting the carrier rocket's final stage), was put into orbit by Soviet scientists July 16 by means of what Tass described as "a new powerful booster." *Proton 1*, "equipped [according to Tass] with special scientific equipment for study of cosmic particles of super-high energies," achieved an orbit with an apogee of 627 kilometers (389.367 miles), perigee of 190 kilometers (117.99 miles), 92.45-minute period and 63.5° angle of inclination to the equatorial plane.

In a July 30 announcement, Tass reported that *Proton 1*'s mission included (a) "the study of solar cosmic rays and their radiation hazard," (b) "the study of the energy spectrum and chemical composition of the particles of primary cosmic rays in the energy interval of up to 100 trillion electron volts," (c) "the determination of absolute intensity and energy spectrum of galactic electrons," (d) "the determination of intensity and energy spectrum of galactic gamma rays with energies exceeding 50 million electron volts."

Prof. Nikolai L. Grigorov, a Proton project official and head of the Cosmic Ray Laboratory at Moscow University's Institute of Nuclear Physics, reported in a *Pravda* interview Aug. 21 that particles with energy of 100 trillion electron volts had been trapped for the first time by *Proton 1* and that the satellite had measured the energy spectrum of cosmic rays containing such particles. Grigorov said that the orbiting of heavy equipment such as the ionizing calorimeter needed for such studies had been impossible with previously used rocket carriers and had been made possible only by the development of the USSR's new booster.

A 2d Proton satellite, *Proton 2*, described as a "heavy scientific space station," was sent Nov. 2 into an orbit with an initial apogee given as 637 kilometers (395.577 miles), perigee of 191 kilometers (118.611 miles), period of 92.6 minutes and inclination of 63° 30'. The announcement of *Proton 2*'s launching, made after *Proton 2* was successfully in orbit, said that the satellite's "overall payload," "exclusive of the last stage of the carrier rocket," was the same as that of *Proton 1*, a record "12.2 tons." The announcement said *Proton 2* carried equipment "to study solar cosmic rays and their radiation danger; to study the power spectrum and chemical composition of cosmic rays in the interval of energies of up to 100 trillion electron volts; to study the nuclear interaction of cosmic particles of superhigh energies of up to one trillion electron volts; to determine the absolute intensity and power spectrum of electrons of galactical origin; to determine the intensity and power spectrum of the galaxy's gamma rays with energies exceeding 50 million electron volts."

Prof. Georgi V. Petrovich (believed to be the penname of a veteran Soviet rocket scientist) reported in the November issue of *Aviation & Cosmonautics* that the Proton satellites had been sent into orbit by means of booster rockets capable of more than 60 million horsepower—roughly equivalent to some 3 million pounds of thrust. By comparison, the U.S.' Titan-3C, previously considered the world's most powerful booster, generated 2,400,000 pounds of thrust. (Petrovich had reported in *Komsomolskaya Pravda* Oct. 5 that the USSR's first experimental liquid rockets had been launched "in the latter half" of 1933. He said: "The experimental work on Soviet electric and liquid rockets was started in May 1929 at the gasdynamic laboratory of the Leningrad Military Research Committee"; "a jet research institute set up at the end of 1933 . . . developed in the 40's a series of aircraft liquid rocket engines with the pump-type supply of nitric acid and kerosene, with chemical ignition, an unlimited number of repeated, fully automated starts, controlled thrust and the maximum near-earth thrust of 300 to 900 kilograms [661,500 to 1,984,500 pounds]"; the first engine in the series "passed ground and flight tests in 1943," and "about 400 firing tests" of 2 of the engines were conducted on various aircraft by 1946).

The Soviets sent into orbit at least 36 unmanned Cosmos satellites during the latter half of 1965. Dr. G. A. Skuridin reported in the Bul-

INTERNATIONAL DEVELOPMENTS 33

letin of the Soviet Academy of Sciences that Cosmos satellites had helped map the earth's upper atmosphere, magnetic field and radiation belts, had helped determine the atmosphere's density and the chemical composition of the outer ionosphere and had provided data used by Soviet designers of manned spaceships. Skuridin said that *Cosmos 5* had gathered data on the U.S.' high-altitude atomic explosion of July 9, 1962.

Cosmos launching data reported by the Soviets:

Cosmos 67—May 25. Apogee, 350 kilometers (217.35 miles). Perigee, 207 kilometers (128.547 miles). Period, 89.9 minutes. Angle of inclination to the equatorial plane, 51.8°.

Cosmos 68—June 15. Apogee, 334 kilometers (207.414 miles). Perigee, 205 kilometers (127.305 miles). Period, 89.77 minutes. Inclination, 65°.

Cosmos 69—June 25. Apogee, 332 kilometers (206.172 miles). Perigee, 211 kilometers (131.031 miles). Period, 89.7 minutes. Inclination, 65°.

Cosmos 70—July 2. Apogee, 1,154 kilometers (716.634 miles). Perigee, 229 kilometers (142.209 miles). Period, 98.3 minutes. Inclination, 48.8°.

Cosmos 71, 72, 73, 74 and *75*—July 16. Apogee and perigee, 550 kilometers (341.55 miles). Period, 99.5 minutes. Inclination, 56.1°.

Cosmos 76—July 23. Apogee, 530 kilometers (329.13 miles). Perigee, 261 kilometers (162.081 miles). Period, 92.2 minutes. Inclination, 48.8°.

Cosmos 77—Aug. 3. Apogee, 291 kilometers (180.711 miles). Perigee, 200 kilometers (124.2 miles). Period, 89.3 minutes. Inclination, 51.84°.

Cosmos 78—Aug. 14. Apogee, 329 kilometers (204.309 miles). Perigee, 206 kilometers (127.926 miles). Period, 89.8 minutes. Inclination, 69°.

Cosmos 79—Aug. 25. Apogee, 359 kilometers (222.939 miles). Perigee, 211 kilometers (131.031 miles). Period, 90 minutes. Inclination, 64.9°.

Cosmos 80, 81, 82, 83 and *84*—Sept. 3 (all 5 satellites put into orbit by a single rocket carrier). Altitude of their near-circular orbits was given as "nearly 1,500 kilometers" (931½ miles). Period of revolution, 116.6 minutes. Inclination, 56°.

Cosmos 85—Sept. 9. Apogee, 319 kilometers (238.099 miles). Perigee, 212 kilometers (131.652 miles). Period, 89.6 minutes. Inclination, 65°.

Cosmos 86, 87, 88, 89 and *90*—Sept. 19 (all 5 satellites put into orbit by a single rocket carrier). Apogee, 1,690 kilometers (1,049.49 miles). Perigee, 1,380 kilometers (856.98 miles). Period, 116.7 minutes. Inclination, 56°. (The announcement said that one of the satellites was powered "by a device in which the energy is produced by a radioactive substance" but that "measures have been taken to exclude any possibility of the isotope spreading in the atmosphere or on the earth's surface.")

Cosmos 91—Oct. 12. First weather satellite announced by USSR; carried temperature-recording equipment.

Cosmos 92—Oct. 16. Apogee, 353 kilometers (219.213 miles). Perigee, 212 kilometers (131.652 miles). Period of revolution, 89.9 minutes. Inclination, 65°.

Cosmos 93—Oct. 19. Apogee, 522 kilometers (324.162 miles). Perigee, 220 kilometers (136.62 miles). Period, 91.7 minutes. Inclination, 48°24'.

Cosmos 94—Oct. 28. Apogee, 293 kilometers (181.959 miles). Perigee, 211 kilometers (131.031 miles). Period, 89.3 minutes. Inclination, 65°.

Cosmos 95–Nov. 4. Apogee, 521 kilometers (323.541 miles). Perigee, 207 kilometers (128.547 miles). Period, 91.7 minutes. Inclination, 48°25'.

Cosmos 96–Nov. 23. Apogee, 310 kilometers (192½ miles). Perigee, 227 kilometers (140.967 miles). Period of revolution, 89.6 minutes. Inclination, 51°51'.

Cosmos 97–Nov. 26. Apogee, 2,100 kilometers (1,304 miles). Perigee, 220 kilometers (136½ miles). Period, 108.3 minutes. Inclination, 49°.

Cosmos 98–Nov. 27. Apogee, 570 kilometers (154 miles). Perigee, 216 kilometers (134 miles). Period, 92 minutes. Inclination, 65°.

Cosmos 99–Dec. 10. Apogee, 320 kilometers (199 miles). Perigee, 199 kilometers (123.58 miles). Period, 89.6 minutes. Inclination, 65°.

Cosmos 100–Dec. 17. Altitude (of presumably circular orbit), about 650 kilometers (403½ miles). Period of revolution, 97.7 minutes. Inclination, 65°

Cosmos 101–Dec. 21. Apogee, 550 kilometers (342½ miles). Perigee, 260 kilometers (161½ miles). Period, 92.4 minutes. Inclination, 49°.

Cosmos 102–Dec. 28. Apogee, 172 miles. Perigee, 135 miles. Period, 89 minutes.

Cosmos 103–Dec. 28. Altitude (of near circular orbit), 372 miles. Period, 97 minutes.

The Soviets made several series of tests of rocket carriers that came down in the Pacific. Tass announced Oct. 13 that the USSR had completed a series of launchings of "new versions of rocket carriers of space probes" to a previously designated mid-Pacific area and that the area therefore was open to ships and planes. Tass said Nov. 24 that a series of Soviet launchings to a mid-Pacific target area had been completed successfully and that shots to a newly designated mid-Pacific target area would take place Nov. 25–Dec. 25. Successful completion of the latter tests "ahead of schedule" was announced Dec. 15. After the tests the USSR announced the reopening of the target zones (from which planes and ships had been warned away during the launching period). Tass announced Dec. 14 that "tests of a variant of a system of landing space vehicles" would be carried out between Dec. 16, 1965 and June 1, 1966 with "some elements of the booster-rockets" scheduled to fall into a newly designated mid-Pacific target area.

Soviets Display 'Orbital Missiles'

At Red Square ceremonies marking the 48th anniversary of the Bolshevik Revolution, the USSR Nov. 7 displayed 2 apparently operational models of a rocket it said could deliver a nuclear warhead on target from orbit. Tass said the 3-stage 115-foot rockets were "orbital missiles whose warheads can deliver their blow unexpectedly to the aggressor at the first or any other orbit around the earth." The rocket, modified versions of which had orbited the Vostok and Voskhod spaceships, had been displayed previously—May 9—in Red Square. At that time Soviet officials had described it simply as having "orbital capability."

The Soviet Union Nov. 7 also displayed a new intercontinental ballistic missile mounted on a self-propelled transport. Tass said the 72-foot rocket was "solid fuel[ed]" and was "elusive to the enemy's air and space reconnaisance." Several new smaller rockets for tactical use in support of ground troops were also shown Nov. 7. These highly-mobile weapons brought to 13 the number of new missiles systems the USSR had introduced in the past 2 years.

(Col. Gen. Vladimir F. Tolubko said Nov. 13 that the USSR was developing "rockets which can deliver nuclear warheads over ballistic as well as orbital trajectories and are capable of maneuvering on trajectory.")

Activity by Other Nations

4 other nations—France, Canada, Indonesia and West Germany—conducted experiments in space exploration during the latter half of 1965. Some were with U.S. cooperation. Among these developments:

France—The first French earth satellite, a $92\frac{1}{2}$-pound A-1 capsule, was put into orbit Nov. 26 by means of a 62-foot 3-stage Diamant rocket launched by French rocket experts at 2:47 p.m. from the French launching base at Hammaguir in the Algerian Sahara. The satellite went into an orbit with an initial apogee given as $1,098\frac{1}{2}$ miles, a perigee of 328.08 miles and a 108-minute period of revolution. French Gen. Pierre Nardin said that the apogee was 349 miles lower, the perigee 10 miles higher and the period 5 minutes shorter than intended but that the orbit still was "within planned limits."

The A-1, a limited-purpose satellite designed to do little more than emit a radio signal for 2 weeks, carried no sophisticated scientific equipment and had no scientific mission, according to French officials. Its rôle was to test the Diamant. The French Armed Forces Ministry reported Nov. 27 that the A-1 had stopped sending radio signals because of a defective transmitter.

The Diamant was scheduled to serve as the basis for French military rockets, carrying nuclear weapons, that were intended to start replacing France's Mirage-4 atomic bombers in about $2\frac{1}{2}$ years. The Diamant had a thrust reported at 56,000 pounds. Its first stage, the Emeraud, was liquid fueled and fired for one minute 35 seconds in the Nov. 26 shot. The remaining 2 stages, both solid-fueled, fired for 44 seconds and 45 seconds, respectively, to inject the A-1 into orbit at a velocity of 26,000 feet a second.

The launching was the first firing of a complete 3-stage Diamant. All 3 stages of the Diamant had been tested, however, in one- or 2-stage combinations during 22 launchings of Diamant elements at Hammaguir. Critics of French Pres. Charles de Gaulle charged that he had forced a speed-up in the launching schedule (the original launching date had been set for 1966) in order to have a satellite in orbit before the French elections Dec. 5. The Diamant and A-1 were flown to Hammaguir by cargo planes from the French space research center at Saint-Mederd-en-Jalles Nov. 5, but the French rocket experts reportedly failed in an effort to meet a Nov. 22 launching date and postponed the attempt for 4 days.

France was the 3d country to launch a satellite of its own. The first 2: the USSR and the U.S. British, Canadian and Italian satellites had all been put into orbit by U.S. rockets.

Alexander MacLeod had reported in the *Christian Science Monitor* Sept. 10 that "France's space effort is No. 3 in the world, pushing Britain into 4th place." French expenditures on space rose "from an officially admitted $8,120,000 in 1961 to around $56 million"—nearly $\frac{3}{4}$ of it for "strictly national space projects" (the rest for joint European research)—in 1965, MacLeod said.

France's National Center for Space Studies (CNES) in Paris June 8 had announced the successful completion of tests on the Diamant and had scheduled the Hammaguir launching. Since the Hammaguir range was to be returned to Algeria in 2 years, France had begun to develop a better site near Kourou, French Guiana.* CNES' new national space center was also under construction in Toulouse. France had also announced launchings of its Dragon scientific sounding rocket for radiation studies. The use of a site in southern Iceland for such a launching (to an altitude of 255 miles) was reported by the CNES Sept. 7.

---

*At the spring 1965 meeting of the International Committee for Space Research, France had offered to make bilateral arrangements for other nations to use the launching range it was developing near Kourou. By December, only the U.S. had indicated a desire to accept the offer and to open negotiations. It was reported that the USSR had rejected French suggestions that Russia discuss using the range. (The USSR and France, however, had reached agreement on use of the Soviet Molniya-1 communications satellites and on France's SECAM color TV system. An experimental 55-minute color TV program using SECAM was broadcast from Moscow to Paris by means of a Molniya-1 for the first time Nov. 29, and a 75-minute color TV program was broadcast from Moscow to Paris by the same system Nov. 30.)

The U.S. cooperated in a French satellite launching Dec. 6 when France's 135-pound *Fria* (originally designated FR-1A), a scientific satellite, was sent into a near polar orbit Dec. 6 by means of a 4-stage U.S. Scout rocket launched by U.S. personnel at Vandenberg Air Force Base, Calif. at 1:06 p.m. The satellite achieved a nearly circular orbit with an altitude of about 490 miles, a period of one hour 40 minutes and a 76° angle of inclination. Its mission was to study very-low-frequency (VLF) radiowave propagation and to measure electron densities at orbital altitudes. The satellite was launched by the U.S.' NASA under a Feb. 1963 agreement between NASA and France's CNES. Under the agreement, NASA had previously launched 4 French payloads on suborbital investigations by means of Aerobee-150A sounding rockets fired from Wallops Island, Va. Oct. 17 and 31, 1963 and Sept. 17 and 25, 1965.

Fria's radio-wave experiments were designed by France's CNET (Centre National d'Etudes des Telecommunications, or National Telecommunications Center). The electron-density probe experiment was provided by the University of Birmingham, England. Data from the satellite were received by ground stations operated by the U.S. and France at locations around the globe. *Fria* consisted of 2 8-sided prisms attached to an 8-sided main body about 27 inches in diameter and 52 inches long (including a magnetic antenna support boom).

Canada—The Canadian satellite *Alouette 2* and the 218-pound NASA satellite *Explorer 31* were shot into orbit Nov. 28 by means of a single 2-stage Thor-Agena-B rocket launched by U.S. personnel at Edwards Air Force Base, Calif. The 2 satellites, on a mission to study the ionosphere and radio-wave propagation, traveled within a few miles of each other in nearly polar orbits with apogees of about 1,850 miles, perigees of some 310 miles and 80° angles of inclination to the equatorial plane. The 320-pound *Alouette 2* was designed and built for Canada's Defense Research Board (DRB) by the DRB's Defense Research Telecommunications Establishment. *Explorer 31* was designed and built for NASA's Goddard Space Flight Center by the Applied Physics Laboratory of Johns Hopkins University.

Indonesia—2 Kappa-8 rockets were launched by Indonesian technicians from the Indonesian launching site at Pameungpeuk on the south coast of Java Aug. 7 and 12, according to Japanese sources. It was reported that 10 of the rockets, designed by Hideo Itokawa of Japan, had been bought by Indonesia and assembled by Indonesians. The first rocket was said to have achieved an altitude of 197.4 miles, the 2d an altitude of 118 miles.

West Germany—The U.S.' NASA and West Germany's Scientific Research Ministry announced July 19 the signing of an understanding under which a satellite "designed and constructed in Germany" would be launched from the U.S.' "Western Test Range . . . on a Scout vehicle provided by NASA." The launching, to put the German payload into a polar orbit, was scheduled for 1968. The joint program's primary objective was a study of the earth's inner radiation belt.

1966

The U.S. and the Soviet Union took big strides toward exploration of the moon during 1966 although the Soviets made no manned flights. Both "soft-landed" instrument packages on the lunar surface. A Soviet satellite orbited the moon for the first time in March. The U.S.' Gemini series of 2-man spaceship flights ended after 5 flights during 1966, most of them successful in experimental rendezvousing and "docking" in space. The first successful docking in space was achieved in March by *Gemini 8* and a target vehicle; this was seen as an essential step toward the first manned flight to the moon. With the $4.1 billion Gemini program completed, the U.S. ended the year looking forward to Apollo—the program to land men on the moon.

MAN IN SPACE

Gemini 8 'Docks' in Space

Civilian command pilot Neil Alden Armstrong, 35, and Air Force Maj. David Randolph Scott, 33, were shot into orbit aboard the U.S. spaceship *Gemini 8* Mar. 16. Some $6\frac{1}{2}$ hours later they successfully executed the first "docking" (physical joining) of 2 orbiting spacecraft. But a malfunction of a small maneuvering rocket aboard the $3\frac{1}{2}$-ton *Gemini 8* then forced them to cut short their flight. They made an emergency but safe splash-down in the Western Pacific at one of several pre-designated secondary landing areas.

The 2 downed astronauts had to forgo scheduled attempts at 3 more link-ups of *Gemini 8* with the previously launched Agena target vehicle and a planned 2-hour 40-minute "space walk" by Scott. The flight, scheduled to last 71 hours, ended after less than 11 hours, and *Gemini 8* completed fewer than 7 of the planned 44 revolutions around the earth.

Gemini 8 had been launched from Cape Kennedy's Launch Complex 19 at 11:41 a.m. EST. It was carried aloft by a 2-stage Titan-2 launch vehicle that developed 430,000 pounds of thrust. The 26-foot-long Agena target vehicle had been launched from Cape Kennedy 101 minutes earlier—at 10 a.m.—by means of an Atlas rocket (with 2 outboard booster rockets) developing a total thrust of 390,000 pounds.

The Agena went into a circular orbit with an altitude of 185 miles. *Gemini 8* assumed a lower orbit with an initial apogee of 169 miles and perigee of 100 miles. Both orbits were inclined at angles of 28.87° to the equatorial plane. On entering orbit, *Gemini 8* was both lower than the Agena and 1,200 miles behind it.

During the next $6\frac{1}{2}$ hours, while the 2 space vehicles traveled about 105,000 miles each, Armstrong maneuvered *Gemini 8* closer and closer to the Agena. Armstrong, the first American civilian to achieve orbit, piloted the spaceship on the basis of calculations worked out by Scott by means of the onboard computer and with data provided by radio from ground stations.

Armstrong reported the target "in sight"—76 miles away— at 4:21 p.m. EST. He announced at 5:40 p.m. that "we are station keeping"—at a speed of 17,295 mph. and a distance of 150 feet. Slowly the distance was narrowed. Finally *Gemini 8*'s nose was eased into the opening in the Agena's forward end, and docking was completed at 6:16 p.m. with the snapping of the Agena's latches at an altitude of 185 miles over Brazil.

The malfunction in the maneuvering rocket apparently developed during the next 26 minutes. Armstrong found that he was unable to turn off the No. 8 thruster rocket, one of the 16 thrusters attached to *Gemini 8*. Although the defective rocket developed only 25 pounds of thrust, Armstrong reported that the 2 linked spacecraft were "toppling end over end" and that he had therefore "disengaged from the Agena." Armstrong then deactivated the entire maneuvering system in order to shut off the single faulty thruster, and he stabilized *Gemini 8* by firing the re-entry rockets. But the use of the re-entry fuel, coupled with the failure of the thruster, made it necessary to cut short the flight at the earliest practicable moment, and the astronauts were ordered down.

Using data provided by ground stations, the astronauts brought *Gemini 8* out of orbit and down into the Western Pacific about 600 miles east of Okinawa. Their fall slowed by parachute, they splashed down at 10:24 p.m. EST. Frogmen flown to the floating capsule were parachuted into the water. They attached a flotation collar to the spaceship to make sure it did not sink. *Gemini 8* and the 2 astronauts, both in excellent physical condition, were picked up by the U.S. destroyer *Leonard F. Mason* $3\frac{1}{2}$ hours after splash-down.

NASA (National Aeronautics & Space Administration) officials reported Mar. 19 that they had traced the thruster malfunction to a probable short circuit.

The Agena target remained in orbit. It carried bacteria and penicillium mold for later recovery (by a Gemini "space-walking" crewman) and equipment to determine the effects of cosmic rays. NASA officials started up the Agena's rocket engine Mar. 17 by radio signal from the ground; they then made 2 changes in the Agena's orbit and put it into a stable circular orbit at an altitude of 253 miles.

The *Gemini 8* flight had been delayed for one day (it had been scheduled for Mar. 14) because of an oxygen leak in the capsule and an overflowing fuel tank in Agena's Atlas booster. Both troubles were cleared up satisfactorily.

(Charles W. Mathews, manager of NASA's Gemini program office in Houston, Tex., had disclosed Feb. 23 that NASA had discontinued development work on a system for landing Gemini and other spacecraft on solid ground instead of on water. NASA had already spent more than $30 million on the canceled program. Most of the money had been used for a bat-wing "paraglider.")

Gemini 9 Fails in Docking Attempt

The U.S.' *Gemini 9*, with Air Force Lt. Col. Thomas Patten Stafford, 35, and Navy Lt. Cmndr. Eugene Andrew Cernan, 32, aboard, was shot into orbit June 3 by means of a 2-stage Titan-2 booster rocket launched from Cape Kennedy. During the following 3 days *Gemini 9* achieved close rendezvous 3 times with an orbiting target vehicle, and 2d pilot Cernan spent a record 2 hours 9 minutes in airless space outside the capsule. But plans to dock with the target were abandoned because a fouled protective shroud obstructed the target's docking apparatus; and plans for Cernan to fly independently in space with a Buck Rogers-like back pack were also cancelled because his helmet visor became fogged.

As *Gemini 9* was completing its 44th revolution around the earth, command pilot Stafford brought it out of orbit to a safe landing in the Atlantic 345 miles east of Cape Kennedy June 6. Both astronauts, brought quickly aboard the waiting carrier *Wasp*, were found to be in excellent physical condition after 72 hours 21 minutes in space. Pres. Johnson phoned the 2 astronauts aboard the *Wasp* June 6 to congratulate them.

This was Stafford's 2d time in orbit, Cernan's first. Cernan was the 3d person to step out of an orbiting capsule (he had been preceded by a Russian and an American).

The *Gemini 9* launching, originally scheduled for May 17, had been delayed by technical troubles. *Gemini 9*'s target, as originally planned, was to have been a maneuverable Agena rocket, which was to have been sent aloft about $1\frac{1}{2}$ hours before the manned capsule went up. With Stafford and Cernan waiting aboard *Gemini 9* on Cape Kennedy's Launch Pad 19, the Agena was launched from Launch Pad 14 at 11:15 a.m. May 17 atop an Atlas booster rocket. But one of the Atlas' 3 engines malfunctioned, and the Atlas and Agena rockets crashed separately into the Atlantic 185 miles or more from the launching site. Stafford and Cernan were then ordered out of the *Gemini 9*, and their launching was postponed—tentatively to May 31.

The loss of the target rocket resulted not only in the delaying of the *Gemini 9* mission but in a decision to use a non-maneuverable substitute target since a back-up Agena was not available. The substitute, called an ATDA (augmented target docking adapter), had been developed as an alternate target for the *Gemini 8* mission or for some subsequent mission in which an Agena might not be available. The ATDA, put together from existing equipment, was an 11-foot cylinder that weighed about 2,400 pounds at launch and about 1,700 after achieving orbit. A new 24-hour postponement was announced May 27 because more time was needed to prepare and test the replacement Atlas booster for launching the ATDA.

Stafford and Cernan were back aboard *Gemini 9* June 1 when the Atlas was launched from Pad 14 at 11 a.m. EDT. The Atlas, developing

390,000 pounds of thrust, put the ATDA into a circular orbit about 185 miles high and inclined at an angle of 28.87° to the equatorial plane.

Shortly thereafter—within 3 minutes of *Gemini 9*'s scheduled liftoff—electronic trouble forced a new delay in *Gemini 9*'s launching, and the 2 astronauts were again ordered to quit the spaceship. NASA technicians discovered at about the same time that the shroud protecting the ATDA's docking apparatus had apparently failed to fall away from the target and therefore would probably block the physical joining of *Gemini 9* with the ATDA in orbit. They also discovered that 6 unidentified objects were following the ATDA and the burned-out Atlas in their separate orbits.

Gemini 9 was finally launched successfully June 3 at 9:39 a.m. EDT. The Titan-2 rocket, developing 430,000 pounds of thrust, put the spacecraft into an orbit with an apogee of 174 miles, a perigee of 99 miles and a 27.87° angle of inclination.

When *Gemini 9* entered orbit, it was lower than the ATDA target vehicle and some 620 miles behind it. Stafford adjusted *Gemini 9*'s orbit twice by using the spaceship's thruster rockets as he maneuvered to catch up with the target. By *Gemini 9*'s 3d revolution around the earth he had closed the distance to about 25 feet.

The astronauts were then able to see that the shroud was blocking the ATDA docking cone into which they had planned to thrust *Gemini 9*'s nose for docking. Ground controllers rejected suggestions that *Gemini 9* "bump" the shroud loose or that Cernan try to free it when he left the spaceship for his EVA (extra-vehicular activity) experiment.

The astronauts June 3-4 executed 2 more difficult rendezvous exercises—an equiperoid rendezvous using only on-board equipment and a rendezvous-from-above designed to simulate a lunar rendezvous that could take place if a lunar excursion module had to be retrieved after it had started to descend to the moon. The maneuvers ended up with *Gemini 9* in an orbit with a 194-mile apogee and 183-mile perigee. The rendezvous exercises were so difficult and exhausting that the astronauts decided to postpone Cernan's EVA for a day.

The astronauts began preparing for Cernan's space walk at about 8 a.m. EDT June 5. Cernan donned a chest pack and attached the 25-foot "umbilical cord" that carried his communication lines and through which he was to receive oxygen. With both astronauts sealed up in their spacesuits and with their helmets closed, Stafford depressurized the cabin.

Cernan opened the hatch at 11:02 a.m. EDT and stood up on his seat. Slowly he pulled himself out of the capsule and retrieved a micrometeorite detector attached to the outside of the spaceship. Cernan attached a rear-view mirror and 16-mm. camera to the front of the capsule and deployed a handrail for use in moving around the capsule. He rested briefly and then allowed himself to drift away to the full length of the umbilical cord, noting that this maneuver made him float upwards. Cernan experimented with Velcro hand pads that were supposed to engage Velcro handholds on the outside of the spaceship and make it easy for him to move around, but he found the pads ineffective.

Later Cernan tried to don the 166-pound AMU (astronaut maneuvering unit), a back-pack device designed to make it possible for an astronaut to move independently outside his spaceship with his own oxygen supply, propulsion jets and radio. The AMU was attached to the rear of the capsule. It had been planned that Cernan would put on the AMU and then propel himself 150 feet from the spaceship to the end of a 125-foot tether that he was to attach to the 25-foot umbilical cord. Cernan, however, found that detaching the AMU and getting into it in space was more difficult and exhausting than he had found it to be in practice on the ground. He also reported that his visor was fogging up and blinding him, and Stafford discovered that Cernan's voice through the AMU radio transmitter was virtually unintelligible. As a result of these difficulties, Stafford decided against using the AMU and cut short the EVA.

Cernan retrieved his camera, reentered the capsule and closed the hatch at 1:11 p.m.—2 hours 9 minutes after he had opened it but 17 minutes sooner than initially planned.

Stafford began the reentry procedure June 6 while *Gemini 9* was 165 miles above the Pacific on its 44th revolution around the earth. The blunt end of the capsule was turned in the direction of flight, and Cernan activated switches to jettison the rear equipment compartment— including the AMU back pack. The retro (braking) rockets were fired automatically at 9:26 a.m. EDT to reduce the capsule's 17,500-mph. speed, and *Gemini 9* began to drop out of orbit.

Eased downward by parachute, the spaceship came down in the Atlantic at 10 a.m. It fell less than 3 miles from the waiting carrier *Wasp* and less than $\frac{1}{2}$ mile from the target point. The landing set a record for accuracy in descending from orbit, and TV cameras aboard the *Wasp* recorded the descent for immediate broadcast by means of the communications satellite Early Bird. Navy frogmen dropped from helicopters fastened a flotation collar around the capsule to keep it from sinking, and shortly thereafter the spaceship with the 2 astronauts still in it was hauled aboard the *Wasp*.

Gemini 10's Complex Mission

The 8th manned mission in the Gemini series began July 18 when *Gemini 10* was launched from Cape Kennedy with Navy Cmndr. John Watts Young, 35, and Air Force Maj. Michael Collins, 35, aboard. Its mission was one of the most complex flown in the U.S. manned space flight program so far.

Gemini 10 achieved rendezvous with 2 previously launched Agena target rockets and docked with one of them. The astronauts used the motors of one of the Agenas to boost *Gemini 10* into a higher orbit with a 475-mile apogee. During the flight, co-pilot Collins left *Gemini 10* in his space suit and retrieved a meteoroid detection box from one of the Agenas.

Command pilot Young brought *Gemini 10* out of orbit to a safe splash-down in the Atlantic July 21 for pick-up by the helicopter carrier *U.S.S. Guadalcanal*. Medical checks indicated that the astronauts had suffered no ill-effects from their 70 hours 47 minutes in space.

Several "firsts" were claimed by the *Gemini 10* mission. Among them: The first time a spaceship achieved rendezvous with 2 different vehicles; the first time a spaceship linked up with a 2d vehicle in space and used the latter's engines for propulsion; the first time men flew as high as 475 miles above the earth's surface; the first time a man left a spaceship and made physical contact with another body in space; the first time a man stepped out of a capsule into space twice.

The *Gemini 10* mission started at 4:40 p.m. EDT July 18 when a 26-foot-long Agena target vehicle was launched from Cape Kennedy's Launch Complex 14. The Agena, sent aloft in the nose of an Atlas booster rocket, went into an orbit with an apogee of 190 miles, perigee of 183 miles and 28.87° angle of inclination. Then *Gemini 10* was launched from Launch Complex 19 at 6:20 p.m. It was carried up by a modified 2-stage Air Force Titan-2 rocket. Its initial orbit had an apogee of 167 miles and perigee of 100 miles.

On injection into orbit, *Gemini 10* was some 1,160 miles behind its target. But by means of the small maneuvering rockets, Young brought *Gemini 10* to within sight of the Agena target by 10:45 p.m., and the 2 spaceships were flying in close formation by midnight. Shortly thereafter, at 12:18 a.m. July 19, *Gemini 10*'s nose was thrust into the Agena's docking collar; after a chase of 103,000 miles, the 2 space vehicles were joined. The maneuver, however, used up more fuel than had been planned; only 380 pounds of the original 940 pounds of monomethyl-hydrozine fuel remained. 2 additional planned docking exercises were "scrubbed," therefore, to guard against any possibility of the fuel supply becoming dangerously low.

Using the Agena's motors, the astronauts put the linked *Gemini 10*-Agena assembly into a higher orbit with a 475-mile apogee and 184-mile perigee. The 2d Agena, which had been launched Mar. 16 as a target for *Gemini 8*, was in an orbit with an apogee of 250 miles and perigee of 247 miles. To prepare for rendezvous, Young began moving *Gemini 10* down to a lower orbit at 2:46 p.m. July 19 by using the engines of the attached Agena. The new orbit, a near-circular one, was about 245 miles high.

Later the afternoon of July 19, as both astronauts sat in their space suits, air was let out of the cabin. At 5:47 p.m. Collins opened his hatch, hauled the upper part of his body out of the spaceship, and started taking a series of photos. He and Young, however, soon found their eyes watering because of pungent fumes in their helmets. The irritating gas forced them to cut Collins' EVA (extra-vehicular activity) short after about a half-hour, and Collins climbed back into the cabin.

Young cut *Gemini 10* loose from the attached Agena the afternoon of July 20 before bringing *Gemini 10* to rendezvous with the 2d Agena. The astronauts again depressurized the cabin.

Then, while Young kept *Gemini 10* within feet of the 2d Agena, space suit-clad Collins opened his hatch at 7:02 p.m. July 20 and stepped out into space at the end of a 50-foot "umbilical cord" attached to his spaceship. By use of a hand-held nitrogen-jet gun, Collins maneuvered himself over to the Agena. He removed the Agena's micrometeoroid package but could not attach a new one because part of the Agena's nose was loose and dangling dangerously.

Because of the shortage of maneuvering fuel, the astronauts eliminated half an hour of EVA experiments; Collins returned to the capsule and shut the hatch at 7:40 p.m. While he was out of the cabin, Collins lost his $470 Hasselblad camera, which drifted away in space.

Leaving both Agenas in orbit, Young fired *Gemini 10*'s 4 retrorockets at 4:31 p.m. EDT July 21 over the South Pacific to bring the spaceship out of orbit. Eased down by parachute, it splashed into the Atlantic at 5:07 p.m. about 500 miles east of Cape Kennedy and less than 3 miles from the recovery carrier *Guadalcanal*. Navy frogmen, brought by helicopter, quickly secured a flotation collar to the capsule to keep it from sinking, and the astronauts were flown to the *Guadalcanal* by helicopter.

This was the 2d space flight for Young, who had been co-pilot of *Gemini 3*. It was Collins' first space flight.

Gemini 11 Sets Records

Navy Cmndr. Charles (Pete) Conrad Jr., 36, and Lt. Cmndr. Richard Francis Gordon Jr., 36, achieved an unprecedented first-orbit "docking" with a target vehicle in space Sept. 12 and then flew their spaceship to a record altitude of 849.85 miles Sept. 14. Both feats were achieved during a record-making flight in *Gemini 11* Sept. 12-15. The capsule was brought down safely Sept. 15 after 44 revolutions around the earth, and neither astronaut suffered any physical harm from their 71 hours 17 minutes in space flight.

The mission had started at 9:05 a.m. EDT Sept. 12 when an Atlas rocket was sent up from Cape Kennedy's Launch Complex 14 with an Agena-D target vehicle. The 26-foot Agena went into a near circular orbit about 185 miles high and inclined at an angle of 28.87° to the equatorial plane.

Gemini 11 went aloft 97 minutes $24\frac{1}{2}$ seconds later in the nose of a 2-stage Titan-2 rocket launched from Launch Complex 19 at 10:42 a.m. *Gemini 11* achieved an initial orbit with an apogee of 174 miles, perigee of 100 miles and angle of inclination identical to that of the Agena's orbit.

Gemini 11 entered orbit about 250 miles behind its higher-flying Agena target. Speeding after the Agena at a velocity of about 17,500 mph., the 2 astronauts used their on-board computer to calculate the 3 maneuvers that brought them to within 50 feet of the Agena at 12:08 p.m. and enabled command pilot Conrad to thrust the nose of *Gemini 11* into the Agena's docking collar at 12:16 p.m. *Gemini 11* thus set a record by docking 94 minutes after lift-off, and it became the first spaceship to link up with a target in space during its first revolution around the earth.

After the first-orbit docking, Conrad detached *Gemini 11* from the target and began additional docking practice. He docked a 2d time, and co-pilot Gordon, in the right-hand seat, also docked twice. Gordon thus became the first co-pilot to dock in space. After the 4th docking, the Agena was left attached to *Gemini 11*.

107 minutes of EVA (extra-vehicular activity) had been scheduled for Gordon for Sept. 13, but the EVA was cut short after 44 minutes because Gordon became tired very fast and because his right eye was blinded by sweat.

The Sept. 13 EVA began during the 15th revolution after the space-suited astronauts depressurized the cabin. Gordon, wearing a chest pack that carried emergency oxygen and equipment to clear carbon dioxide and moisture from his space suit, opened his hatch door at 10:40 a.m. EDT. He then had to untangle himself from his life-support "umbilical cord" before he could climb out.

Linked to the spaceship by the 30-foot umbilical cord, Gordon set up a 16-mm. movie camera to follow his movements, and he retrieved a micrometeorite recording device from the spaceship's outer surface. He then moved to the Agena and uncoiled a 100-foot dacron tether that had been rolled up in a compartment in the Agena. Straddling the nose of the Agena, Gordon leaned over and attached the free end of the tether to *Gemini 11*'s nose. The other end remained attached to the Agena. Gordon reported that he was tired, and Conrad directed him to rest there.

Because of Gordon's exhaustion and his sweat-blinded eye, Conrad "scrubbed" an experiment calling for Gordon to simulate working with a power wrench especially designed for use under zero-gravity conditions. Instead, Gordon was directed to return to the spaceship, and the hatch was fastened at 11:28 a.m. About an hour later Gordon opened the hatch for 5 minutes to jettison the chest pack, umbilical cord and other items no longer needed.

The 2 astronauts Sept. 13 spotted and photographed, from a distance of 279 miles, a satellite later identified as the USSR's *Proton 3*.

The ascent to a record altitude, accomplished by using the Agena's rocket power, was begun at 3:12 a.m. Sept. 14, when Conrad positioned the linked Agena-*Gemini 11* assembly by firing 2 of the Agena's small maneuvering rockets. The Agena's main 16,000-pound-thrust engine was then fired for 23 seconds, and the 2 joined space vehicles rose to an orbit with an apogee of 849.85 miles and a perigee of about 175 miles. 2 revolutions around the earth were made in this record-high orbit. The Agena engine was then fired again in retro position, and the linked vehicles dropped to an orbit with an apogee of some 190 miles and perigee of 180 miles.

2 hours later, with the cabin again depressurized and the 2 astronauts breathing oxygen fed into their helmets, Gordon opened the hatch and leaned out for a 128-minute session of scientific photography.

At 12:30 p.m. Sept. 14 *Gemini 11* began a successful experiment in tandem flying by means of the 100-foot dacron tether joining the spaceship with the Agena. Conrad first "undocked"–detached *Gemini 11*'s nose from the Agena's docking device. He then used *Gemini 11*'s thrusters to move the 2 spaceships apart. Conrad reported, however, that the tether was acting like "a skip rope"–"rotating and making a big loop"–and that the 2 vehicles were alternately drawing together and then separating until brought to a jerking stop when the tether was stretched taut. *Gemini 11*'s rockets were therefore used to put the 2 linked vehicles into a gentle rotation, which created enough centrifugal force to keep them apart to the full extent of the tether. This slow movement, at the rate of one complete rotation in 9 minutes, produced the effect of a slight artificial gravity (estimated at .00015 that of the earth's gravity) in the 2 vehicles. As a result, loose objects in *Gemini 11* would move slowly to the spaceship's blunt end—the end furthest from the tether—instead of floating almost motionlessly in the cabin. The chief purpose of the tether experiment was to show whether a tether was a good method of keeping spaceships in formation without using fuel.

After circling the earth twice in tethered formation, the 2 vehicles were separated when Conrad, by closing a switch, détached the tether from *Gemini 11*'s nose.

An extra rendezvous was ordered Sept. 15, and the astronauts successfully brought *Gemini 11* from a distance of about 25 miles to a distance of about 40 feet from the Agena. The Agena was then left in orbit.

Re-entry activity started at 9:21 a.m. Sept. 15 when the astronauts jettisoned the equipment section and thus exposed the retro-rockets. The on-board computer was then given the job of bringing the capsule safely back into the atmosphere and down to the target area in the Atlantic where the *U.S.S. Guam* was waiting to recover it. This was the first automatic computer-piloted landing in the U.S.' manned space program, and it was executed successfully. The astronauts monitored the computer as it started the retro-rocket sequence at 9:24 a.m. and continued the landing procedure.

Eased down by parachute, *Gemini 11* splashed into the Atlantic at 10 a.m., 700 miles southeast of Cape Kennedy, less than 2 miles from the aiming point and 3 miles from the *Guam*. The astronauts, both in fine physical condition, and the capsule were quickly brought aboard the ship.

The mission was the first space flight for Gordon and the 2d for Conrad, who had been co-pilot of *Gemini 5*.

Gemini 12 Mission Ends Series

Gemini 12 circled the earth 59 times Nov. 11-15 in a successful conclusion to the Gemini manned space-flight program. 10 2-man crews had gone safely into orbit in 20 months in the $1.4 billion Gemini project.

Records for individual space flight were set by both of *Gemini 12*'s astronauts during their 4-day flight: Navy Capt. James Arthur Lovell Jr., 38, the command pilot, raised his total time in space to a record 425 hours 8 minutes, or more than $17\frac{1}{2}$ days (he had spent 330 hours 35 minutes in space as co-pilot of *Gemini 7* in Dec. 1965); and Air Force Maj. Edwin Eugene (Buzz) Aldrin Jr., 36, the co-pilot, making his first space flight, set a record total of 5 hours 32 minutes for EVA (extravehicular activity).

The *Gemini 12* mission had started at 2:08 p.m. EST Nov. 11 when an Agena-D target rocket was launched from Cape Kennedy's Launch Pad 14. The Agena, to be used for rendezvous and docking practice, went up in the nose of a $1\frac{1}{2}$-stage Atlas launching rocket powered by 2 outboard booster engines and a sustainer engine. The Agena achieved a nearly circular orbit with an apogee of 188 miles and perigee of 184 miles.

Gemini 12, with Lovell and Aldrin aboard, was launched at 3:47 p.m. Nov. 11 in the nose of a 2-stage Titan-2 booster rocket fired from Launch Pad 19. *Gemini 12*, sent up as the target Agena passed overhead at the completion of its first revolution around the earth, went into an initial orbit with an apogee of 168 miles and perigee of 100 miles. (The launching had been postponed twice—Nov. 8 and Nov. 9—because of technical troubles in the guidance system.)

Gemini 12 started off about 575 miles behind the Agena. After it had achieved orbit, command pilot Lovell fired 3 posigrade blasts of the maneuvering rockets to close the distance. The astronauts were hampered in their rendezvous attempt by the failure of the Agena to return their radar signals. By 6:40 p.m., however, they reported that they could see the target.

The first docking—in which *Gemini 12*'s nose was thrust into the Agena's docking collar—took place about an hour later. The astronauts then practiced docking and undocking, completed a total of 4 successful dockings and kept their spaceship attached to the Agena overnight. But a scheduled plan to use the Agena's engine for a flight to a 460-mile-high orbit was "scrubbed" because of a faulty Agena fuel pump.

Aldrin's first EVA, a record 2 hours 28 minutes in duration, took place between 11:16 a.m. and 1:44 p.m. EST Nov. 12. It started after the 2 astronauts had closed their space helmets and depressurized the cabin. Aldrin then opened his hatch, stood up and thrust his head and shoulders outside. *Gemini 12* circled the earth twice during this EVA, and Aldrin worked without fatigue. During the 2 "night-time" periods, Aldrin took photos of the ultra-violet emanations from predetermined young "hot" stars for use by astronomers in research on the creation of such stars. He also photographed the U.S., he retrieved an experiment package (containing a meteoroid collection box and a container of bacteria and molds) from the outer surface of the spaceship, and he attached to the outside a telescoping handrail for use in his next EVA.

Aldrin made a true "space walk" of 2 hours 9 minutes between 10:33 a.m. and 12:42 p.m. Nov. 13. He performed all the tasks and practice work assigned to him without exhaustion or excessive perspiration (fatigue and blinding perspiration had cut short EVAs of his predecessors). His success was attributed to periodic 2-minute rests—which he took even when he felt no need for them—and to the use of handholds, foot restraints and waist-lines to keep him close to his work without physical effort.

Attached to the spacecraft by a 25-foot umbilical cord through which he received oxygen and communicated with his command pilot, Aldrin emerged for his Nov. 13 space walk as *Gemini 12* hurtled over the Pacific Ocean on its 27th revolution around the earth. He asked for and received permission to release 2 pennants reading "November 11, Vets Day" and "Go Army, Beat Navy." Aldrin moved hand-over-hand along his handrail to the spaceship's nose, which was still latched to the Agena target, and fastened his waist-lines—one to the end of his handrail, the other to the Agena. Then he took a 100-foot tether that was linked to *Gemini 12*'s nose and attached the other end of it to the Agena's docking apparatus. Unfastening his waist-lines, Aldrin moved back across his spaceship, set up 2 cameras and used Velcro pads to bring himself hand-over-hand to the back of the spaceship. He fastened his waist-lines there and put his feet—with some difficulty—into a pair of "foot restraints" attached to the vehicle. These aids helped make it possible for him to perform 17 specified tasks—attaching and detaching electrical plugs, cables, bolts and other items. Before reentering the cabin Aldrin examined 2 of *Gemini 12*'s 8 thruster rockets that had failed to operate, took photos, wiped the capsule's windows and threw away his handrail, waist-lines and other items no longer needed. (The inoperative thrusters posed no hazard to the mission.)

Aldrin's 3d and final EVA was a 55-minute stand-up exercise between 9:55 and 10:50 a.m. Nov. 14. He thrust his head and shoulders out of his hatch, performed some light calisthenics, took photos and threw overboard additional unneeded equipment.

The photography mission provided the first photographs of an eclipse ever taken from above the atmosphere. A total eclipse of the sun took place Nov. 12, and *Gemini 12* was in direct line with the eclipse for 7 seconds at 7:49 a.m. Before the eclipse took place, Lovell had used his maneuvering rockets to point the spaceship's nose toward the sun. An opaque screen was placed over Aldrin's window and a polarized filter over Lovell's. Lovell took the eclipse photos with a 16-mm. Maurer movie camera.

An exercise in "tether flying" was attempted with limited success Nov. 13. For this task the astronauts used the 100-foot tether with which Aldrin had fastened the Agena to *Gemini 12*. The exercise consisted of flying in orbit with the tether extended full length between the 2 vehicles. When done right, the pull of gravity stabilized the 2 spacecraft and there was little need to use rocket fuel for stabilization. Because of the 2 inoperative thruster engines (a 3d became inoperative later), Lovell had difficulty in getting the tether sufficiently taut, and it took 3 hours before the exercise was performed satisfactorily.

At 7:48 p.m. Nov. 13 Lovell fired an explosive bolt that released the tether. Shortly thereafter he fired a maneuvering rocket to put *Gemini 12* into an orbit slightly above the Agena's. The 2 vehicles then quickly separated.

During 2 passes over the Sahara Nov. 14 the astronauts turned their cameras in the direction of 2 Centaure rockets launched from the French space center at Hammaguir, Algeria. On each of the 2 French shots the rockets released sodium clouds at altitudes of 38 to 111 miles. The astronauts took photos of the areas the clouds were supposed to have been in, but they said they had been unable to see the clouds either time.

The return to earth Nov. 15 was managed by the on-board computer, which handled both guidance and rocket firing. The retrorockets were fired at 1:47 p.m. EST to bring the capsule out of orbit. Eased down by parachute, *Gemini 12* splashed into the Atlantic at 2:21 p.m., only 3.85 miles northeast of its aiming point and about 3 miles from the carrier *Wasp*, which was waiting some 700 miles southeast of Cape Kennedy.

The 2 astronauts were picked up by helicopter and brought aboard the *Wasp* less than $\frac{1}{2}$ hour after splash-down. Initial medical exams showed them both to be in excellent shape after 94 hours 33 minutes in space and a trip of about 1,600,000 miles.

Budget Cut Threatens Moon Landing

NASA Director James E. Webb told the House Science & Astronautics Committee Mar. 10 that the tight $5.012 billion budget imposed on NASA for fiscal 1967 might be sufficient to put an American on the moon by the target date of 1970—or even by 1969—but that it "provides no margins of time or of resources to counter the effects of setbacks or failures."

NASA had requested $5.6 billion for fiscal 1967, but nearly $600 million had been pared from this request by the Budget Bureau. "We are as much as 2 years behind the Soviet Union in certain aspects of space power," Webb testified, and the $5.012 billion budget "will not close" this gap. He said there was "more chance than I thought a year ago" that the Russians would make a manned landing on the moon "before 1969." The Soviet space program advanced "very much more rapidly" in 1965 than the U.S. program did, Webb declared. He added that NASA had to plan post-lunar landing goals and programs soon or lose momentum and fall behind in space.

Webb repeated the substance of this testimony May 25 when he presented NASA's budget arguments at a hearing of a Senate Appropriations subcommittee.

Expanding on his assertion that post-lunar landing planning must be started soon, Webb said in an interview published in the *N.Y. Times* May 30 that (a) the $20 billion Apollo program was already beginning to slow down; (b) no Apollo capsules or Saturn-1B or Saturn-5 rockets had been ordered for missions to follow the manned lunar landing; (c) 40,000 to 60,000 Apollo employes would leave the program in fiscal 1967, and those going included men needed to modify the capsules and rockets for future assignments. Webb said that Apollo hardware could be used, after minor changes, to put men into 45-to-90-day orbits and to carry large scientific payloads into orbit or to nearby planets.

LUNAR & INTERPLANETARY PROBES

Soviet & U.S. Craft 'Soft Land' on Moon

The unmanned Soviet lunar probe *Luna 9* Feb. 3 became the first man-made object to land on the moon without being destroyed. Subsequently, the U.S. achieved a "soft-landing" with an instrument package in May, and another Soviet probe soft-landed in December. Both the U.S. and the Soviets placed satellites in moon orbit, with the Soviets Mar. 31 having the first success.

Luna 9, launched Jan. 31, made its successful soft landing on the moon at 9:45 p.m. Moscow time Feb. 3. The 3,482-pound vehicle transmitted TV photos of the lunar landscape that were picked up by Britain's Jodrell Bank radio-telescope as well as by Soviet receiving stations. The first photos, of excellent quality, were released by the Jodrell Bank observatory to the world's press and TV broadcasters

Feb. 4 as soon as they were processed. The Soviet press and TV, however, received no photos from Soviet authorities until Feb. 5; Soviet citizens, therefore, did not see the photos taken by the Soviet probe until one day after they were shown throughout the rest of the world.

Sir Bernard Lovell, director of Jodrell Bank, asserted Feb. 4 that the photos disproved a theory that the moon was covered with loose dust to a depth of many feet. He said the photos showed instead that the lunar surface in the vicinity of *Luna 9* appeared "more like pumice stone." Lovell lauded the quality of the pictures, which showed what he called "small stones on the surface of the moon which may be less than one inch in diameter."

U.S. Pres. Johnson Feb. 3 sent Soviet Pres. Nikolai V. Podgorny a cablegram congratulating Podgorny and the Soviet people on *Luna 9*'s "great success." Similar messages were sent by other world leaders.

The *Luna 9* shot was the 5th acknowledged attempt by the USSR at a soft landing on the moon. There had been 4 known Soviet failures. U.S. sources, however, speculated that there may have been unacknowledged failures as well.

The first Soviet announcement of the launching was made by Tass Jan. 31 after *Luna 9* had headed into what appeared to be a successful trajectory toward the moon. The initial announcement said *Luna 9* was "moving toward the moon" with "all apparatuses . . . functioning normally," but it did not even hint that a soft landing would be tried.

Western observers said the probe had probably been launched by means of a multi-stage rocket from the Soviet cosmodrome at Baikonur, Kazakhstan and placed in "parking orbit" around the earth. At a calculated point in orbit, a rocket attached to the probe was ignited to accelerate it to the escape velocity of about 25,000 mph. and to send it on course toward the moon. A course-correction maneuver was made Feb. 1 to put *Luna 9* directly on target.

Retro-rockets, which acted as brakes, were ignited Feb. 3 when *Luna 9* was about 75 kilometers ($46\frac{1}{2}$ miles) from the moon's surface. The probe's speed was reduced "from about 2,600 meters [1.6 miles] a second to several meters a second" before impact 48 seconds later; the retro-rocket was detached just before the probe landed, its instruments protected by a shock absorbing system.

An hour after *Luna 9* ended its journey, Tass announced that the probe had "accomplished a soft landing on the surface of the moon in the area of the [dry] Ocean of Storms to the west of the craters Reiner and Marius." This was the first Soviet admission that the shot was a soft-landing attempt. Tass reported Feb. 5 that the lunar coordinates of the landing point were $7°$ 8 minutes N., $64°$ 22 minutes W. (The exact distance traveled by *Luna 9* could not be computed until Soviet space officials released the details of the probe's trajectory; the mean distance of the moon from the earth is 238,857 miles.)

The Jodrell Bank pictures were obtained by feeding the TV transmission signals from *Luna 9* into wirephoto equipment borrowed from the *London Daily Express.* Lovell, commenting Feb. 4 on the Soviet delay in publishing the photos, called it "slightly strange in that they have taken the trouble to announce the frequency and transmitted on standard scanning lines." He speculated that "it may possibly be an indirect indication that they would be interested in us obtaining photographic material with the high sensitivity of our [radio-] telescope." He added that "the Russians appear to have made it possible in every way for us to receive these pictures without ever asking for us to do so."

Soviet space authorities began releasing their *Luna 9* photos Feb. 5, and they simultaneously criticized the Jodrell Bank scientists for rushing to publish what they called distorted photos. Prof. Anatoly A. Blagonravov, chairman of the Soviet Commission for Exploration & Use of Outer Space, reported that the Jodrell Bank photos were $2\frac{1}{2}$ times too narrow. He said the British scientists should have released the photos only after applying "the information necessary for correct reproduction of the image." "Apparently some motives of a sensational nature played a role in this case," he said. Lovell replied at Jodrell Bank that the photos had too much international importance to be withheld. He said that he had "allowed for a wide margin for error" and had "stressed . . . that there was no indication of scale."

In a special Moscow TV program Feb. 5, Soviet scientists showed 2 of *Luna 9*'s photos and commented: "The surface [of the moon] in the vicinity of the station [*Luna 9*] is sufficiently hard [to support a spaceship] because the pictures show that the station did not sink into the soil to any substantial degree. No visible traces of dust are found on the lunar surface. In the pictures one can see details one to 2 millimeters [.04 to .08 inch] in size."

Soviet Academician Nikolai P. Barabashov, chairman of the Soviet Committee for the Study of the Physical Conditions of the Moon and chief of the Kharkhov University Observatory, said Feb. 6 that the *Luna 9* photos confirmed his 30-year-old theory that the moon's surface was "highly pitted, porous and probably covered with numerous rocks and fragments." He asserted that the photos "proved beyond doubt that the upper layer of lunar soil is a sponge-like rough-textured mass scattered with individual sharp-edged fragments of various sizes." "Simultaneously," he continued, "it has become clear that this layer is strong enough to support more or less heavy objects."

Dr. Gerard P. Kuiper, director of the Lunar & Planetary Laboratory at the University of Arizona and chief of the photo evaluation group of the U.S.' Ranger lunar probe program, had suggested Feb. 4 that "the Russian pictures show a lava field." He said: "You can also see the depressions made by impact of meteorites. It is of the consistency of crunchy snow that has been slightly pressed down."

Tass announced Feb. 6 that a final radio contact with *Luna 9* had been ended that evening at 8:41 p.m. Moscow time and that *Luna 9*'s research program "has been fulfilled and successfully concluded." But Jodrell Bank scientists, who had already received 10 photos from *Luna 9,* said they had picked up transmissions from the probe after the Tass-reported conclusion of the program. Tass explained Feb. 7 that a final radio session with *Luna 9* had been held for 2 hours beginning at 11:37 p.m. Moscow time Feb. 6 because the probe was found to have "power left in excess of the rated level."

U.S. *Surveyor 1* Soft Lands on Moon

The U.S.' TV-equipped *Surveyor 1* was launched toward the moon May 30 by a 2-stage Atlas-Centaur booster rocket fired from Cape Kennedy at 10:41 a.m. EDT. The unmanned probe traveled 231,483 miles in 63 hours 36 minutes and came down softly on the moon at 2:17 a.m. EDT June 2. It landed almost exactly on its designated target point in the dry Ocean of Storms.

Telemetry signals indicated that the spacecraft and its equipment had landed intact, and *Surveyor 1* began radioing back high-quality photos of the lunar surface that showed details as small as a fraction of an inch. Ultimately *Surveyor 1* sent back 11,237 closeup photos of its landing area.

Pres. Johnson lauded the Surveyor team June 2 on the achievement. "We can be as proud of the openness of our space program as we are of its successes," he declared. (As with all other U.S. civilian space shots, details of the planned lunar mission had been publicized months in advance of the launching, and every possible step of the experiment took place in view of the press and TV cameras.) "As the day approaches when men land on the moon," Johnson said, "it is of the greatest importance that we agree to exchange openly all information that could affect their safety and welfare. It is equally important that we preserve the regions [of space] for peaceful, scientific activities."

According to NASA scientists, initial examination of *Surveyor 1* photos and telemetry data indicated that the surface of the moon was firm enough to support a space vehicle and astronauts. The data apparently helped disprove theories that either (a) the moon was covered to a depth of several feet by fine dust or (b) its surface was a fragile froth, and in either case a spaceship or astronauts would sink in deeply and find exploration of the moon extremely difficult. 2 of the spacecraft's 3 landing feet were photographed, and they appeared to have made only shallow indentations in the lunar surface. In the vicinity of *Surveyor 1*, the surface seemed to be relatively flat and strewn with "rubble"—small and large rocks and "projections."

A carefully prepared test failed to show any indication of loose dust on the moon. A small gas jet on the lower end of one of *Surveyor 1*'s landing legs was fired 7 times June 4 in the direction of the lunar surface. Sharp photos of the landing foot and the surrounding lunar surface were made immediately before, after and during the experiment, but no cloud of gas or plume of disturbed dust was detected by the camera.

The launching May 30 had put *Surveyor 1* into a direct-ascent lunar trajectory. The parking-orbit technique was not used. The Atlas first stage, with its 3 main engines and 2 vernier (control) engines burning, provided 388,000 pounds of thrust for lift-off. It boosted the Atlas-Centaur-Surveyor combination to a velocity of 7,600 mph. and an altitude of 98 miles. It then separated from the assembly and fell back into the Atlantic. The Centaur stage, developing 30,000 pounds of thrust, accelerated the spacecraft to a speed of 23,500 mph., put it into lunar trajectory and then separated from the probe. (The Centaur stage remained in a highly ellipitical orbit around the earth with an initial apogee of more than 250,000 miles, a period of about 11 days and an inclination of about 33.6°.)

Surveyor 1, a 10-foot-high instrument-laden framework with 3 landing legs, had been folded up in the nose of the Centaur stage on launching. After separation from the Centaur, the outboard elements of the probe were deployed, but telemetry indicated that one of the 2 antennas had failed to unfold. As it sped away from the earth, *Surveyor 1*'s first maneuver after separation was to align its solar panel with the sun to obtain solar power. It did this by turning on small nitrogen gas jets that turned the entire probe in controlled roll and yaw movements. The spacecraft's next step was to establish communications with the ground control station at Johannesburg, South Africa. The final orientation maneuver was to lock the star sensor on the star Canopus as a fixed inertial reference during the trip to the moon.

The launching was so accurate that the probe would have landed only 215 miles southwest of the designated target even without a scheduled mid-course correction. NASA officials decided, however, not only to make a mid-course maneuver but to direct *Surveyor 1* to a slightly smoother landing point 9 miles away from the originally chosen point. The course-changing maneuvers took place between 2:30 a.m. and 4 a.m. EDT May 31. They were made in compliance with radioed orders from the Goldstone space communications center in the Mojave Desert in California. As ordered, the gas control jets turned the spacecraft until a liquid-fueled vernier rocket motor was aligned in a freshly-calculated direction. The firing of the motor caused the necessary adjustment in trajectory and decreased the speed by about 40 mph. The tiny control jets then turned the spacecraft back to its proper cruise attitude with the solar panel locked on the sun and the star sensor locked on Canopus.

The landing sequence started when *Surveyor 1* was about 1,000 miles from the moon. Discarding its locks on the sun and Canopus, the probe turned until its main retro-rocket faced the direction of flight. It continued its approach to the moon under the direction of its "closed-loop system," which developed altitude, attitude, flightpath and velocity data, fed the data into a computer and used the computer-produced result to control the braking and attitude rockets.

When the spacecraft was about 200 miles from the moon, a radio signal from the ground turned on the attitude-marking radar. At a slant range of about 60 miles from the surface, the marking radar turned on the flight control programmer, which ignited the 3 liquid-fueled vernier engines and the solid-propellant main retro-rocket. The main retro, burning out in about 40 seconds, reduced the speed of *Surveyor 1* from about 6,000 mph. to about 250 mph. at an altitude of about 25 miles; the main retro then was separated from the probe.

A few seconds later the RADVS (radar altimeter and doppler velocity sensor) took over, bouncing signals off the surface to provide data that was computer processed to control the vernier engines. The verniers slowed *Surveyor 1* further until they were cut off at an altitude of about 14 feet. The spacecraft then dropped to the moon's surface at a speed of 8 mph. (The balky antenna apparently deployed properly with the jar of landing.)

Surveyor 1 settled down upright on its 3 landing legs just south of the moon's equator in a position designated as $2°\ 49'$ S. and $43°\ 32'$ W. The landing point, 60 miles north of the crater Flamsteed, was 10 miles from the original target.

Surveyor 1, complete with retro-rockets and fuel, had weighed 2,194 pounds on launching. By the time it had used up most of its fuel and jettisoned its retro-rocket before landing, the probe's weight was down to 620 pounds (earth weight).

LUNAR & INTERPLANETARY PROBES

(In a preparatory shot Apr. 7, an Atlas-Centaur assembly carrying a dummy Surveyor was launched from Cape Kennedy at 8 p.m. EST into a 100-mile-high parking orbit. The Centaur stage, however, failed to complete a planned reignition; the Surveyor model, therefore, remained in the initial orbit instead of being sent out on a transfer trajectory toward a hypothetical landing on an "imaginary moon" about 236,000 miles out in space. The main purpose of the shot was to test the Centaur's ability to restart its hydrogen-fuel engines in orbit.)

A 2d soft-landing attempt by the U.S. failed when *Surveyor 2* crashed on the moon's surface Sept. 23. The 2,204-pound unmanned *Surveyor 2* had been sent up from Cape Kennedy's Launch Complex 36 at 8:32 a.m. EDT Sept. 20. The launching, by means of a 2-stage Atlas-Centaur, put the probe into a near-perfect trajectory toward the moon. But the failure of one of *Surveyor 2*'s 3 vernier engines during a delicate mid-course correction Sept. 21 caused the probe to tumble out of control. As the probe approached the moon Sept. 22, flight controllers at the Jet Propulsion Laboratory in Pasadena, Calif. sought to stabilize it by sending a radio signal to fire the probe's main braking rocket. This maneuver failed to save the $65 million probe, however, and *Surveyor 2* crashed into the moon southeast of the crater Copernicus at about 11:18 p.m. EDT Sept. 23.

The TV-equipped *Surveyor 2* had been assigned a difficult mission—a tricky 23° approach to the surface of the Sinus Medii (Central Bay), a potential manned landing site, and the transmission of photos of the immediate vicinity of the landing area.

USSR's 2d Soft Landing on Moon

Before 1966 had ended, a 2d Soviet unmanned spacecraft had made a soft landing on the moon. The probe, dubbed *Luna 13,* was launched from a Soviet site toward the moon at 1:17 p.m. Moscow time Dec. 21. After a radio-controlled mid-course correction at 9:41 p.m. Moscow time Dec. 22, the probe made its soft landing at 9:01 p.m. Dec. 24 on the surface of the moon's arid Ocean of Storms. Tass said the selenographic coordinates of the landing point were Latitude 18° 52′ N. and Longitude 62° 3′ W.

Tass reported that in preparation for the landing, *Luna 13*'s retrorockets had been switched on at an altitude of 70 kilometers ($43\frac{1}{2}$ miles) from the lunar surface. The probe landed 15 minutes later. Like *Luna 9, Luna 13* ejected its instrument capsule just before touchdown, and the capsule landed near the rocket carrier.

Tass reported that *Luna 13*, on radio command from the earth, had started transmitting a good-quality "lunar panorama" at 3:15 p.m. Moscow time Dec. 25. This was the first disclosure that the probe was equipped for photography. (Similarly, the USSR had withheld the information that a lunar launching was planned until *Luna 13* had successfully achieved a lunar trajectory.) Tass said that the probe was also transmitting "scientific data and the readings of sensors characterizing the functioning of its systems and equipment." 2 of the lunar photos were shown on Moscow TV the same day. Discussing the photos, Tass said that "there is no dust visible either on the ground or on parts of the station [the probe]."

Tass reported Dec. 26 that *Luna 13* was "testing the firmness and density of the lunar soil by its mechanical manipulators." Disclosing the probe's possession of such "manipulators" for the first time, Tass quoted Soviet lunar scientist Aleksandr Lebedinsky as saying: *"Luna 13* is equipped with instruments that can feel the moon. This set of equipment is a sort of miniature laboratory. One of the instruments ... [a device] on a long arm outside the station, forced a test rod into the soil .... The depth to which it penetrated ... will serve as an indication of the firmness of the surface layer." He added that the probe measured the density of the same layer by testing its resistance to radiation.

Luna 10 Orbits Moon

The first man-made object to achieve a selenocentric (near-lunar) orbit was the Soviets' *Luna 10,* which was launched from Soviet territory at 1:47 p.m. Moscow time Mar. 31 and put into orbit around the moon at 9:44 p.m. Apr. 3.

As usual with Soviet space projects, the planned launching was not disclosed until *Luna 10* had first been put into parking orbit around the earth and then had been sent off into what appeared to be a successful trajectory toward the moon. The attempt at a lunar orbit, however, was revealed in the first announcement of the launching. This first announcement was made about 5 hours after the launching but still 3 days before a lunar orbit was achieved. The announcement, made by Tass, said: "The main purpose ... is to test a system ensuring the setting up of an artificial moon satellite with the object of exploring near-lunar outer space and also testing the systems installed on board for putting the station [*Luna 10*] in a selenocentric orbit."

Tass disclosed that the lunar orbit had an initial apocynthion (maximum distance from the lunar surface) of about 632 miles, a pericynthion (minimum distance from the lunar surface) of 217 miles, a period of 2 hours 58 minutes 15 seconds and a 72° angle of inclination to the lunar equator.

It was reported that *Luna 10* weighed 1,600 kilograms (about 3,530 pounds) and consisted of "2 main parts: a man-made moon satellite ... and the propulsion facilities with instrument compartments"; after lunar orbit was achieved, the satellite separated from the "propulsion facilities" and began "to conduct scientific exploration." Tass said that all of *Luna 10*'s instruments worked well, that the satellite's temperature was in the range of 75.2° to 78.8° F. and that internal pressure was 850 millimeters (compared with 760 millimeters on the earth's surface). *Luna 10* was reported to be carrying (a) a micrometeorite-particle recorder, which had been in operation for the entire flight from earth, (b) a gamma spectrometer to study gamma radiations of the lunar surface and (c) a magnetometer to measure the moon's magnetic field.

In achieving lunar orbit, it was reported, a course correction was made Apr. 1; a retro-rocket was then fired just before *Luna 10* entered its orbit Apr. 4 to reduce the probe's speed from 2.1 kilometers (1.3 miles) a second to $1\frac{1}{4}$ kilometers (about .78 miles) a second. (Observers at Britain's Jodrell Bank radio-astronomy observatory had reported that after the retro-rocket was fired, *Luna 10* appeared to be "tumbling" in orbit.)

(*Luna 10* "successfully completed its research program" May 30 after its 219th communication session with the earth and its 460th revolution around the moon, according to a report of the USSR Academy of Sciences published in *Pravda* June 3. As of May 30 *Luna 10*'s orbit had an apocynthion of 985.3 kilometers [611.9 miles], pericynthion of 378.7 kilometers [235.2 miles], period of 2 hours 58 minutes 3 seconds and 72° 2′ angle of inclination.)

First U.S. & 2d Soviet Probes in Moon Orbit

2 probes—one American, the other Soviet—were sent into lunar orbit in August.

The first U.S. spacecraft to orbit the moon was placed in lunar orbit Aug. 14. It was followed within 2 weeks by *Luna 11*, the 2d Soviet lunar orbiting craft. The U.S. spacecraft, whose mission was to send back photos of the surface, was called *Lunar Orbiter 1*. Launched from Cape Kennedy Aug. 10, it went into orbit around the moon after a 236,319-mile trip that took 92 hours.

Luna 11, a 1,640-kilogram (3,616-pound) spacecraft whose mission appeared to be similar to *Lunar Orbiter 1*'s, was launched from a Soviet site Aug. 24. It achieved orbit around the moon Aug. 28.

NASA had publicized full details of the U.S. Lunar Orbiter plans months before the launching, and it provided full facilities for press, radio and TV coverage of the launching and of the events that followed. The USSR, as usual, gave no hint that a launching was planned until after the probe seemed to be successfully on its way. Its early announcements thereafter omitted major details of the experiment.

Lunar Orbiter 1 rose from Cape Kennedy's Launch Complex 13 at 3:26 p.m. EDT Aug. 10 in the nose of a 2-stage Atlas-Agena-D rocket. After the Atlas stage separated, the Agena stage ignited and put the Lunar Orbiter-Agena assembly into a 115-mile-high parking orbit around the earth. 28 minutes later the Agena stage ignited a 2d time and pushed the probe into a trajectory designed to bring it to the vicinity of the moon. The Agena stage then separated, and *Lunar Orbiter 1* continued on alone.

Despite initial inability to find and "lock on" Canopus, its reference star, the probe executed a successful mid-course correction maneuver on ground-based radio command at 8 p.m. EDT Aug. 11 by using the moon as a reference point. (The probe's star-tracker finally located Canopus Aug. 13.) Pitching and rolling maneuvers were started on ground command at 11:28 a.m. Aug. 14, the retro-rocket was fired at 11:34 a.m., and *Lunar Orbiter 1* went into orbit around the moon at 11:44 a.m. Aug. 14. The initial orbit had a pericynthion of 119 miles, an apocynthion of 1,160 miles, a period of 3 hours 37 minutes 36 seconds and a $12.16°$ angle of inclination to the moon's equator.

Lunar Orbiter 1 began photographing the lunar surface and transmitting the photos back to earth on ground-based command Aug. 18. Photos taken through the probe's medium-resolution lens proved to be sharp and of high quality, but those taken with the high-resolution lens —the lens that had been expected to produce the most useful photos— turned out to be blurred. After experimenting with shutter speeds, NASA technicians Aug. 20 finally obtained their first good photo through the high-resolution lens—a high-quality picture of part of the hidden "back" of the moon. NASA officials speculated that changes in timing might still bring good results with the high-resolution lens, so they decided to go on with the probe's principal mission—the photographing of 9 areas selected as potential landing sites for Surveyor and Apollo spaceships.

To put *Lunar Orbiter 1* in a suitable position for the planned landing-site photography, the probe's braking rocket was fired for 24 seconds at 5:49 a.m. EDT Aug. 21, and the orbit was thus lowered. The new orbit had an initial pericynthion of 36 miles, an apocynthion of 1,148 miles and a period of 3 hours 20 minutes.

Lunar Orbiter 1 Aug. 22 began its task of taking 32 photos—16 through each lens—of each of the 9 potential landing sites. After taking photos of the earth Aug. 22 and 25, the probe fired its retro-rocket for 3 seconds Aug. 25 to bring its orbit's pericynthion down to 24.7 miles.

In addition to its photography mission, *Lunar Orbiter 1*'s instruments were at work on selenodesy (the study of gravitational field and shape of the moon), meteoroid measurements and radiation measurements. William H. Michael Jr. of the NASA's Langley Research Center reported Aug. 26 that "from what *Orbiter* had told us so far, we figure that the moon bulges out about a quarter mile at the north pole and sinks in about a quarter of a mile at the south pole."

Britain's Jodrell Bank radio-telescope observatory reported Aug. 28 that *Luna 11* had gone into orbit around the moon, and Tass confirmed the Jodrell Bank report Aug. 29. Jodrell Bank added Aug. 29 that the Soviet probe had transmitted photos for 30 minutes that day starting at 4 p.m. EDT, but Tass made no immediate comment on this report.

Tass' Aug. 29 announcement said that *Luna 11*'s orbit had a pericynthion of 99 miles, apocynthion of 745 miles, period of 2 hours 58 minutes and 27° angle of inclination. Tass reported that *Luna 11*'s retro-rockets had been fired at 12:49 a.m. Moscow time Aug. 28 after a trip of 85 hours 46 minutes. A course-correcting maneuver had been made at 10:02 p.m. Aug. 26. Tass scientific commentator Valentin Alekseyev said the Soviet probe had "brought new data about radiations of different kinds, the properties of near-lunar plasma, the density of micrometeoritic matter and magnetic and gravitational fields."

Pravda reported Sept. 28 that *Luna 11* carried "instruments [to] analyze the gamma- and X-ray emissions from the lunar surface" to "make it possible to ascertain the chemical composition of the lunar rock." "The measurements of the evolution of the sputnik orbit will help to get a more accurate idea of the features of the gravitational field of the moon," *Pravda* said. It reported that *Luna 11* carried "special devices" to investigate "the concentration of meteorite showers and their distribution and the intensity of the hard corpuscular radiation near the moon." "Similar investigations have already been done" by *Luna 10*, *Pravda* said. Tass reported Oct. 4 that *Luna 11* had "completed its research program" Oct. 1 after its 137th communication session with the earth. Still in selenocentric orbit, it had made 277 revolutions around the moon during its active phase.

(A 3-stage thrust-augmented Delta rocket had been launched from Cape Kennedy's Launch Pad 17A at 12:02 p.m. EDT June 30 in an unsuccessful attempt to send an unmanned 206-pound U.S. probe, later designated as *Explorer 33*, into orbit around the moon. The rocket, however, gave the probe a velocity 36 miles faster than the 21,135-mph. speed required, and NASA scientists concluded that it would be impossible to achieve a lunar orbit. By ground-based radio signal, therefore, the probe's retro-rocket was fired at 6:32 p.m. July 1, and *Explorer 33* was put into an earth orbit with an apogee of 279,630 miles and perigee of 18,642 miles. In its new orbit, the satellite was assigned to measure the Van Allen radiation belts and the earth's magnetic field.)

The USSR achieved its 3d success of the year with a moon-orbiting vehicle in October. *Luna 12* was launched from the Baikonur cosmodrome at 11:42 a.m. Moscow time Oct. 22. On the basis of signals intercepted by Britain's Jodrell Bank radio-telescope, Sir Bernard Lovell reported that *Luna 12* had gone into orbit around the moon at 8:47 p.m. Greenwich Mean Time Oct. 25, and Tass confirmed his report Oct. 26.

The first disclosure that *Luna 12* had a TV camera aboard was made Oct. 29 after Soviet space authorities had received and processed high-quality photos from the probe. Less than an hour after the announcement, 2 of the photos were shown on Moscow TV. It was also announced that *Luna 12*'s orbit had an apocynthion of 940 miles, a pericynthion of 58 miles.

The announcement of the *Luna 12* shot had been made by Tass nearly 5 hours after the launching, when it was possible to report that the probe had achieved a successful lunar trajectory. Tass scientific commentator Valentin Alekseyev reported Oct. 22 that *Luna 12*'s main tasks were (1) "to improve the systems of artificial satellites of the moon" and (2) "a complex of [scientific] research in near-moon space."

Soviet Probe Hits Venus

The unmanned Soviet spaceship *Venera* (*Venus*) *3*, launched Nov. 16, 1965, reportedly crashed into the planet Venus Mar. 1 at 9:56 a.m. Moscow time. The one-ton probe, which carried a Communist hammer-&-sickle pennant, was the first man-made object to touch another planet.

The Soviet news agency Tass, announcing the Soviet achievement, revealed simultaneously that the 2,123-pound Soviet *Venus 2*, launched Nov. 12, 1965, had passed Venus at a distance of 14,912 miles Feb. 27 and had continued on in independent orbit around the sun. *Venus 2* made its entire trip without course correction, but the course of *Venus 3* had been altered Dec. 26 by a mid-course correction.

Tass reported that Soviet scientists had maintained regular radio communication with the 2 probes throughout most of their 175 million-mile journeys but that radio contact with *Venus 3* was lost at the end of the trip and that the last scheduled communication period before impact "did not take place."

Soviet sources gave no indication initially of any scientific findings the 2 probes may have reported. Neither had the Soviets told what specific data was being sought.

Soviet disclosures on some aspects of the mission seemed contradictory. Prof. Vladimir Ivanchenko, a Soviet space expert, had said shortly after the launchings, that the 2 probes would pass on opposite sides of the planet. Dr. Mstislav V. Keldysh, president of the Soviet Academy of Sciences, had told questioning newsmen Feb. 10 that the 2 probes would "approach" Venus about Mar. 1 but would not attempt a "soft" landing.

Sir Bernard Lovell, director of Britain's Jodrell Bank radio-telescope, expressed fear Mar. 1 that the probe might have contaminated the planet with earthly microbes and thus "endangered the future biological assessment of Venus."

Mariner 4 Reports

The U.S. interplanetary probe *Mariner 4*, in orbit around the sun after flying past Mars July 14, 1965, reestablished telemetry communication with the earth May 21 by radioing back from a record distance of $197\frac{1}{2}$ million miles. NASA reported that the probe's signal, picked up in the Mojave Desert (Calif.) by the Goldstone station's new 210-foot dish antenna, "indicated that all spacecraft systems are operating properly." The *Mariner 4* signal "was received at the time the spacecraft was nearly on the opposite side of the sun from the earth," the NASA said. "This was the first time in history that an earth station received a spacecraft radio signal after it [the signal] had passed deep into the solar corona." The ground station, however, had sent a half-dozen signals "blind" (without possibility of answer) through the sun's corona to *Mariner 4* to make changes in the "look angle" of the probe's star tracker.

U.S. PROGRAM

NASA & Military Launchings

NASA and the U.S. Air Force launched many earth orbital probes and unmanned scientific packages during 1966. Most were reported to be successful in their missions. The launchings, in chronological order:

Feb. 3—Essa 1, the first operational weather satellite, was sent into near polar orbit by means of a 3-stage Thor-Delta rocket booster launched by NASA from Launch Pad 17A at Cape Kennedy. The 305-pound satellite, named for the Commerce Department's new Environmental Science Services Administration, went into a sun-synchronous orbit with an apogee given as 523 miles, perigee of 433 miles, period of about 100 minutes and 81° angle of inclination to the equatorial plane.

Essa 1, which began transmitting cloud-cover photos to U.S. receiving stations Feb. 4, was designed to provide complete global weather coverage daily. Actually the 11th satellite in the successful Tiros (TV infra-red observation satellite) program, *Essa 1* was the first of a new Tos (Tiros operational satellite) system financed, managed and operated by the Environmental Science Services Administration. It cost 2\frac{1}{2}$ million to build, $3,600,000 to launch. Like its experimental Tiros predecessors, *Essa 1* was an 18-sided hatbox-shaped container, 22 inches high and 42 inches in diameter. Its power came from 9,100 solar cells encrusting its top and sides. It took pictures by means of 2 TV cameras with wide-angle (104°) lenses.

Feb. 28—The 285-pound *Essa 2,* the U.S.' 2d operational weather satellite, was launched from Cape Kennedy by means of a Delta rocket. The satellite achieved a sun-synchronous, near-polar orbit with an apogee of 885 miles, perigee of 843 miles, period of 113.57 minutes and 79° angle of inclination. *Essa 2*'s 2 camera systems were switched on Mar. 2, and the satellite began transmission of cloud-cover pictures. *Essa 2,* unlike *Essa 1,* was designed for use by simple, relatively inexpensive ground stations. About 50 such stations were already being operated by the U.S. in various parts of the world, and about 30 were being operated by foreign countries. After $2\frac{1}{2}$ weeks of tests, the NASA turned over *Essa 2* for operation to the Environmental Science Services Administration.

Apr. 8—A 3,900-pound OAO (orbiting astronomical observatory) was launched from Cape Kennedy at 2:36 p.m. by means of a 2-stage Atlas-Agena-D rocket. The OAO, initially designated *OAO-A1,* achieved a near-circular orbit with an apogee of 502 miles, perigee of 496 miles, period of about 101 minutes and 35° angle of inclination. NASA spokesmen said that the spacecraft's sensors then detected and "locked" onto pre-determined stars to stabilize *OAO-A1*'s orientation. But a power failure reported Apr. 10 resulted in the total failure of *OAO-A1* to provide astronomical data.

OAO-A1, the first of 4 projected OAOs in a $250 million astronomical research program, was described as the most complicated unmanned spaceship launched by the U.S. so far. Its nearly 500,000 parts included 10 telescopes with which astronomers had hoped to get a clear look at about 150,000 stars from above the earth's obscuring and distorting atmosphere. *OAO-A1* carried 4 experiments to study the nonvisible ultra-violet, X-ray and gamma-ray emissions from stars and nebulae that cannot be observed from the ground.

Apr. 22–A 132-pound satellite designated *OV3-1* was shot into polar orbit by means of a 4-stage solid-fueled Scout booster rocket launched from Vandenberg Air Force Base, Calif. *OV3-1* was the first in a projected series of 6 satellites designed to check on radiation trapped in the earth's magnetic field.

May 15–The 912-pound experimental *Nimbus 2* weather satellite was launched from Vandenberg Air Force Base at 6:56 a.m. atop a thrust-augmented Thor-Agena-B rocket. The satellite achieved a sun-synchronous near-polar orbit, inclined at an 80° angle, with an apogee of 726 miles, perigee of 687 miles and period of 108 minutes. It began transmitting weather photos the day it went into orbit.

Nimbus 2 was the heaviest and most complex weather satellite orbited by the U.S. so far. It carried infra-red cameras to enable it to take night pictures as well as the already familiar daytime photos–a total of 3,000 pictures a day. Its polar orbit made it possible for the satellite to photograph the earth's entire 200 million-square-mile surface nearly twice daily and thus provide instantaneous, around-the-clock weather photo service twice a day to 150 simple, inexpensive APT (automatic picture transmission) ground stations around the world–including 44 stations in 26 foreign countries. The satellite also carried infra-red heat sensors to measure the earth's "heat balance"–the difference between the amount of solar radiation absorbed by the earth and the amount reflected into the atmosphere. This measurement was expected to help explain how storms are created and dissipated.

(*Nimbus 1,* launched Aug. 28, 1964, had stopped operating Sept. 23, 1964 because of power failure.)

May 25–The 495-pound *Explorer 32* was launched from Cape Kennedy by means of a 3-stage Thor-Delta booster rocket. Its mission, aeronomy, was to measure the temperatures, compositions, densities and pressures of various parts of the upper atmosphere. Because of an extra 8-second burn of the 2d booster stage, the satellite went into an orbit with an apogee of 1,629 miles instead of the 750 miles programmed. The 173-mile perigee was close to the 170 miles planned, however. The angle of inclination was 64°. NASA reported that all 8 experiments aboard were operating normally and that *Explorer 32* would be able to deliver not only the data planned but–because of the higher apogee–additional useful scientific measurements. The satellite, initially designated Atmosphere Explorer-B, had the secondary mission of studying the effects of short-term atmospheric disturbances caused by radiation from solar storms.

June 6—A 1,135-pound scientific satellite dubbed *Ogo 3* (for orbiting geophysical observatory) was sent into orbit by means of a 2-stage Atlas-Agena-B booster rocket launched from Cape Kennedy at 10:48 p.m. EDT. The orbit, almost precisely as planned, had an apogee of 75,768 miles, perigee of 170 miles, period of 48 hours 37 minutes and 31° angle of inclination.

Ogo 3 carried a record 21 scientific experiments contributed by 10 U.S. universities, 2 NASA field centers and 4 other U.S. government agencies. The overall objective of the Ogo series was to provide data on the relationship between the sun and the earth's environment. *Ogo 3*'s scientific mission was: (a) to "investigate the charged particle population and energy spectra of trapped radiation and the interrelationship with magnetic activity"; (b) to continue *Ogo 1*'s observations of particle and electro-magnetic fluctuations; (c) to measure the interplanetary solar plasma and magnetic field; (d) to study galactic and solar cosmic rays and "the modulation mechanisms associated with solar activity"; (e) to analyze "the charged portion of the atmospheric composition"; (f) to observe "the correlation between ion distribution, magnetic fields and electron density"; (g) to measure "very low frequency noise of natural and man-made origin, of galactic emissions and of planetary and solar bursts for correlation with magnetic field measurements, solar flare activity and electro-magnetic radiation"; (h) to study the geocorona, gegenschein (faint, nebulous light opposite the sun), and micrometeoroid distributions.

NASA reported June 17 that all 21 experiments were working. But NASA disclosed July 26 that *Ogo 3*'s attitude control system had failed after the satellite had "operated almost flawlessly in an earth-stabilized mode for more than 6 weeks, thus proving the engineering feasibility of the 3-axis stabilization system." According to NASA, "operation in this mode for 30 days was one of the [satellite's] 2 primary mission objectives." Following the failure of the 3-axis system, the satellite was put into a "spin-stabilized" mode July 25 by ground-based radio command.

June 10—A satellite carrying imitation human flesh was launched by the Air Force from Wallops Island, Va. into an orbit that took it into the Van Allen radiation belts girdling the earth. Its mission was to provide data on the effect this radiation might have on astronauts.

June 16—An Air Force 3-stage Titan-3C rocket, launched from Cape Kennedy at 10 a.m. EDT, put 8 satellites into orbit around the earth at an altitude of 20,941 miles.

The first of the satellites was a General Electric-built research package testing a device that used gravity to keep satellites oriented. The remaining 7 satellites were 100-pound radio relay stations, built by the Ford Motor Co.'s Philco Western Development Laboratories at a cost of $1\frac{1}{2}$ million each, to give the armed forces a jam-proof global communications system.

The rocket's 2 initial stages put the satellite package into 2 preliminary "aiming" orbits before the 3d stage ignited, some 6 hours 3 minutes after launching and carried the 8 satellites into the final orbit. The research satellite was then ejected, followed by the communications satellites at 20–25-second intervals. Since the 3d Titan stage was accelerating slightly as the satellites were ejected, each satellite went into a slightly different orbit at a slightly different speed. The purpose of this difference was to allow the satellites to spread out around the globe over a period of 3 months. The 7 satellites were moving in orbit in the same direction as the earth's rotation and at such a speed that it took about 12 days for each satellite to move completely around the earth.

The satellites had an expected life of $1\frac{1}{2}$ to more than 2 years, but each carried an automatic device to turn it off after 6 years to make sure it did not interfere with later communications satellites.

The Air Force planned to put up 15 more of the communications satellites in 2 additional 1966 launchings. But the first of the planned additional shots failed Aug. 26 when a Titan-3C, launched from Cape Kennedy at 10 a.m. EDT with 8 communications satellites aboard, exploded 80 seconds after lift-off. The cause of the failure was said to be a rocket cover that broke away prematurely, shook the rocket off course and activated the automatic destruction system.

June 23–Pageos 1, a 125-pound aluminum-coated plastic balloon satellite, was launched from Vandenberg Base at 11:12 p.m. PDT by means of a 2-stage thrust-augmented Thor-Agena-D rocket assembly. Its mission was to assist mapmakers. *Pageos 1* (for passive geodetic earth-orbiting satellite), 100 feet in diameter after inflation in orbit, achieved a near-polar orbit with a 2,649-mile apogee, 2,616-mile perigee and 3-hour period. Sunlight glinting from its aluminum surface made the satellite a bright reference point for the mapping project.

July 5–A 92-foot Saturn-4B (S4B) rocket stage weighing 58,537 pounds (including about 22,000 pounds of hydrogen fuel) was sent into orbit. The successfully tested S4B was a working model of the 3d (final) stage of the Saturn-5 rocket assembly, the booster scheduled for use in the U.S.' first manned trip to the moon. The 29-ton satellite, the heaviest ever orbited by the U.S. (although believed to have been outweighed by the USSR's *Proton 1*), demonstrated that NASA had solved the problem of keeping hydrogen fuel liquid and in proper position for the reignition in space required in the Apollo lunar project. The solution was venting jets. The constant venting of a small amount of hydrogen in space provided a small amount of forward thrust, which created enough artificial gravity to keep the liquid hydrogen in the bottom of the tank ready for use.

The Saturn-4B had been sent up as the 2d stage of an uprated 2-stage Saturn-1 assembly launched from Cape Kennedy's Launch Complex 37 at 10:53 a.m. EDT July 5. The 173-foot Saturn-1B first stage, generating 1,600,000 pounds of thrust, carried the S4B to an altitude of 50 miles; at this height the S4B ignited and continued upward into an orbit with an apogee of 117 miles, perigee of 115 miles, period of 88 minutes and $32°$ angle of inclination to the equatorial plane.

After the S4B had been in orbit for 4 revolutions and the experiment completed, NASA officials blew it up by shutting the vents (by radio signal from the ground); the shut-off caused the hydrogen to build up pressure until it exploded.

Aug. 4—In another study of the Van Allen radiation belts, the Air Force used a Scout rocket to launch a satellite from Vandenburg Base into a polar orbit with a 2,776-mile apogee and 220-mile perigee.

Aug. 17—The 140-pound *Pioneer 7* space probe was sent into orbit around the sun by means of a 3-stage thrust-augmented Improved Delta rocket launched at 11:28 a.m. EDT from Launch Pad 17A at Cape Kennedy. The orbit's aphelion (furthest point from the sun) was about 102 million miles, its perihelion (closest point to the sun) about 93 million miles and its period 404 days. On radio orders from the Goldstone Tracking Station in the Mojave Desert Aug. 19, the scientific probe turned itself exactly $31°$ to focus its high-gain antenna precisely on the earth to maintain strong 2-way radio communication.

Pioneer 7's mission was to perform 6 experiments to produce data on magnetic fields, cosmic rays and the solar wind. The probe's predecessor, *Pioneer 6*, had helped show that the "solar wind," a supersonic flow of ionized gas moving constantly outward from the sun's surface, establishes the interplanetary magnetic field by drawing out magnetic fields near the sun. These fields and the solar wind's direction are changed by the collision of solar-wind "beams," masses of gas that move away from the sun on separate paths. As a result, according to NASA interpretations, the field lines of force, rooted in the sun and stretched through interplanetary space, are twisted about each other "like many strands of spaghetti in boiling water." High-energy particles generated by solar explosions travel down the twisted magnetic field lines, often at speeds of hundreds of millions of miles an hour, in the well-defined, clearly-separated "streams" known as solar cosmic rays. Knowledge of the mechanics of these streams was needed to assure the protection of astronauts.

Aug. 18, Oct. 5, Oct. 28, Nov. 8 & Dec. 14–2 secret satellites were launched by the Air Force into polar orbit by means of Scout rockets fired from Vandenberg Base Aug. 18 and Oct. 28. Air Force-industry teams at Vandenberg Base used an Atlas-Agena assembly Oct. 5, a Thor-Agena-D assembly at 11:52 a.m. Nov. 8 and a Titan-3B-Agena assembly Dec. 14 to send secret satellites into polar orbits.

Aug. 25–A 2-stage uprated Saturn-1, consisting of a Saturn-1B first stage and Saturn-4B (S4B) 2d stage, was used in a 17,625-mile suborbital test of a 2-module Apollo, the spacecraft designed to carry the first American astronauts to the moon. The launching, from Cape Kennedy Launch Complex 34, took place at 1:16 p.m. EDT; splash-down was in the South Pacific southeast of Wake Island at 2:49 p.m. EDT.

The 53-foot Apollo, whose command module, service module and adapter weighed a total of 48,400 pounds on launching (including 23,000 pounds of service module propellants), fired its service-module rocket engine 5 times after separation from the Saturn-4B. The first ignition brought the spacecraft to its peak altitude of more than 750 miles (the goal had been only 703 miles). The next 3 ignitions accelerated the spacecraft back toward earth and were used to help simulate the speed of an Apollo re-entering the atmosphere after a trip back from the moon. The 5th ignition separated the heat-shielded command module from the service module. On plunging into the atmosphere at a speed of more than 19,000 mph., the command module fired its reaction control rockets to make the module skip back up again and increase the duration of the entry test.

Oct. 2–The 32-pound *Essa 2*, the 3d weather satellite in a series launched by NASA for the Environmental Science Services Administration, was sent up from Vandenberg Air Force Base at 3:39 a.m. PDT by means of a 3-stage thrust-augmented improved Delta rocket. The hatbox-shaped satellite went into a near-polar orbit with an initial apogee given as 923 miles, perigee as 859 miles and period as 114 minutes. It was then put into a sun-synchronous circular orbit with an altitude of about 865 miles and period of $113\frac{1}{2}$ minutes. The 18-sided polygon replaced *Essa 1* in providing global weather pictures for the Tos (Tiros operational satellite) system. The satellite, 22 inches high and 42 inches in diameter, "rolled" in orbit like a wheel; each of the 2 cameras on its rim pointed directly to the earth (for picture-taking) once during each revolution.

Dec. 6–A satellite described by NASA as "one of the most versatile spacecraft ever developed" was shot into space by means of an Atlas-Agena-D rocket combination launched from Launch Complex 12 at Cape Kennedy at 9:12 a.m. EST. *ATS-1* (for applications technology satellite) was the first of 5 satellites planned in the ATS program. The satellite, carrying communications, meteorology, control technology and scientific experiments, achieved a transfer orbit with an apogee of about 23,000 miles, perigee of just over 100 miles and inclination of about 31°.

At 1:42 p.m. EST Dec. 7, on radioed command from the Goldstone tracking station in the Mojave Desert, Calif., the satellite fired its apogee rocket to transfer to a synchronous orbit in which it appeared to hang motionless 22,300 miles over the Pacific Ocean east of Christmas Island at the equator. NASA authorities reported later that *ATS-1* was on station at 151° W. Longitude and was operating properly. *ATS-1* weighed about 1,550 pounds at launching but was down to 775 pounds in orbit. It was designed and built by the Space Systems Division of the Hughes Aircraft Co. in El Segundo, Calif.

Dec. 11 – 2 radiation research satellites dubbed *OV 1-9* and *OV 1-10* (for orbiting vehicle) were sent into polar orbit by means of a single Atlas-D rocket launched by the Air Force at Vandenberg Base.

Dec. 14 – A space biology package carrying more than 10 million microbes, insects, mold spores and plants was shot into a 195-mile-high circular orbit by means of a 2-stage thrust-augmented improved Delta rocket launched from Cape Kennedy. The space vehicle, *Biosatellite 1*, was the first of about 6 such satellites planned in a $2\frac{1}{2}$-year $80–$100 million program. The initial shot, however, produced disappointment, *Biosatellite 1*'s 280-pound specimen-laden capsule failed to come out of orbit as planned after its 47th revolution around the earth Dec. 17. The satellite's retro-rocket failed to fire on radioed signal from the earth, and the specimens continued on in orbit. Had the recovery system operated, the capsule would have re-entered the atmosphere, made a parachute-slowed descent and been caught in mid-air by a Hawaii-based Air Force C130 cargo plane carrying equipment for seizing the capsule's parachute.

INTERNATIONAL DEVELOPMENTS

U.S. Urges Moon Treaty

Pres. Lyndon B. Johnson announced May 7 that the U.S. would seek through the UN a treaty barring any nation from asserting sovereignty over the moon or other celestial bodies. In a statement released from the LBJ Ranch in Texas, Johnson said the purpose of the treaty would be to "ensure that explorations of the moon and other celestial bodies will be for peaceful purposes" and to guarantee that astronauts of the U.S. and other nations "can freely conduct scientific investigations of the moon." "We want the results of these activities to be available for all mankind," he declared. The "essential elements" of the proposed treaty:

> The moon and other celestial bodies should be free for exploration and use by all countries. No country should be permitted to advance a claim of sovereignty.
> There should be freedom of scientific investigation, and all countries should cooperate in scientific activities relating to celestial bodies.
> Studies should be made to avoid harmful contamination.
> Astronauts of one country should give any necessary help to astronauts of another country.
> No country should be permitted to station weapons of mass destruction on a celestial body. Weapons tests and military maneuvers should be forbidden.

Johnson said the treaty would be aimed at preventing "serious political conflicts . . . [from] aris[ing] as a result of space activities." He disclosed that U.S. Amb.-to-UN Arthur J. Goldberg had been instructed to "seek early discussion of such a treaty in the appropriate United Nations body."

Goldberg presented the President's proposal to UN Secy. Gen. U Thant at UN headquarters May 9 and asked that it be circulated to all UN members.

USSR Orbits 2 Dogs

The USSR sent 2 male dogs named Veterok (Breezie) and Ugolyok (Blackie, or Little Coal Nut) into orbit Feb. 22 aboard a satellite designated *Cosmos 110*. The announcement of the launching, made after Soviet authorities were sure that the 2 dogs were alive and well in the spaceship, said that *Cosmos 110* had achieved an orbit with an apogee of 562 miles, perigee of 116 miles, period of 95.3 minutes and a 51°54' angle of inclination to the equatorial plane.

Tass reported that the dogs were participating in a "new biological experiment" and added that such biological studies "will precede every new important [Soviet] step of man into space." Tass did not even hint at the specific mission of the flight. U.S. experts pointed out, however, that *Cosmos 110* was the first satellite to carry animals into the Van Allen radiation belts, which girdle the earth in space.

Pictures of the dogs in orbit in their spaceship were shown on Moscow TV Feb. 26. This was the first disclosure that *Cosmos 110* was

TV equipped. Soviet Dr. Boris B. Yegorov, reporting Feb. 27 that the experiment was "proceeding successfully and the animals are feeling well," disclosed that the satellite was "equipped for a wide range of radiobiological studies."

Soviet cosmonaut Gherman S. Titov had been quoted by Tass Feb. 4 as saying the USSR might send dogs to the moon before attempting a manned lunar landing. Dr. Mstislav Keldysh, president of the Soviet Academy of Sciences, was questioned later about Titov's remark. Keldysh answered that such use of dogs was one of several proposals being debated.

Soviet scientists Yegorov, V. N. Pravetsky and N. M. Sisakyan disclosed in a *Pravda* article Mar. 1 that Veterok was subjected to various tests whereas Ugolyok was used as a control. The 2 dogs made their flight strapped in place in separate containers in the spaceship. During flight food was forced by tube directly into the dogs' stomachs. Medical and biological preparations were injected directly into the bloodstream. Reactions of the dogs were checked by telemetry.

The 2 dogs were brought back to earth Mar. 16 after circling the earth 330 times during 22 days in orbit. Tass reported Mar. 16 that the 2 dogs apparently had needed 8 or 9 days to adjust to weightlessness in space but that they had suffered no apparent ill effects.

Vasily V. Parin, representative of the Soviet Academy of Medicine Institute of Biomedicine, said Mar. 16 that various irritants had been introduced into the blood of one of the dogs (presumably Veterok) during flight to determine the effects on the cardiovascular system. Yegorov and other Soviet sicentists reported Mar. 18, after preliminary examination of the dogs, that they had found "no symptoms of disturbances due to cosmic radiation" although the dogs had passed repeatedly through the Van Allen radiation belts surrounding the earth.

At the concluding session of the International Astronautical Congress in Madrid Oct. 15, Parin showed a short black-and-white film of Veterok and Ugolyok after their space flight. They could barely stand the first day, could stand on their hind legs to reach for food the 2d day but seemed to be active and normal after 4 months.

Eldo Fires First Rocket

The 7-nation European Launcher Development Organization (Eldo) fired its first rocket, the 105-ton Europa-1, from the Woomera launching site in Australia at 9:05 a.m. May 24. Signals from a radar station, however, indicated incorrectly that the rocket was veering off-course. The range safety officer, therefore, cut off the rocket's engines

2 minutes $15\frac{1}{2}$ seconds after blast-off—or 20 seconds early—and the rocket came down 270 miles north of the launching site. Plans had called for a $6\frac{1}{2}$-minute 430-mile flight.

After the flight was cut short, it was discovered that the rocket had been operating properly and was on course. Eldo officials expressed disappointment at the premature ending of the flight but said that the flight had been largely successful and had provided most of the information sought.

The only stage fired was the British-built Blue Streak first stage. The rest of Europa-1 consisted of dummy elements—a dummy French-built Coralie 2d stage, a dummy West German 3d stage and a dummy Italian satellite. The purpose of the dummies was to provide full orbital lift-off load and to carry telemetry equipment.

Meeting in Paris July 7–8, cabinet ministers of the Eldo nations agreed to proceed with plans to build a rocket designed to place a communications satellite in orbit by 1970. The ministers simultaneously agreed to reduce Britain's share of Eldo's expenses from its current 38.79% to 27%. The British Foreign Office June 4 had issued a statement threatening to quit Eldo unless the UK's share of the costs was reduced.

Under the July 8 agreement Eldo set an expenditure ceiling of $330 million with an additional $8,400,000 for experimental studies for the period 1967–71. The 3-stage Europa rocket to be built, an improved perigee-apogee system (ASP), would be capable of placing a 200-kilogram (440-pound) communications satellite in a stationary (synchronous) equatorial orbit. The launching site would be French Guiana, where France was building its new launch complex. At the meeting, the Australian supply minister, Sen. Norman H. D. Henty, said his government had abandoned its claim to provide the launching site. The ministers agreed, however, that Australia's Woomera launching site would continue to be used for research and development firing even after the French Guiana site was completed.

Under the revised budgetary division, France's share rose from 24% to 25%, Germany's from 22% to 27%, Italy's from 10% to 12% and Belgium's and the Netherlands' from 2.85% and 2.64%, respectively, to 9% together. Australia's contribution consisted of placing the Woomera site at Eldo's disposal.

Following Britain's June 4 threat to leave Eldo, the Eldo ministers had reached tentative agreement at a meeting in Paris June 9–10 to reduce Britain's share of the costs. A State Department spokesman said in Washington June 9 that State Secy. Dean Rusk had told the British that the U.S. favored continued UK participation in Eldo.

International Congress

The 17th International Astronautical Congress was held in Madrid Oct. 10-15 with more than 1,000 scientists and engineers from 30 countries attending. The congress was sponsored by the International Astronautical Federation, a non-governmental organization with about 50,000 members in 33 countries. At the concluding business session Oct. 14, Dr. Luigi G. Napolitano of the University of Naples was elected to succeed Dr. William H. Pickering of the U.S.' Jet Propulsion Laboratory as federation president.

Dr. Hillard W. Paige, general manager of the General Electric Co.'s Missile & Space Division (Valley Forge, Pa.), reported at the opening session Oct. 10 that 2 U.S. Naval Research Laboratory experimental satellites had collided in space but had separated with no apparent damage and had continued working. The 2 satellites were among 8 strung out in similar orbits by a single Air Force rocket assembly Mar. 9, 1965, and they were traveling at an altitude of 560 miles and a speed of some 20,000 mph. when they touched.

Soviet scientists V. E. Belai, P. V. Vassilyev and G. D. Glod indicated at the congress' Oct. 10 session that Soviet research had showed an "urgent" need for the development of new drugs and medicines to help overcome the effects of long space flight. Soviet studies apparently had disclosed that current drugs and medicines sometimes have different effects when used in space conditions than when used on earth.

Richard S. Johnston, chief of crew systems at the U.S.' Manned Spacecraft Center (Houston, Tex.), told the delegates Oct. 11 that U.S. EVAs (extravehicular activities) had disclosed a need for more flexible space suits, better individual propulsion and "a stable work platform" if astronauts were to work outside their spaceships effectively.

Prof. Leonid I. Sedov, 58, University of Moscow aerodynamicist and leader of the Soviet delegation, denied in an Oct. 12 interview at the congress that he had predicted (as had been reported) that the USSR would put a man on the moon in 1969. He said Soviet space scientists had "difficult problems" to solve before they could launch a manned lunar mission. Sedov, ex-president of the International Astronautical Federation, was currently the editor of the Soviet space bimonthly *Cosmic Research* and had been chairman of the Interplanetary Travel Commission of the Soviet Academy of Sciences 1956-61 (current chairman: Prof. Anatoly Blagonravov). Sedov agreed that there were good economic and scientific reasons for the U.S. and USSR to cooperate in joint space activities. The reason "we don't have any plans for joint launchings... is not technical but political," he declared, "and I think in this international situation in these days we cannot change this." He said the USSR had not let Western newsmen see Soviet launchings or space facilities because of the political situation. "We can open all our works only when the international situation will be better," he said.

Prof. Oleg G. Gazenko, the USSR's leading space medicine expert, said at a press conference at the congress Oct. 14 that the USSR was planning a "serious new step" in manned space exploration that "requires a lot of preparatory work." He reported that the USSR planned to test several new crew propulsion methods, "including man's ability to move in space by his own resources without use of supplementary equipment."

Spain, the congress' host country, fired its first rocket, a British-built Carabela-4, on the final day of the congress Oct. 15. The 9-foot 88-pound projectile, launched at 11:56 a.m. at the Campo de Arenosillo rocket base near Huelva, carried a 12-pound meteorological payload up 40 to 50 miles and then plunged into the Atlantic. Spain had bought 6 of the rockets, known in Britain as the Skua, at a cost of $1,689 each. (A 2d Spanish launching attempt at the Huelva site failed Oct. 20 when the guidance system malfunctioned. But Spanish scientists successfully fired a U.S.-made Judy-Dart meteorological rocket Oct. 21.)

Soviets Launch Satellites

The USSR's 3d Molniya-1 communications satellite was launched Apr. 25 into what Tass described as "a high elliptical orbit in the Soviet Union." The apogee was given as "39,500 kilometers [about 24,500 miles] in the Northern hemisphere" and the perigee as "499 kilometers [310 miles] in the Southern hemisphere." The period was listed as 11 hours 50 minutes and the angle of inclination as 64.5°.

Tass disclosed May 18 that the new Molniya-1 carried "experimental instruments for observing the earth from outer space." "The experiment in observing the earth and taking TV pictures of our planet was carried out for the first time [earlier May 18]," Tass reported. "Pictures were taken from the height of 30,000 to 40,000 kilometers.... The lenses and light filters were changed [during the photographing], which made it possible to get images of the earth on different scales and to observe the elements of the earth's surface in the conditions of different illumination." Soviet experts disclosed that the new satellite broadcast at a peak power of 40 watts, compared with the 4 watts of such U.S. communications satellites as *Syncom* and *Early Bird*.

The Paris TV Center May 28 successfully transmitted a color TV program from Paris to Moscow by way of the new Molniya-1.

Proton 3, described by Tass as "a heavy scientific space station," was launched July 6 into an orbit with an apogee of 630 kilometers ($391\frac{1}{2}$ miles), perigee of 190 kilometers (118 miles), inclination of $63\frac{1}{2}°$ and period of $92\frac{1}{2}$ minutes. The disclosure of the launching was made after the Soviet satellite was safely in orbit. Tass said *Proton 3's* equipment was designed to continue cosmic-ray studies. Specific experiments included (1) the "study of the power spectrum and chemical composition of cosmic rays in the power interval of up to 100 trillion electron volts"; (2) the "study of nuclear interdependence of cosmic particles in the sphere of power up to 1 trillion electron volts"; (3) "measuring the absolute intensity and power spectrum of electrons of galaxy origin"; (4) "search in primary cosmic rays for particles with a fractional electrical charge."

Non-Soviet Reds View Launchings

Officials of 8 non-Soviet Communist countries reportedly were present at the Baikonur cosmodrome in Kazakhstan Oct. 20 to witness the launching of 2 unmanned Soviet earth satellites—*Cosmos 130* and the USSR's 4th Molniya-1. The only foreigner known to have been permitted at Baikonur previously was French Pres. Charles de Gaulle, who saw the launching of *Cosmos 122* June 15.

It was reported that non-Soviet Communists had complained mildly that de Gaulle, a Westerner, had been permitted to see Soviet space facilities that were kept off-limits to heads of Communist countries. The complaints apparently led to the invitations to Communist Party leaders, premiers and defense ministers of Bulgaria, Cuba, Czechoslovakia, East Germany, Hungary, Mongolia, Poland and Rumania to attend the Oct. 20 launchings.

Communist sources in Moscow had been cited Oct. 15 as authority for conflicting rumors that the foreign Communist leaders might (or might not) have been slated to see a spectacular launching that was to have orbited as many as 8 to 10 men—cosmonauts from Bulgaria, Czechoslovakia, East Germany, Hungary, Poland and Rumania as well as from the USSR—in the biggest spaceship fabricated so far. At least one other Communist source was cited the same day, however, as saying that such a multiple launching was planned for later in 1966, not for Oct. 20.

The Tass announcements of the Oct. 20 launchings said that *Cosmos 130*'s orbit had an initial apogee of 340 kilometers (211 miles), perigee of 211 kilometers (131 miles), period of 89.8 minutes and 65° angle of inclination. The Molniya-1's orbit was reported to have an initial apogee of 39,700 kilometers (24,654 miles) "in the Northern hemisphere," perigee of 458 kilometers (284 miles) "in the Southern hemisphere," period of 11 hours 53 minutes and 64.9° angle of inclination. The Molniya-1, according to Tass, was a communications satellite that carried "rebroadcasting apparatuses to transmit TV programs and long-distance multi-channel radio communications." Its "main task" was given as "the further testing of the system of long-range 2-way television and telephonic-telegraphic radio communication and its experimental exploitation." *Krasnaya Zvezda (Red Star)*, the Soviet Defense Ministry newspaper, reported Oct. 22 that the satellite's TV camera was expected to provide photos to be used in accurately mapping the earth's surface. The satellite also had a weather mission—the observation of cloud movements, warm- and cold-air masses and other atmospheric phenomena.

3d Soviet Space Base Reported

Evart Clark reported in the *N.Y. Times* Dec. 20 that "a secret new space launching base at Plesetsk, 400 miles north of Moscow" and 140 miles south of Archangel, had been used for the launching of at least 5 Cosmos satellites since Mar. 17. Previously it apparently had been used only for intercontinental ballistic missiles. 2 other Soviet space bases were known to the West: Kapustin Yar and Tyuratam (Baikonur). Although the U.S. government had known of the Plesetsk base's use for months, the knowledge did not become public until students at the Kettering Grammar School in Northamptonshire, England had calculated the base's approximate latitude and longitude by means of war-surplus radio equipment with which they had been tracking Cosmos satellites.

(The USSR had conducted a series of rocket tests Apr. 25-May 25 with the final stages landing in 2 mid-Pacific target areas west of Midway Island and north of Wake Island. It was reported that equipment tested was designed for landing spaceships at sea. A series of Soviet rocket tests was conducted Apr. 24-July 4 with the final stages landing in 2 mid-Pacific areas west and northwest of Midway Island. A Soviet announcement said that carrier rockets "to develop new space systems" had been tested.

(Tass announced Aug. 24 that Soviet "booster rockets" would be test-fired to a mid-Pacific impact area between Aug. 26 and Oct. 25, but a Sept. 6 announcement said the tests had been completed in 11 days. A mid-Pacific impact area was also used for a new series of tests of "carrier rockets" announced by Tass Sept. 24 for the Sept. 26-Oct. 25 period.)

Cosmos Launchings

In addition to those already mentioned, the Soviets reported launchings of the following Cosmos satellites during 1966:

Cosmos 104 –Jan. 17.   Apogee (maximum altitude), 401 kilometers (249 miles).   Perigee (minimum altitude), 204 kilometers (126.67 miles).   Period, 90.2 minutes. Angle of inclination to equatorial plane, 65°.

Cosmos 105 –Jan. 22.   Apogee, 324 kilometers (201.2 miles).   Perigee, 204 kilometers (126.7 miles). Period, 89.7 minutes. Inclination, 65°.

Cosmos 106 –Jan. 25.   Apogee, 564 kilometers (350.2 miles).   Perigee, 290 kilometers (180.1 miles). Period, 92.8 miles. Inclination 48.4°.

Cosmos 107 –Feb. 10.   Apogee, 322 kilometers (200 miles).   Perigee, 204 kilometers (126.8 miles). Period, 89.7 minutes. Inclination, 65°.

Cosmos 108 –Feb. 11.   Apogee, 865 kilometers (537.4 miles).   Perigee, 227 kilometers (141.1 miles). Period, 95.3 minutes. Inclination, 48.9°.

Cosmos 109 –Feb. 19.   Apogee, 309 kilometers (192 miles).   Perigee, 209 kilometers (129.9 miles). Period, 89½ minutes. Inclination, 65°.

Cosmos 111 –Mar. 1.   Apogee, 226 kilometers (140½ miles).   Perigee, 191 kilometers (118⅔ miles). Period, 88.6 minutes. Inclination, 51° 51 minutes.

Cosmos 112 –Mar. 17.   Apogee, 565 kilometers (351 miles).   Perigee, 214 kilometers (133 miles). Period, 92.1 minutes. Inclination, 72°.

Cosmos 113 –Mar. 21.   Apogee, 327 kilometers (203 miles).   Perigee, 210 kilometers (130½ miles). Period, 89.6 minutes. Inclination, 65°.

Cosmos 114 –Apr. 6.   Apogee, 374 kilometers (232¼ miles).   Perigee, 210 kilometers (130½ miles). Period, 90.1 minutes. Inclination, 73°.

Cosmos 115 –Apr. 21.   Apogee, 294 kilometers (182½ miles).   Perigee, 190 kilometers (118 miles). Period, 89.3 minutes. Inclination, 65°.

Cosmos 116 –Apr. 26.   Apogee, 478 kilometers (297 miles).   Perigee, 294 kilometers (182½ miles). Period, 92 minutes. Inclination, 48° 25 minutes.

Cosmos 117 –May 6.   Apogee, 308 kilometers (191¼ miles).   Perigee, 207 kilometers (128½ miles). Period, 89½ minutes. Inclination, 65°.

Cosmos 118 –May 11.   Altitude of circular orbit, "about 640 kilometers" (397½ miles). Period, 97.1 minutes. Inclination, 65°.

Cosmos 119 –May 24.   Apogee, 1,305 kilometers (810½ miles).   Perigee, 219 kilometers (136 miles). Period, 99.8 minutes. Inclination, 48.5°.

Cosmos 120 –June 8.   Apogee, 300 kilometers (186 miles).   Perigee, 200 kilometers (124 miles). Period, 89.4 minutes. Inclination, 51.8°.

Cosmos 121 –June 17.   Apogee, 354 kilometers (220 miles).   Perigee, 210 kilometers (130 miles). Period, 89.9 minutes. Inclination 72.9°.

INTERNATIONAL DEVELOPMENTS

Cosmos 122—June 25. Circular orbit about 625 kilometers (388 miles) above the earth's surface. Period, 97.1 minutes. Inclination, 65°.*

Cosmos 123—July 8. Apogee, 529 kilometers (328½ miles). Perigee, 263 kilometers (163 miles). Period, 92.2 minutes. Inclination, 48.8°.

Cosmos 124—July 14. Apogee, 303 kilometers (188 miles). Perigee, 208 kilometers (129 miles). Period, 89.4 minutes. Inclination, 51.8°.

Cosmos 125—July 20. Circular orbit about 250 kilometers (155 miles). Period, 89½ minutes. Inclination, 65°.

Cosmos 126—July 28. Apogee, 359 kilometers (223 miles). Perigee, 212 kilometers (131½ miles). Period, 90 minutes. Inclination, 51.8°.

Cosmos 127—Aug. 8.

Cosmos 128—Aug. 27. Apogee, 364 kilometers (226 miles). Perigee, 212 kilometers (131½ miles). Period, 90 minutes. Inclination, 65°.

Cosmos 129—Oct. 14. Apogee, 307 kilometers (191½ miles). Perigee, 202 kilometers (125½ miles). Period, 89.4 minutes. Inclination, 65°.

Cosmos 131—Nov. 12. Apogee, 360 kilometers (223½ miles). Perigee, 205 kilometers (127.3 miles). Period, 89.9 minutes. Inclination, 72.9°.

Cosmos 132—Nov. 19. Apogee, 280 kilometers (173.9 miles). Perigee, 207 kilometers (128½ miles). Period, 89.3 minutes. Inclination, 65°.

Cosmos 133—Nov. 28. Apogee, 232 kilometers (144 miles). Perigee, 181 kilometers (112.4 miles). Period, 88.4 minutes. Inclination, 51.9°.

Cosmos 134—Dec. 3.

Cosmos 135—Dec. 12. Apogee, 662 kilometers (411.1 miles). Perigee, 259 kilometers (160.8 miles). Period, 93.5 minutes. Inclination, 48.5°.

Cosmos 136—Dec. 19. Apogee, 305 kilometers (189.4 miles). Perigee, 198 kilometers (123 miles). Period, 89.4 minutes. Inclination, 64.6°.

Cosmos 137—Dec. 21. Apogee, 1,720 kilometers (1,068 miles). Perigee, 230 kilometers (143 miles). Period, 104.3 minutes. Inclination, 48.8°.

Space Shots by Other Nations

A 42-pound French satellite called *Diapason* (previously designated D-1A) was launched into orbit Feb. 17 by means of a 3-stage Diamant rocket fired by French personnel at 8:33 a.m. (GMT) from France's Saharan base near Hammaguir in Algeria.

Diapason, an 8-inch-high cylinder 20 inches in diameter, achieved an orbit with an apogee of 2,753.46 kilometers (1,710.92 miles), perigee of 503.06 kilometers (312.59 miles), 118.64-minute period and 34.04° angle of inclination. Plans for the launching, announced in detail

---

*The USSR disclosed Aug. 17 that some of its Cosmos satellites, "including *Cosmos 122*," were equipped to collect weather data. Tass said that "*Cosmos 122* carried instruments for taking TV pictures of the clouds, cameras to photograph clouds in infra-red rays on the day-and-night sides of the earth and instruments to measure radiation in the earth-atmosphere system." The USSR Aug. 18 began transmitting *Cosmos 122*-secured weather data to the U.S. on its weather communications link with the U.S. This was the first time satellite-obtained weather data had been transmitted on the network.

months earlier, had called for a 2,700-kilometer (1,678-mile) apogee, 505-kilometer (314-mile) perigee, 118.14-minute period and 34° angle of inclination. A 42-pound battery-powered equipment case accompanied the satellite in orbit.

Diapason was the 3d satellite built by France but only the 2d launched by means of a French-made rocket. *Diapason* was a technical research satellite. Its $4\frac{1}{2}$ watts of solar power was produced by 2,304 silicon cells mounted on 4 paddles. The satellite's mission was to continue the testing of France's 62-foot Diamant booster rocket and French equipment, to furnish geodetic measurements (of the earth's shape) and to provide information on the behavior of solar cells in space. Ground stations reported Feb. 17 that all equipment aboard the satellite was working perfectly.

The Diamant rocket consisted of a solid-fueled Emeraude first stage, solid-fueled Topaze 2d stage and liquid-fueled Rubis 3d stage. Its total weight: 18.4 metric tons (20.3 short tons).

Mechanical failures, revealed to the public as they occurred, had thwarted 2 attempts at launching the D-1A Feb. 11 and 12. The Feb. 17 shot was the 3d attempt.

A $36\frac{1}{2}$-kilogram ($80\frac{1}{2}$-pound) West German scientific payload was shot to an altitude of 2,035 kilometers (1,261 miles) Apr. 22 by means of a French-built Rubis rocket fired by France's CNES space authority at the Hammaguir launching site. The shot completed a 2-nation project that included several French launchings of German sounding rockets in 1965. The instruments were provided by the Max Planck Institute for Extraterrestrial Research in Garching, near Munich; the institute's director, Prof. Reimar Lust, supervised the program. In the Apr. 22 launching, the German probe released 2 clouds of barium ions that spread out along the earth's magnetic field and made it possible to observe the Van Allen radiation belt surrounding the earth.

Japanese scientists launched a 4-stage Lambda-4S1 rocket from a mobile truck on an ocean-side cliff in Uchinoura, Japan at 11:58 a.m. Sept. 26 in an unsuccessful attempt to put the rocket's 57-pound burned-out 4th stage into orbit. Because of 4th-stage failure, an orbit was not achieved. The Japanese-made Lambda-4S1, 55 feet high and weighing 16,960 pounds, was a solid-fueled booster capable of 80,000 pounds of thrust. The Sept. 26 shot was conducted primarily to test altitude control prior to the first launching of Japan's first Mu-1, a 4-stage solid-fueled rocket weighing 84,000 pounds and capable of 400,000 pounds of thrust. Dr. Hideo Itokawa of Tokyo University, head of the Japanese space program, had said that the Sept. 26 shot had less than a 25% chance of putting the 4th stage into orbit.

A 2d Japanese attempt to orbit a satellite ended in failure Dec. 20. The 4-stage Lambda-4S2 rocket was launched from the Tokyo University space center at Uchinoura, but the final stage did not ignite. The 57-pound prospective satellite presumably crashed into the ocean.

An Arctic launching site 30 miles north of Kiruna, Sweden was opened officially Sept. 24 by the 10-nation Esro (European Space Research Organization). The $7.6 million base, 100 miles north of the Arctic Circle, was to be used for an 8-year study of the aurora borealis. Plans called for the firing of about 400 British-made Skylark and French-made Centaure rockets. The first launching, which took place Nov. 20, was of a Centaure, which went up 125 kilometers (78 miles).

1967

A flash fire in their capsule killed the 3 astronauts who had been named to fly the first Apollo spaceship. The disaster took place in late January as the astronauts rehearsed for the scheduled launching. The fatalities put a damper on the Apollo program, a series of missions designed to be climaxed by landing men on the moon. In April, a Soviet cosmonaut was killed when a spaceship he was testing crashed during the first Soviet manned space flight in 2 years. The U.S. continued unmanned probes to the moon, and the U.S. and USSR launched probes of Venus as well as earth orbiting missions. International concern over the problems of space exploration led to a treaty signed by 60 nations, including the U.S. and the Soviet Union, banning mass destruction weapons in space.

DEATHS IN U.S. & USSR PROGRAMS

3 Astronauts Killed in 'Rehearsal'

The 3 astronauts who had been designated to fly the first Apollo spaceship into orbit were killed by a flash fire in the spacecraft at Cape Kennedy, Fla. Jan. 27 while rehearsing for the scheduled Feb. 21 launching. The victims were:

- Air Force Lt. Col. Virgil Ivan (Gus) Grissom, 40, one of the 7 original Mercury astronauts, who had been the pilot aboard the Mercury capsule *Liberty Bell 7* in a suborbital flight and the command pilot of *Gemini 3*, the first 2-man spaceship orbited by the U.S.
- Air Force Lt. Col. Edward Higgins White 2d, 36, co-pilot of *Gemini 4* and the 2d person (the first American) to execute an EVA (extravehicular activity, or space walk).
- Lt. Cmndr. Roger B. Chaffee, 31, who had not yet made a space flight.

The fire started at 6:31 p.m., and the spacesuit-clad astronauts apparently died in about 13 frantic seconds as they fought unsuccessfully to open the capsule's single hatch. The fire took place 218 feet above Cape Kennedy's Launch Complex 34 in an Apollo capsule (the command module) atop an unfueled 2-stage Saturn-1 rocket.

The astronauts had entered the module at 1 p.m. to begin the tests. They presumably were reclining in their flight positions when they detected the fire. According to a *N.Y. Times* report Jan. 31, a tape recording of their intercom conversation indicated that the first sign of anything wrong was this almost calm announcement by one of the astronauts, whose voice could not be identified: "Fire. I smell fire." 2 seconds later White shouted: "Fire in the cockpit!" 3 more seconds passed, and an unidentifiable voice declared in what was described as "an hysterical shout": "There's a bad fire in the spacecraft!" About 6 seconds later there were sounds described as frantic scurrying, poundings, clawings and unintelligible shouts. The last word from the capsule came 4 seconds later when Chaffee cried: "We're on fire! Get us out of here!"

According to early reports, the astronauts had no chance at all to unscrew the hatch and get out. The fire flared too rapidly in the pure-oxygen atmosphere of the closed module, and the heat was too intense. It took 5 minutes before members of the launch pad crew could open the hatch, and 27 of the pad-crew members later required treatment for smoke inhalation. The capsule's interior was described as damaged beyond repair. The rocket, however, seemed untouched except for some burned and blistered paint.

In a statement issued later Jan. 27, Pres. Johnson paid tribute to the 3 astronauts. "3 valiant young men have given their lives in the nation's service," the President said. "We mourn this great loss, and our hearts go out to their families." Vice Pres. Hubert H. Humphrey, chairman of the National Aeronautics & Space Council, also expressed his feeling of "personal loss." But, he added, the U.S. "will push ever forward in space, and the memory of these men will be an inspiration to all future spacefarers."

A 7-man inquiry board was appointed Jan. 28 by Dr. George E. Mueller, associate NASA (National Aeronautics & Space Administration) administrator for manned spaceflight, to try to find out the cause of the fire. Dr. Floyd L. Thompson, director of NASA's Langley Research Center (Hampton, Va.), was named chairman of the inquiry board. The membership of the board was increased to 12 by the appointment of 5 additional members Jan. 29.

While the Soviet press and radio paid tribute to the 3 U.S. astronauts, Soviet science writer T. Borisov asserted Jan. 28 in *Trud*, the USSR trade-union newspaper, that "this tragedy is far from being a pure accident." He charged that "the astronauts fell victim to the space race, which the men in charge of the U.S. space program launched." "Recently, ... the haste in [U.S.] space flights has continued to grow," he declared. "There were a number of flaws in the Apollo system."

(The 3 astronauts were the first known to have died in a spaceship, but 3 U.S. astronauts had been killed previously in plane crashes. Rumors of Soviet cosmonauts killed in space had never been confirmed. Later in 1967, 3 other astronauts died in accidents. Marine Maj. Clifton C. Williams Jr., 35, died Oct. 5 in a jet plane crash in Florida; Air Force Maj. Robert H. Lawrence Jr., 31, the first Negro astronaut, was killed in an F-104 crash in California Dec. 8; and Air Force Maj. Edward G. Givens Jr., 37, died June 6 in an auto accident in Texas.)

Equipment Failure Blamed

An 8-man board of review reported Apr. 9 that "some minor malfunction or failure of equipment or wire insulation" was "most probably" the cause of the Apollo spacecraft fire that took the lives of astronauts Grissom, White and Chaffee. The board concluded that "this failure ... most likely will never be positively identified." The investigators, however, reported "many deficiencies in design and engineering, manufacture and quality control" in the Apollo program that might have contributed to the tragedy. They criticized NASA and the chief Apollo contractor, North American Aviation, Inc., for "these deficiencies," which "created an unnecessarily hazardous condition." The board recommended various studies, changes and improvements to correct the defects uncovered.

The findings were made in a 2,375-page report produced after 10 weeks of investigation. Board members besides Dr. Thompson were Col. Frank Borman, an astroanut; Dr. Maxime Faget, director of engineering and development at NASA's Manned Spacecraft Center in Houston; E. Barton Geer, associate chief of the Langley center's flight vehicle and systems division; Col. Charles F. Strang, chief of the Air Force's missiles and space safety division; George C. White Jr., director of Apollo reliability and quality at NASA headquarters in Washington; Dr. Robert W. Van Dolah, research director of the Bureau of Mines' Explosive Research Center in Pittsburgh; John J. Williams, director of spacecraft operations at the John F. Kennedy Space Center at Cape Kennedy.

Chairman Olin E. Teague (D., Tex.) of the House Science & Astronautics Subcommittee on NASA Oversight asserted that "the report is a broad indictment of NASA and North American and the whole program." He said he was "disappointed and surprised" at what he described as "unbelievable" "carelessness and laxity" in the Apollo program.

The report called it "impossible" to determine exactly when the astronauts died. "It is estimated that consciousness was lost between 15 and 30 seconds after the first suit failed [was penetrated by smoke]," the report said. "Chances of resuscitation decreased rapidly thereafter and were irrevocably lost within 4 minutes." The autopsy found that the astronauts had died of "asphyxia due to inhalation of toxic gases due to fire" and that "a contributory cause of death was the thermal burns."

The board, admitting that it "was not able to determine conclusively the specific initiator" of the fire, said it had identified these "conditions which led to the disaster": "(1) A sealed cabin, pressurized with an oxygen atmosphere. (2) An extensive distribution of combustible materials in the cabin. (3) Vulnerable wiring carrying spacecraft power. (4) Vulnerable plumbing carrying a combustible and corrosive coolant. (5) Inadequate provisions for the crew to escape. (6) Inadequate provisions for rescue or medical assistance."

The report noted that "frequent interruptions and failures had been experienced in the overall communication system during the operations preceding the accident." It said "deficiences existed in command module design, workmanship and quality control, such as: (A) Components of the [capsule's] environmental control system ... had a history of many removals and of technical difficulties including regulator failures, line failures and environmental control unit failures.... (B) Coolant leakage at solder joints has been a chronic problem.... (D) Deficiencies in design, manufacture, installation, rework and quality control existed in the electrical wiring. (E) No vibration test was made of a complete flight-configured spacecraft.... (G) No design features for fire protection were incorporated." It was also found that "non-certified equipment items were installed in the command module at time of test."

Congressional Committee Hearings

Hearings on the Apollo disaster were held by the House Science & Astronautics Subcommittee on NASA Oversight Apr. 10-May 11 and by the Senate Aeronautical & Space Sciences Committee Apr. 11-May 9. Among hearing developments:

Apr. 10—NASA Administrator James Edwin Webb, 60, the opening witness, assured the House subcommittee Apr. 10 "that NASA and its contractors have the capability to overcome every deficiency required to proceed on and to successfully fly the Apollo-Saturn system and accomplish its objectives." Webb said he had asked astronaut Frank Borman, a member of the board that had criticized NASA and North American Aviation over "deficiencies" involved in the Apollo fire, "if he [Borman] felt that the deficiencies ... could be overcome in a manner that would give him confidence to fly the Apollo system." Borman's "answer was yes," Webb reported.

Webb asserted that "the capability" NASA had "demonstrated in Project Mercury and in Project Gemini, flying 20 men in 20 months and fully accomplishing the program objectives, has not all been consumed in one Apollo fire." "Whatever our faults, we are an able-bodied team," he insisted. He pointed out that NASA's difficulties "in bringing into play the strengths of over 20,000 contractors and over 400,000 men and women in industrial occupations has not permitted a cleancut or static pattern that could be established in advance and adhered to without change."

Dr. Thompson, the review board chairman, testified that "nothing is farther from the board's intent" than for its report to be interpreted as "an indictment of the entire manned space flight program and a castigation of the many people associated with the program." "Our review has been directed at faults, not the many things that have been done well," he declared.

Reps. Donald Rumsfeld (R., Ill.), John W. Wydler (R., N.Y.) and Edward J. Gurney (R., Fla.) were among the subcommittee members who criticized NASA for conducting the investigation instead of having the inquiry made by disinterested outside experts. Rumsfeld held in a statement that the board had "failed to examine, or at least report on, the fundamental conditions which permitted the accident to occur." He expressed a belief that such conditions were "the direct result of serious and fundamental management defects within NASA."

Apr. 11–J. Leland Atwood, president of North American Aviation, conceded to the House subcommittee that there might have been some relaxation of standards as a result of "overconfidence" produced by previous space successes. But he denied that the accident was caused by pressure and shortcuts taken in an effort to meet the goal of landing Americans on the moon by 1970. Although he termed the review board's report on the accident generally "excellent," Atwood disagreed with some of its criticism of his company. He said that although "some deficiencies did in fact exist," extensive review procedures had been instituted at the very beginning to eliminate or to identify and correct them. On such specific criticisms as the report's charge that the coolant used was corrosive and combustible, he pointed out that the report was referring to "the fluid added to prevent the water from freezing." "This fluid is not easily ignited," he declared, and "is not combustible below a temperature of 250°F., which is above the boiling point of water." He expressed the belief that his company had chosen "the best coolant" in view of the fact that "all possible coolants presented some problems." He similarly defended the company against other specific criticisms made in the report.

Dr. John F. McCarthy Jr., the company's vice president for research and engineering, told the House subcommittee that he and "the top technical men" in NASA and North American Aviation "feel . . . [that] one of the gravest errors we ever made" was the failure to design protection against fire in the capsule on the ground.

Astronaut Frank Borman, testifying before the Senate committee, blamed NASA and the company about equally for not evaluating the chances of such an accident. But he asserted that he would be "willing and eager" to fly in an early Apollo flight if the board's recommendations for improvements were followed.

Apr. 12–Maj. Gen. Samuel C. Phillips, NASA's Apollo program director, testified before the House subcommittee about the so-called "Phillips report"–variously described as memoranda or notes he had made criticizing North American Aviation procedures. The notes were based on several months of investigation by Phillips and a NASA group in 1965 and were turned over to the review board after the accident. Phillips said that his notes, in which he recommended engineering, manufacturing and quality-control improvements, were presented to North American Aviation Pres. Atwood Dec. 19, 1965 while Phillips was reporting orally to Atwood on his findings. Phillips said the company had accepted his report, "and a very aggressive effort . . . was applied immediately to these problems." On the advice of NASA's counsel, Phillips at least tentatively declined to give the subcommittee a copy of the "Phillips report."

Apr. 13–Dr. Robert C. Seamans Jr., deputy NASA administrator, told the Senate committee that prior to the production of the "Phillips report," "North American had not . . . shown sufficient dedication to the engineering design or workmanship on the job" and "did not address itself properly to the training and supervision of its personnel and the adequate inspection of the work that was done." He reported that "there had been marked improvement at North American in the last year and a half," but, he said, "it should not have been necessary for the government to take this kind of strenuous action."

Apr. 17–Webb denied to the Senate committee that unnecessary risks had been taken in order to meet the goal of getting Americans on the moon before 1970. But he questioned whether "we should have ever flown the Block 1 spacecraft [the one in which the 3 astronauts had been killed] in the first place since we know there are many deficiencies which we've already fixed in Block 2 [the next series of Apollo spaceships]." One of the major changes permitted much quicker entry and exit through the Block 2 hatch since the Block 2 craft, unlike the Block 1 vehicle, were to be used for rendezvous and docking in space.

Apr. 18—Webb gave the House subcommittee an 8-page digest of the "Phillips report." Webb reported in his summary: "North American Aviation experienced a trend of slippages in key milestone accomplishments, shortcomings in equipment performance, and increasing costs. By the fall of 1965 it was apparent that the rate of progress [in spaceship and rocket booster projects] ... was not sufficient to meet the requirements of the Apollo program." Following the Phillips inquiry, North American Aviation "organized an extensive effort" to improve. In Apr. 1966 "we determined that the contractor had made substantial progress to correct many of the deficiencies noted and that the planned future course of action provided confidence that the required rate of improvement ... would be achieved. ... I do not intend, however, to leave the impression that we were completely satisfied with North American's performance."

Apr. 21—Thomas R. Baron, who had been dismissed as a North American Aviation quality control inspector, told the House subcommittee at a session at Cape Kennedy that a former co-worker, Mervin Holmburg, a North American electrical technician, had told him Feb. 2 that the astronauts had smelled smoke about 12 minutes before they died and had tried to escape for 5 minutes. Holmburg appeared before the subcommittee a few minutes later and denied Baron's story. Baron, who had written 3 long reports accusing North American Aviation of mismanagement, "gets all his information from anonymous phone calls and people dropping a word here and there," Holmburg said. Rep. Ken Hechler (D., Va.), a subcommittee member, said the taped conversation of the astronauts before the accident also refuted the assertion that they had detected the fire 12 minutes before they died. (Baron was killed Apr. 28 when his car collided with a train near Titusville, Fla. His wife, Marlene, 27, and a stepdaughter, Penny Frye, 4, also died in the accident. A 2d stepdaughter, Robin Frye, 7, was injured.)

Apr. 26—Rep. William F. Ryan (D., N.Y.), a member of the House subcommittee, announced that he had obtained a copy of the "Phillips report" and that it indicated "incredible mismanagement" of the Apollo program. (Ryan made parts of the text public Apr. 29.) Among criticisms of North American Aviation that Ryan said the report contained: (a) "The condition of hardware shipped from the factory, with thousands of hours work to complete, is unsatisfactory to NASA." (b) "North American Aviation quality is not up to NASA required standards." A "large number of discrepancies ... escape North American inspectors but are detected by NASA inspectors." (c) "Effective planning and control from a program standpoint does not exist." (d) "There is little confidence that North American will meet its schedule and performance commitments within the funds available for this portion of the Apollo program."

May 4—North American Aviation Pres. Atwood told the Senate committee that improvements and changes his company was making should make it possible for the U.S. to achieve its goal of landing Americans on the moon "in this decade."

May 9—Webb told the Senate committee that the fire would add at least $75 million to the cost of the lunar program. He confirmed that a 190-member source evaluation board that had checked the bids for the Apollo program in 1961 had named the Martin Marietta Corp. as its first choice on the basis of technical superiority. (Webb had told the committee Apr. 17 that North American Aviation had been the board's first choice.) Webb said that he and 3 other senior officials (Dr. Robert Seamans, Dr. Robert F. Gilruth and the late Dr. Hugh L. Dryden) had awarded the contract to North American Aviation instead "for 2 reasons"—"because their proposal was lowest in cost and... because they had the most experience." Webb said that the job of integrating and checking out Apollo spacecraft before launching was being shifted from North American Aviation to the Boeing Co. He indicated that North American Aviation would be excluded from the job of preparing spacecraft for missions that were to follow the initial lunar landing.

May 10—Webb told the House subcommittee that after the source evaluation board had recommended Martin Marietta, he and the other senior NASA officials had ordered a re-evaluation because they had found that the board "had not done a full and complete job." The re-evaluation showed North American Aviation to be "superior," Webb testified.

Shake-Up in Top Echelon

One of the first major results of the Congressional investigation of the fatal Apollo fire was the signing by NASA June 16 of a contract for the Boeing Co. to help manage the Apollo project. Boeing said that one of its tasks would be to integrate the Saturn-5 rocket with the Apollo spacecraft and lunar landing vehicle.

North American Aviation, Inc., builder of the Apollo spacecraft, had announced May 1 that Harrison A. Storms Jr., 51, a North American vice president who had headed the company's Apollo program as president of North American's Space & Information Division, had been replaced in the job by William B. Bergen, 52. Bergen, a former president of the Martin Marietta Co., had joined North American Apr. 17 as a vice president.

NASA had announced Apr. 5 that Dr. Joseph F. Shea, 40, who had been manager of the Apollo spacecraft office in Houston since 1963, had been transferred to NASA headquarters in Washington, where he was to serve as deputy associate NASA administrator for manned space flight. He was replaced by Vienna-born George M. Low, 41, who had been deputy director of the Manned Spacecraft Center in Houston. (Shea resigned in July to become a Polaroid Corp. vice president.)

Maj. Gen. John G. Shinkle resigned May 18, effective May 31, as Apollo program manager for NASA's Kennedy Space Center. NASA officials said his resignation was not connected with the Apollo fire. Rear Adm. Roderick O. Middleton was named July 18 to succeed Shinkle.

NASA July 26 picked the Martin Marietta Co. as engineering and scientific coordinator for major space projects that would follow the Apollo lunar landing. William D. Smith, who had been project manager of Martin Marietta's Titan-2 rocket development, was named general manager of the program, which NASA had designated the Apollo Applications Program.

NASA Administrator James Webb announced Oct. 2 the resignation of Dr. Robert C. Seamans Jr., 48, as NASA deputy administrator (effective Jan. 1, 1968). NASA also announced that Air Force Brig. Gen. Carroll Bolender was replacing William A. Lee as manager of the Lunar Excursion Module program. NASA said Lee had resigned for "personal reasons."

The redesigned Apollo command module was pronounced "quite fire-safe" Dec. 6 after a series of exhaustive tests at the NASA Space Center in Houston, Tex. The oxygen-filled capsule withstood 41 attempts to set it afire by overloading its wiring and heating its solder at danger points.

Cosmonaut Killed in Crash

Col. Vladimir Mikhailovich Komarov, 40, a Soviet cosmonaut, was killed Apr. 24 in the crash of a new-type Soviet spaceship he was testing. Komarov, who had just brought his *Soyuz* (*Union*) *1* out of orbit after 18 revolutions around the earth, was the first man definitely known to have been killed in a space flight.*

Soyuz 1, with Komarov aboard as the sole crew member, had been launched at 3:35 a.m. Moscow time Apr. 23 from the Baikonur cosmodrome at Tyuratam in Kazakhstan. Tass reported that the multi-stage launching rocket had put the spaceship into an orbit with an apogee of 224 kilometers (139.1 miles), perigee of 201 kilometers (124.8 miles), period of 88.6 minutes and inclination of 51° 40' to the equatorial plane. *Soyuz 1* traveled in orbit at a speed of 18,641 mph. According to Western authorities, *Soyuz 1*, the heaviest manned Soviet spaceship sent up so far, weighed about 15,000 pounds.

Western sources reported that trouble apparently had developed aboard *Soyuz 1* early in the flight and that the spaceship was tumbling dangerously toward the end of the flight. They expressed the conviction that the Russians had cut the flight short because of the control failure. According to Western experts, Komarov tried unsuccessfully to

---

*Julius Epstein, a research associate at the Hoover Institution on War, Revolution & Peace (at Stanford University), asserted Apr. 30 that U.S. authorities knew of at least 11 other Soviet cosmonauts killed in space-flight accidents. Epstein indicated that the alleged victims probably included Piotr Ivanovitch Dolgov, Serenty (or Terentiy) Shiborin, Vassilyevitch Zovodovsky, Aleksei Belokonev, Ivan Kachur (or Kascheur), Aleksis Gratzev (or Aleksei Grachev) and Gennady Michailov.

stabilize the spaceship sufficiently for re-entry into the atmosphere during the 16th and 17th revolutions around the earth. He finally succeeded during the 18th revolution and brought the *Soyuz 1* out of orbit shortly after 6 a.m. Apr. 24 after 26 hours 35 minutes in space. Tass reported, however, that when the main parachute of the descending spaceship was opened at an altitude of 7 kilometers (4.3 miles), the parachute lines "got snarled." The *Soyuz 1* crashed in an undisclosed location in the USSR, and Komarov was killed.

Messages of sympathy were sent to the Soviet people and to Komarov's family by Pres. Johnson and other world leaders and by the U.S.' 47 astronauts. An urn containing Komarov's ashes was buried in the Kremlin wall in Moscow Apr. 26 following a state funeral.

Komarov, who died on his 2d space flight, was the first Soviet cosmonaut to make more than one trip in space. He had been the pilot of the 3-man *Voskhod 1* in its 16-orbit flight Oct. 12-13, 1964.

Komarov's flight, the first manned Soviet space flight in 2 years, had been preceded by unconfirmed reports from Moscow and elsewhere that the USSR was readying a space "spectacular." The Soviet government gave no hint that a space shot was planned until more than an hour after Komarov was in orbit. The Russians then continued to withhold most data on the flight except for items such as orbital measurements that Western experts would have derived anyway from their own observations. (Komarov's wife told a *Pravda* reporter that she had not known of his mission until after the launching.)

Col. Yuri A. Gagarin, the first man to fly in orbit, had hinted in the magazine *Ogonyok* Apr. 8 that the launching of a large manned space station for a flight of "long" duration in orbit was "not far off." "It will be necessary to supply them [the station's crew members] and change crews with the use of simplified types of space vehicles," Gagarin said. Reuters had reported from Moscow Apr. 21 that, according to "informed sources," "Russia's 11 astronauts and top scientists" had gone to the Baikonur cosmodrome. Other reports said the Soviet space experts probably were planning to put at least 2 spaceships into orbit with multi-manned crews, to attempt the first Soviet rendezvous and "docking" in space and the first exchange of crew members in space.

(Warsaw Radio had reported Mar. 9 that Lt. Gen. N. P. Kamanin, commander of the cosmonauts' training school and a member of the USSR's State Commission for Space Exploration, had denied Western speculation that Russia was getting ready to orbit a spaceship with a crew of 6 to 8. Kamanin said the West credited the USSR with a greater capacity than it had and thus reduced the impact of Russia's actual exploits. But he asserted that the USSR did have "tremendous capacity" and was planning a complex 1967 flight program whose preparation needed the more than 2 years that had expired since the conclusion of the Voskhod flights in Mar. 1965.

(Western experts speculated that the spaceships the USSR planned to use in the new program had been tested without crews in launchings announced as Cosmos shots Nov. 28, 1966 and Feb. 7, Mar. 10 and Apr. 8, 1967. Soviet space scientist Mstislav V. Keldysh, president of the Soviet Academy of Sciences, confirmed in a speech at Komarov's funeral Apr. 26 that the *Soyuz 1* had been tested in "unmanned flights" before Komarov's disastrous flight. Tass had said in its Apr. 23 announcement that "the aims" of Komarov's flight were: "testing of the new piloted spaceship, checking of the ship's systems and elements in conditions of space flight, . . . expanded scientific and physical-technical experiments and studies in conditions of space flight, the further continuation of medical and biological studies and studies of the influence of various factors of space flight on human organism.")

James E. Webb, administrator of the U.S. National Aeronautics & Space Administration, suggested in his statement of sorrow Apr. 24 that U.S.-Soviet cooperation might prevent space accidents. Asserting that "we at NASA want to make every realistic effort" toward such cooperation, Webb asked: "Could the lives already lost have been saved if we had known each other's hopes, aspirations and plans? Or if full cooperation had been the order of the day?"

LUNAR PROBES

3d U.S. Probe Orbits Moon

The 850-pound unmanned U.S. *Lunar Orbiter 3* was launched from Cape Kennedy Feb. 4 on a 231,600-mile journey that brought it into orbit around the moon Feb. 8. The mission of the TV-equipped probe was to send back photos of the lunar surface both for scientific purposes and to help select a site for a manned landing on the moon.

Lunar Orbiter 3 rose from Cape Kennedy's Launch Complex 13 at 8:17 p.m. Feb. 4 in the nose of an Atlas-Agena-D rocket carrier. After the burned-out Atlas stage was jettisoned, the Agena engine ignited and carried the spacecraft into "parking orbit" around the earth. The Agena engine was shut off temporarily for a coasting period and then reignited to thrust the probe into a lunar trajectory. The Agena then separated from the spacecraft. A NASA spokesman said the launching was so "fantastically accurate" that *Lunar Orbiter 3* would have gone into orbit around the moon even without the usual mid-course correction. The course-correction maneuver, however, was executed by radio signal from the ground at 10 a.m. EST Feb. 6 to give the probe a higher initial orbit.

Lunar Orbiter 3 was put into orbit around the moon by means of a 9-minute 1-second firing of its rocket at 4:54 p.m. Feb. 8. Its initial orbit had an apocynthion (maximum altitude above the lunar surface) given as 1,118 miles, a pericynthion (minimum distance from the surface) of 131 miles, a period of 3 hours 35 minutes and a 21° angle of inclination to the plane of the moon's equator. By a 4-second rocket firing Feb. 12, the orbit's pericynthion was reduced to 32 miles as required for the photo mission (the apocynthion became 1,147 miles), and picture taking was started Feb. 15.

NASA (National Aeronautics & Space Administration) announced Apr. 6 that 8 possible sites for manned lunar landings had been selected on the basis of the photos provided by the 3 Lunar Orbiters.

NASA destroyed *Lunar Orbiter 3* Oct. 9 and *Lunar Orbiter 2* Oct. 11 by sending them crashing into the moon. This was done because their attitude control gas supplies were nearly depleted and neither had enough power to survive the lunar eclipse of Oct. 18. NASA said that by destroying them, it freed their radio frequencies for use by future probes. *Lunar Orbiter 2* was believed to have crashed at a point at Longitude 98° E. and Latitude 4° S. The impact point of *Lunar Orbiter 3* was believed to be at Longitude 91.7° W. and Latitude 14.6° N.

Surveyor 3 Digs Into Moon

The U.S. lunar probe *Surveyor 3* was launched from Cape Kennedy Apr. 17 on a successful mission to "soft-land" on the moon, "sample" the lunar surface with a small mechanical shovel and send back photos.

The lunar probe was sent up from Launch Complex 36B by means of a 2-stage Atlas-Centaur rocket at 2:05 a.m. EST Apr. 17. After the Atlas-D first stage was jettisoned, the Centaur 2d stage put *Surveyor 3* first into parking orbit and then into lunar trajectory. The Centaur was then jettisoned. The trajectory was so accurate that the probe would have landed only 295 miles from its aiming point without course correction. But a correction maneuver, directed by radio signal from the ground, took place at midnight Apr. 17 to put *Surveyor 3* directly on course.

As *Surveyor 3* approached the moon Apr. 19, its retro-rocket was fired at 7:01 p.m. EST to reduce its nearly 6,000-mph. speed. This took place at an altitude of 55 miles from the moon's surface. 3 small engines were ignited at an altitude of 35,000 feet to slow the 3-legged probe further and to put it in landing position. The engines, directed by on-board radar, were supposed to shut off at an altitude of 14 feet. Due to a minor radar failure, however, the shut-off was delayed slightly, and the probe bounced 3 times before it came to rest at 7:04 p.m. on the eastern edge of the dry Sea of Storms. The landing spot was given as Latitude 2° 98′ S. and Longitude 23° 40′ W. The site was 2.4 miles from the aiming point. The 620-pound spacecraft apparently suffered no damage. It rested at an angle of 10° on the side of a lunar crater about 150 feet in diameter and 20 feet deep.

Surveyor 3 began transmitting photos of its surroundings within an hour of landing, and it began digging with its mechanical scoop early Apr. 22.

Dr. Robert F. Scott, chief scientist on the digging project, told reporters after examining the first digging photos Apr. 22 that the moon's dry surface "is like ordinary soil," "a lot like fine-grained sod or damp beach sand." He said that the 2-by-5-inch "claw left a clear impression" and that "this means the soil must be sticky and cohesive, not dusty or sandlike." Calculations showed that the soil "can hold about 6 pounds per square inch on the surface," he reported, "and we have some evidence it can hold more just below the surface." (6 pounds per square inch was more than strong enough to hold men and spaceships.) The mechanical claw, operated by 4 electric motors, was capable of reaching out 5 feet and digging to a depth of 18 inches. It could press the surface with a maximum force of 8 pounds, dig and pull back with a force of 26 pounds, pick up rocks with 300 pounds of force and be dropped from a height of 40 inches with a force of almost 300 pounds. It operated on signal from earth-based stations.

During an eclipse of the moon by the earth Apr. 24, *Surveyor 3* photographed the earth and Venus and took photos of the lunar surface through red, blue and green filters.

Surveyor 3 dug 4 small trenches, made 7 bearing-strength tests and 13 penetration tests and transmitted 6,315 pictures back to earth before the coming of the lunar night May 3 permanently ended contact. At a press conference in Washington May 23, Dr. Eugene Shoemaker of the U.S. Geologic Survey reported that *Surveyor 3* photos had shown the moon to be gray in color, with the shade varying "from a pale to a very dark gray."

4th Lunar Orbiter Circles Moon

The 860-pound *Lunar Orbiter 4* was launched from Cape Kennedy at 6:25 p.m. EDT May 4 on a 245,519-mile trip that put it into a high elliptical orbit around the moon May 8. Its mission was to provide photos of the moon's surface and to relay other scientific data.

The unmanned probe rose from Cape Kennedy's Launch Complex 13 May 4 in the nose of a 2-stage Atlas-Agena-D rocket. After the spent Atlas stage was jettisoned, the Agena stage put *Lunar Orbiter 4* into a parking orbit around the earth and then ignited a 2d time to inject the probe into its lunar trajectory. After ground-directed firing of its retro-rocket, *Lunar Orbiter 4* completed its trip at 8:08 a.m. May 8 and began circling the moon in an orbit with a 3,790-mile apocynthion, 1,677-mile pericynthion, 12-hour period of revolution and 85.8° angle on inclination to the plane of the moon's equator.

Lunar Orbiter 4 began taking photos of the moon's surface May 11 and transmitting them to the earth. A photo of the moon's south polar area revealed a 200-mile-long fissure similar to such terrestrial earthquake faults as the San Andreas fault in California. The photo mission was ended May 26 because of a failure of a switch in the picture readout system. Before the failure, *Lunar Orbiter 4* took and processed 163 of the 180 pictures planned for the mission. The NASA reported June 2 that the probe had provided telephoto pictures of 99% of the moon's "front" face and that the photos "provide scientists 10 times finer resolution than best existing telescope views." "In conjunction with its 3 predecessors, *Lunar Orbiter 4* raised to more than 75% the total coverage of hidden side features." (By 2 earth-directed engine ignitions June 5 and 8, the apocynthion of *Lunar Orbiter 4*'s orbit was lowered from 3,850 miles to 2,450, the pericynthion was dropped from 1,625 miles to 48 and the period was decreased to 5 hours 44 minutes.)

5 More U.S. Lunar Probes

During the remainder of 1967 NASA sent 5 more U.S. probes toward the moon. All but one were successful in their orbital or soft-landing missions.

The 2,290-pound *Surveyor 4*, equipped with TV cameras and a digging tool, was launched from Cape Kennedy at 7:53 a.m. EDT July 14 on a mission calling for a soft landing in the moon's Sinus Medii (Central Bay). But at 10:06 p.m. EDT (7:06 p.m. PDT) July 16, as *Surveyor 4* was completing its 225,569-mile journey and was only 3 minutes from landing, the Jet Propulsion Laboratory (JPL) in Pasadena, Calif. lost radio contact with the probe. Efforts to reestablish contact failed, and the JPL stopped trying July 18. JPL scientists had said before the launching that the terrain of the Sinus Medii was so rough that the probe had only a 50% chance of a successful soft landing. The probe's direct-ascent launching, by means of a 2-stage Atlas-Centaur rocket assembly, was so accurate that *Surveyor 4* would have landed within 93 miles of its target site even without a course-correction maneuver, which was executed successfully July 15. The target was in the center of the "front" face of the moon at Longitude 1° 20′ W. and Latitude 0° 25′ N.

The U.S.' 230-pound *Explorer 35* was launched from Cape Kennedy at 10:19 a.m. EDT July 19 into a lunar trajectory so accurate that NASA experts dispensed with the usual mid-course correction. The probe's retro-rocket was fired by radio signal from the earth at 5:19 a.m. July 22, and *Explorer 35* was put into a lunar orbit with an apocynthion of 4,300 miles, pericynthion of 500 miles, 11-hour period and 147° angle of inclination. The probe, carried into its lunar trajectory by a 3-stage Delta rocket, was given the primary mission of studying the solar wind and the interplanetary magnetic field at lunar distances and interactions with the moon. It also collected data on the lunar radiation environment, dust distribution around the moon, the lunar gravitational field and the weak lunar ionosphere.

The 860-pound *Lunar Orbiter 5* was launched from Cape Kennedy the night of Aug. 1 by means of a 2-stage Atlas-Agena rocket assembly. The Agena 2d stage put the probe into a 115-mile-high parking orbit and then ignited a 2d time to send it into a lunar trajectory. After a mid-course correction, *Lunar Orbiter 5* completed its 250,187-mile trip to the vicinity of the moon Aug. 5, and its retro-rocket was fired to put the probe into its initial orbit around the moon. *Lunar Orbiter 5* began its photo mission Aug. 6 while in an orbit with an apocynthion of 3,740 miles, pericynthion of 122 miles, 85° angle of inclination to the moon's equator and period of 8 hours 22 minutes. By signal from the earth Aug. 8, the probe's rocket was fired to reduce the apocynthion to 931.7 miles and the period to 3 hours 11 minutes. Picture-taking was resumed Aug. 9.

The 5th and final probe of the Lunar Orbiter program, *Lunar Orbiter 5* successfully completed its assigned tasks of re-photographing 5 potential Apollo landing sites and photographing (a) possible sites for Surveyor landings, (b) the remaining 40% of the "hidden" side of the moon and (c) sites of scientific interest such as the "hot spots" of the Aristarchus Crater. *Lunar Orbiter 5* was then directed to concentrate on its additional scientific missions of meteoroid and radiation measurement and of selenodesy (the study of the moon's shape and gravitational field).

The 616-pound *Surveyor 5* was launched before dawn Sept. 8 on a successful mission in which it made a soft landing on the moon Sept. 10

and then chemically analyzed the moon's soil. It found that the most abundant elements on the moon were oxygen—as oxides in combination with other elements—and silicon.

Rising from Cape Kennedy's Launch Complex 36B in the nose of a 2-stage Atlas-Centaur booster rocket, the lunar probe first went into a parking orbit around the earth and then into a lunar trajectory so accurate that it would have landed within 37 miles of its target in the moon's dry Sea of Tranquility even without a course-correcting maneuver executed Sept. 9. Despite a leaking helium valve in *Surveyor 5*'s onboard rocket system, Jet Propulsion Laboratory (JPL) experts, by means of radioed commands, brought the probe to a soft landing on target on the moon at 8:46 p.m. EDT Sept. 10.

After receiving reassuring radio signals and TV photos from *Surveyor 5*, the JPL scientists Sept. 11 sent radio orders that caused the probe to lower its 6-inch-square chemical analysis device to the moon's surface at the end of a nylon cord. The device, which was then put into operation, produced its data by bombarding the moon's surface with subatomic alpha particles and measuring the resulting scattering of particles. Since scatter characteristics vary according to the composition of the substance bombarded, this technique enabled the scientists to determine the composition of the moon's soil.

Surveyor 5 sent to earth an impressive amount of scientific data and a record 18,006 photos before its power was shut off for the lunar night Sept. 24. With the return of lunar day Oct. 15, the probe's instruments were reactivated and communication with earth renewed.

Surveyor 6 was launched toward the moon Nov. 7 by means of a 2-stage Atlas-Centaur booster rocket fired from Cape Kennedy at 2:39 a.m. EST. After a course-correction maneuver executed at 6:20 p.m. Nov. 7, *Surveyor 6* made a successful soft landing at 8:01 p.m. EST Nov. 9 in Sinus Medii in the center of the face of the moon. *Surveyor 6* landed only 3 miles from its aiming point and 5 miles from *Surveyor 4*. 50 minutes after completing its 253,922-mile flight, the spacecraft transmitted to earth high-quality TV photos of its surroundings. On radio signal from earth Nov. 17, *Surveyor 6* ignited its rockets briefly to hop about 8 feet sideways. The change in position enabled it to transmit photos from a new angle. The new photos, paired with the 12,000 taken before the hop, made stereoscopic effects possible.

Surveyor 6 was essentially identical to *Surveyor 5* and also carried an alpha-particle soil-analysis kit, which was put into operation Nov. 11. The spacecraft had weighed 2,223 pounds at launch. But it jettisoned its 1,395-pound retro-rocket and used up enough of its liquid propellants and attitude control gas to bring its weight down to about 656 pounds (as measured on earth) by the time it landed.

Lunar Findings

The moon is a "cold, non-magnetic, non-conducting sphere," Dr. Norman F. Ness reported in the December issue of the *Journal of Geophysical Research*. Ness, *Explorer 35* project scientist at the NASA Goddard Space Flight Center in Greenbelt, Md., came to this conclusion on the basis of data provided by the U.S.' *Explorer 35*, which was circling the moon in orbit. According to these findings, the moon has practically no magnetic field and therefore is not surrounded by radiation belts. No evidence of a lunar ionosphere was found. Based on the findings of a low average electrical conductivity, calculations indicated that the moon's internal temperature was less than $1,800°$ F.

According to NASA, *Lunar Orbiter 1,* launched Aug. 10, 1966, had photographed about 2 million square miles of lunar surface and "provided the first detailed scientific knowledge of the lunar gravitational field and topographic and geological information of direct benefit to the Apollo program and to scientific knowledge of the moon." *Lunar Orbiter 2*, launched Nov. 6, 1966, "provided wide-angle and telephoto coverage of more than $1\frac{1}{2}$ million square miles of the moon's surface not covered by *Orbiter 1*" and added to the data on the moon's gravitational field. *Lunar Orbiter 3*, launched Feb. 4, 1967, made it possible for NASA to announce "the selection of 8 sites suitable" for the manned Apollo landing. (The NASA July 6 gave the Boeing Co. a $1,053,405 incentive award in recognition of *Lunar Orbiter 3* achievements.) *Lunar Orbiter 4,* launched May 4, 1967, "acquired telephoto pictures of 99% of the moon's front face" with "10 times better resolution than the best existing telescopic views and, for most of the area covered . . . , a hundredfold increase in discernible detail."

Dr. Anthony Turkevich of the University of Chicago, who directed the lunar soil analysis experiment of *Surveyor 5*, reported at a press conference at NASA headquarters in Washington Sept. 29 that, according to preliminary conclusions from the experiment, "the most abundant elements on the lunar surface are the same as the most abundant elements making up the surface of the earth." Much of the moon's surface seemed to be made of basaltic soil and volcanic rock. The soil was reported to consist 53%–63% of oxygen, $16\frac{1}{2}$%–$21\frac{1}{2}$% of silicon, 10%–16% of sulphur, iron, cobalt and nickel, $4\frac{1}{2}$%–$8\frac{1}{2}$%, of aluminum and, probably, of lesser quantities of magnesium, carbon, sodium and elements heavier than nickel. (The U.S. Oct. 28 filed with the UN a 60-page report on *Surveyor 5*'s scientific findings. The treaty banning mass-destruction weapons in space called for the sharing of such knowledge through the UN. This was the first report filed under the treaty.)

USSR & U.S. Send Probes to Venus

The USSR and the U.S. sent successful, unmanned probes to the planet Venus. The Soviet's instrument package made a parachute landing on the surface, and the U.S.' probe passed within 2,480 miles of Venus' surface.

The USSR's 2,437-pound *Venera (Venus) 4* was launched at 5:40 a.m. June 12. It was first put into a parking orbit around the earth and was then sent on toward Venus. *Venera 4* was the 4th Venus probe acknowledged by the USSR, but Western sources reported at least 7 others, all failures. Of the 3 acknowledged previous probes, *Venera 1* flew past Venus in 1961, *Venera 2* flew past in 1966, and *Venera 3* crashed into the planet in 1965. Communications difficulties apparently kept any of these probes from providing data. The USSR made no disclosure that it was planning a 1967 Venus probe until *Venera 4* seemed successfully embarked on its 212 million-mile trip. The announcement gave few details. *Venera 4* was the heaviest Venus probe known to have been launched by the USSR or the U.S.

An instrumented capsule ejected from *Venera 4* landed by parachute on Venus shortly after 9 a.m. Moscow time Oct. 18. During its 90-minute descent to the surface, the egg-shaped instrument-laden capsule radioed to earth data on Venus' environment. The 842-pound capsule reported temperatures of up to 536° F. in Venus' atmosphere, but it stopped transmitting on landing (or shortly before or after landing), and Soviet spokesmen said the probe's mission had been successfully completed.

Venera 4, launched June 12, took 128 days to travel the curved 350 million-kilometer ($217\frac{1}{3}$ million-mile) path that brought it to Venus. Venus was about 49 million miles from the earth when the meeting took place. After its 35.4-inch capsule was ejected, the remaining carrier section of the probe was burned "to ashes" by friction with Venus' dense atmosphere, according to Tass, the Soviet news agency.

Tass announced the Venus landing Oct. 18 about 7 hours after it had taken place. Tass reported that *Venera 4* had "entered the atmosphere of Venus" at 7:34 a.m. Moscow time and that the "landing apparatus—a scientific laboratory—was jettisoned." This capsule, protected by a heat-resistant coating, had a weighted bottom so that its antenna would point straight up—toward the earth.

"After the landing apparatus decelerated ... through an aerodynamic braking in the planet's atmosphere," Tass reported, "a special-purpose parachute system operated and the apparatus continued a smooth descent in Venus' atmosphere." During the 90 minutes of the 25-kilometer ($15\frac{1}{2}$-mile) descent, "scientific instruments ... were continuously and steadily taking measurements ... of the atmosphere and transmitting the data to the earth," Tass said. "The apparatus landed on the planet's surface and delivered the 2d pennant with the national emblem of the Soviet Union [the first USSR pennant had crashed into the planet with *Venera 3* in 1965]."

Tass reported that during the descent, "the temperature of the atmosphere changed from 40° to 280° C. [104° to 536° F.] and the atmospheric pressure from one to about 15 atmospheres [15 times the earth's atmospheric pressure at sea level]." It was indicated later that the atmospheric pressure at the surface of Venus might be as high as 22 times the earth's sea-level atmospheric pressure.

According to later calculations, *Venera 4* found Venus' atmosphere to be about 90%–95% carbon dioxide, 0.4% oxygen and 1.6% water vapor. "No noticeable traces of nitrogen were discovered," Tass revealed. But the Russians conceded later that the probe's instruments were not very sensitive and could have failed to detect existing nitrogen. *Venera 4* reported that Venus had no magnetic field or radiation belts (a phenomenon created by a magnetic field), but it also disclosed that a slight increase in magnetism had been detected as the probe approached the planet. A weak hydrogen corona, its strength about 1/1,000 of that of the earth's corona, was also detected.

The USSR had not revealed in advance that a landing on Venus was planned, and a Soviet spokesman had said specifically, shortly after the launching, that *Venera 4*'s mission called for the probe to pass the planet. Prof. Mstislav V. Keldysh, president of the Soviet Academy of Sciences, telegraphed Oct. 16 to request that Britain's Jodrell Bank radio-telescope observatory monitor the probe as it approached Venus and record the data it sent. Even then, however, Keldysh did not say the probe was to land on the planet. In answer to a press conference query in Moscow Oct. 17, however, Keldysh admitted that a landing apparently had been scheduled although, he pointed out, "it is impossible to be certain that everything will work out in such a way as to make a soft landing possible." While other Soviet scientists had hinted that a Venus landing might be accomplished by parachute, Keldysh did not disclose that *Venera 4* had such landing equipment.

Jodrell Bank reported the *Venera 4* landing 90 minutes before the actual touch-down; Sir Bernard Lovell, the observatory's director, explained later that he had mistaken the ejection of the capsule for the landing.

Tass reported that *Venera 4* had landed on the night side of Venus about 1,500 kilometers (930 miles) from the terminator (the line on the surface that separates day from night).

The U.S.' 540-pound *Mariner 5* was launched from Cape Kennedy at 2:01 a.m. EDT June 14, 2 days after *Venera 4*'s launching, by means of a 3-stage Atlas-Agena rocket assembly. Its course was corrected June 19 by radio signal from the JPL (Jet Propulsion Laboratory). The probe's primary mission was "to obtain scientific information on the structure of the planet's atmosphere and on its radiation and magnetic environment." *Mariner 5* was the 3d Venus probe launched by the U.S. Of the first 2, both sent up in 1962, *Mariner 1* veered off-course and was destroyed shortly after launching, but *Mariner 2* flew successfully past Venus and provided a large amount of scientific data.

Mariner 5 flew within 2,480 miles of the surface of Venus at 10:35 a.m. PDT Oct. 19 and then continued on in orbit around the sun. For more than 2 hours, while in the vicinity of the planet, *Mariner 5*'s instruments collected data on Venus' environment. A small amount of the data was radioed immediately to earth, but most of the information was recorded, and it took 34 hours for the probe to transmit the full record to earth stations.

Preliminary findings produced by *Mariner 5* were disclosed at the JPL in Pasadena, Calif. Oct. 23 by 8 of the scientists involved in the experiments. These findings indicated that Venus' atmosphere was composed 72%–87% of carbon dioxide, had some nitrogen but had no detectable oxygen. This conclusion contradicted the findings produced by the USSR's *Venera 4*. The U.S. scientists said *Mariner 5* had detected some magnetic activity but no magnetic field around Venus; the probe's instruments, however, would not have been able to detect a magnetic field weaker than 1/300 of the earth's. A halo of hydrogen was found around Venus, and the scientists reported that the planet's night side emits a faint ultra-violet glow. A radio occultation test indicated that the atmosphere of Venus might be as dense and have as high a pressure as the *Venera 4* findings indicated.

INTERNATIONAL DEVELOPMENTS

Arms-Ban Treaty Signed

A treaty to ban mass-destruction weapons from outer space was signed Jan. 27 by representatives of 60 countries. The "Treaty on Principles Governing the Activities of States in the Exploration & Use of Outer Space, Including the Moon & Other Celestial Bodies," was signed at separate ceremonies in Washington, London and Moscow. Representatives of the treaty's depository countries (Britain, the USSR and the U.S.) signed copies at all 3 locations. The treaty, which had been approved by the UN General Assembly Dec. 19, 1966, did not go into effect with the Jan. 27 signings since it required the ratification of at least 5 states, including the 3 depository countries.

Speaking at the Washington ceremony, Pres. Johnson hailed the signing as an "inspiring moment in the history of the human race." He voiced the hope that U.S. and Soviet astronauts "will meet someday on the surface of the moon as brothers and not as warriors," and he added that the treaty represented a "very hopeful sign" for further East-West agreements. Soviet Amb.-to-U.S. Anatoly F. Dobrynin declared: "Let us hope we shall not wait long for similar approval of other urgent problems." State Secy. Dean Rusk and U.S. Amb.-to-UN Arthur J. Goldberg signed for the U.S., Dobrynin for the USSR and UK Amb.-to-U.S. Sir Patrick Dean for Britain. Among states not signing were France, Communist China, Cuba and Albania. (Hsinhua, the official Communist Chinese press agency, accused the USSR Jan. 28 of betraying the Vietnamese people by cooperating with the U.S. in the treaty.)

The treaty was ratified by the U.S. Apr. 25 and by the Soviet Union May 19. (The U.S. Senate voted unanimously [88–0] Apr. 25 to ratify the treaty. Military leaders had assured Senators earlier that the pact would not endanger national security.) The treaty came into force Oct. 13 when 13 nations, including Britain, the U.S. and the USSR, the treaty's depository nations, deposited their ratifications with the UN. By that date, 84 nations had signed the treaty. In addition to the depository nations, the following states deposited their ratifications Oct. 10: Bulgaria, Niger, Czechoslovakia, Hungary, Finland, Sierra Leone, Denmark, Canada, Japan and Australia. Speaking at a Washington ceremony marking the event, Pres. Johnson expressed the hope that the treaty would help end the wasteful "competitive spacemanship" between the U.S. and the Soviet Union. Alluding to the USSR's launching of *Sputnik 1* Oct. 4, 1957, Johnson said that the first decade of space exploration had been "a kind of a contest" but the 2d should be a "partnership."

(The 28-member UN Scientific & Technical Subcommittee on the Peaceful Uses of Outer Space reported Sept. 6 that it had failed in efforts to produce an agreed definition of the term "outer space." The subcommittee concluded that it was "not possible at the present time to identify scientific or technical criteria which would permit a precise and lasting definition of outer space." The subcommittee had agreed Feb. 13 to recommend at the special General Assembly session in April that a proposed UN conference on the exploration and the peaceful uses of space be postponed for one year. The conference was to have begun in September in Vienna. The postponement had been called for by the USSR Feb. 6 to allow more time for preparation. A 13-nation panel of experts on the proposed conference had elected Vikram A. Sarabhai of India Feb. 8 as chairman. The subcommittee's Legal Subcommittee reached tentative agreement in Geneva July 17 on an accord under which all nations would help and return astronauts of other countries who came down accidentally [or mistakenly] in their territory. They also reached preliminary agreement on an accord requiring a launching country to pay for any damages done in other countries by objects it may have launched. U.S. and Soviet representatives participated in both tentative agreements.)

World Weather Watch Established

The World Meteorological Organization (WMO) Apr. 27 approved the establishment of a World Weather Watch (WWW) for the years 1968-71. Its findings, it was anticipated, would make possible the forecasting of weather conditions "for significantly longer periods," perhaps 7 days in advance. The program was decided on in Geneva at the WMO's 5th world congress. It would utilize a network of artificial satellites, drifting balloons, ocean buoys and electronic computers. The system would be coordinated from 3 centers in Washington, Moscow and Melbourne, Australia but would require a large degree of international cooperation. The WMO Apr. 27 appealed to all non-member countries to "accept the concept of the" WWW program. It would be possible for any country to get data from WWW satellites.

USSR Hints Eventual Collaboration

Soviet space scientist Leonid I. Sedov hinted Sept. 25, in an address opening the 18th congress of the International Astronautical Federation, that the USSR would be willing, eventually, to collaborate with other nations on major space projects. Speaking in Belgrade, where the congress was held Sept. 25-29, Sedov noted that "in order to put the great interplanetary expeditions and other major projects into practice, vast material and creative forces will be called for. The problem of international cooperation acquires decisive significance in relation to this." But he said at a press conference Sept. 28 that U.S.-Soviet collaboration in such projects could come "only after the disarmament problem is solved, completely or partially."

(Sedov, interviewed in Moscow before leaving to head the Soviet delegation to the Belgrade congress, had said the USSR planned to build as interplanetary spaceship weighing more than 1,000 tons. In the interview, published in the Belgrade newspaper *Borba* Sept. 23. Sedov said the projected flights could be made only after the USSR had solved the problems of building a laboratory in space and of returning men safely to earth after space flight beyond earth orbit.)

At a symposium held during the congress Sept. 28, delegates of the various nations agreed on the desirability of establishing an international laboratory on the moon for work on physical and chemical projects requiring vacuum.

Consortium Launches 3 Satellites

3 communications satellites were launched during 1967 for the International Telecommunications Satellite Consortium (Intelsat), a global group whose members included more than 50 nations. (2 others, *Intelsat 1* and *Early Bird*, had been launched earlier.) The Communications Satellite Corp. (Comsat) was U.S. member of Intelsat and manager of the program.

The first of the consortium's satellites of 1967, *Intelsat 2*, was shot into orbit Jan. 11 by means of a TAID (thrust augmented improved Delta) booster rocket launched by the NASA from Cape Kennedy. The 3-stage Taid put the satellite, also known as *Lani Bird 2*, into an initial orbit with an apogee of 22,897 miles and perigee of 184 miles. A signal radioed from Paumalu, Hawaii Jan. 14 ignited a maneuvering rocket aboard *Intelsat 2* and put the satellite into a circular orbit 22,300 miles high. Refining maneuvers put it over the equator at the international dateline. In synchronous orbit then, it appeared to be suspended without motion because its speed in orbit exactly matched the speed of the earth's rotation.

The new satellite, built by the Hughes Aircraft Co., had weighed $357\frac{1}{2}$ pounds at lift-off, but the expenditure of its maneuvering fuel brought its weight down to 192 pounds by the time it was in synchronous orbit. The drum-shaped *Intelsat 2* was 56 inches in diameter. $26\frac{1}{2}$ inches long. NASA launched the satellite under contract to Cosmat for $\$3\frac{1}{2}$ million.

The consortium's 2d satellite of 1967, *Canary Bird,* was orbited Mar. 22 by a 3-stage TAID launched by NASA from Cape Kennedy at 8:30 p.m. The satellite achieved an initial orbit with an apogee of 23,105 miles and perigee of 141 miles. By radio signal from Andover, Me., *Canary Bird*'s apogee rocket was fired at 9:45 a.m. Mar. 25 to put the satellite into a 22,300-mile-high circular orbit. Minor orbital adjustments were then made to put the satellite in a synchronous orbit in which it appeared to remain motionless at Longitude 5° W. over the equator.

The satellite's function was to expand transatlantic communications service and also to act as a communications link for manned expeditions to the moon. *Canary Bird* had weighed 357 pounds on launching, but it was down to 192 pounds by the time it had exhausted its fuel and reached its station. Comsat paid NASA about $\$3\frac{1}{2}$ million for putting the satellite into orbit.

Intelsat's 192-pound *Pacific 2* was launched from Cape Kennedy by a Delta rocket at 8:45 p.m. EDT Sept. 27. It went into an elliptical transfer orbit before its on-board apogee motor was ignited by ground radio signal Sept. 30 to move it to its near synchronous orbit over the Pacific.

(Comsat had asked the FCC [U.S. Federal Communications Commission] Apr. 1 for permission to finance, build and operate a domestic U.S. communications satellite system. It offered to provide free service—one channel in each time zone—for educational broadcasting. It said the system could be in operation by late 1969. The Ford Foundation, which had proposed a competing plan, attacked Comsat's proposal in a brief prepared for the FCC Apr. 3 and urged the FCC not to let Comsat expand into the U.S. domestic field lest it become an unprecedented "world monopoly.")

2 International Launchings Fail

A 4-stage U.S. Scout rocket was launched by the U.S. Air Force's 6595th Aerospace Test Wing at Vandenberg Air Force Base, Calif. May 29 in an unsuccessful attempt to put the European international *Esro 2* into polar orbit. An apparent malfunction while the rocket's 3d stage was in operation was believed to have prevented the 4th stage from igniting. The 163-pound prospective satellite was lost in the Pacific.

INTERNATIONAL DEVELOPMENTS 109

The unsuccessful satellite had been designed and built by the 10-nation European Space Research Organization (Esro). It was launched under a cooperative agreement with the U.S.' NASA to study solar and and cosmic radiation. The unsuccessful project cost Esro about $3½ million, the U.S. $1½ million. (The 10 Esro nations were Belgium, Britain, Denmark, France, Italy, the Netherlands, Spain, Sweden, Switzerland and West Germany.) Under the agreement, *Esro 1* was to be launched later (out of sequence because its experiments were still imcomplete).

(Esro Director Gen. Pierre Auger and NASA Deputy Administrator Robert C. Seamans Jr. in Paris Mar. 8 had signed a contract for Esro's purchase of a U.S. Delta rocket to launch the Esro spacecraft Heos-A [for highly eccentric orbit satellite] from Cape Kennedy in 1968's 2d half. The 230-pound satellite was to be launched during a period of maximum solar activity for a study of the interplanetary magnetic field and of solar and cosmic ray particles outside the magneto-sphere.)

Missile scientists of the 6-nation European Launcher Development Organization (Eldo) launched a 2-stage version of the organization's Europa-1 rocket from the Woomera, Australia rocket center Aug. 4 on what at first appeared to be a successful flight. The rocket, however, crashed short of its target because of failure of the French-made Coralie 2d stage to ignite and separate. The British Blue Streak first stage apparently worked as scheduled, as it had done in a previous one-stage launching. The launching had been postponed 10 times because of bad weather and technical difficulties.

ORBITAL & OTHER MISSIONS

Unmanned Soviet Spacecraft 'Dock'

The unmanned Soviet satellites *Cosmos 186* and *Cosmos 188* were successfully linked in space Oct. 30. The feat was the USSR's first space "docking" and the first docking of 2 unmanned spaceships by any nation. (The U.S.' 10 space dockings were achieved by 4 Gemini capsules with 2-man crews.) The USSR did not reveal in advance that a docking would be attempted. The disclosure was made after the successful docking had been achieved.

Cosmos 186 had been launched Oct. 27 and placed in an orbit with an apogee of 235 kilometers (146 miles), perigee of 209 kilometers (130 miles), period of 88.7 minutes and 51.7° angle of inclination to the equatorial plane. During the next 3 days, according to official Soviet sources, *Cosmos 186*'s orbit "was corrected," and by Oct. 30 the orbit had an apogee of 260 kilometers (161 miles), perigee of 180 kilometers (112 miles), period of 88.64 minutes and 51.68° angle of inclination.

Cosmos 188 was then launched Oct. 30 and put into an initial orbit with an apogee of 276 kilometers (171 miles), perigee of 200 kilometers (124 miles), 88.64-minute period and 51.68° angle of inclination. *Cosmos 188* was about 24 kilometers (15 miles) from *Cosmos 166* when it went into orbit.

Cosmos 186, the "active" satellite, "found" *Cosmos 188*, the "passive" or "target" satellite, drew up to the latter and succeeded in thrusting its docking rod into the latter's docking cone during *Cosmos 188*'s first revolution around the earth and its own 49th revolution. The docking was completed shortly after noon—at 12:20 p.m. Moscow time Oct. 30. The 2 satellites remained linked in orbit for $3\frac{1}{2}$ hours until, at 3:50 p.m., they separated in obedience to an order radioed from the earth.

Cosmos 186, brought out of orbit Oct. 31 on its 86th revolution (on a radioed command from earth), "made a soft landing in the given area at 11:20 a.m. Moscow time," according to Tass. *Cosmos 188* was brought down Nov. 2.

Before the announcement at the docking, space observers had pointed out that *Cosmos 186* was in an orbit similar to the one used by the USSR's *Soyuz 1*, in which Col. Vladimir M. Komarov, a Soviet cosmonaut, had been killed. Heinz Kaminski, director of the Space & Satellite Research Institute at Bochum, West Germany, reported Oct. 29 that *Cosmos 186* was sending an "unusually large amount of information" on the same radio frequency as *Soyuz 1* had used. English physics teacher Geoffrey Perry said similar findings had been made by his student monitoring team at Kettering Grammar School in Northants.

Prof. Mstislav V. Keldysh, president of the Soviet Academy of Sciences, was asked at a Moscow press conference Oct. 30 whether he could give details of *Cosmos 186*'s mission. The question was asked before the announcement of the docking but more than an hour after the docking had taken place. Keldysh replied that there were too many Cosmos satellites in orbit for him to remember details of any particular satellite's mission.

USSR Lofts 3 Communications Satellites

The 5th, 6th and 7th Molniya-1 communications satellites were sent up by the USSR during 1967. (NASA's Goddard Space Flight Center had confirmed Apr. 3 that the USSR's 2d Molniya-1 had dropped out of orbit about Mar. 17.)

The first Molniya-1 of the year was launched May 25. Tass said the satellite's orbit had an apogee of 39,810 kilometers (24,737 miles) "in the Northern hemisphere," perigee of 460 kilometers (286 miles) "in the Southern hemisphere," period of 11 hours 55 minutes and 64.8° angle of inclination.

2 more Molniya-1 satellites were shot into orbit Oct. 3 and 22. The orbit of the first had an initial apogee given as "39,600 kilometers [24,592 miles] in the Northern hemisphere," perigee of "465 kilometers [289 miles] in the Southern hemisphere," period of 11 hours 52 minutes and 65° angle of inclination. The orbit of the 2d had an apogee listed as "39,740 kilometers [24,769 miles] in the Northern hemisphere," perigee of "456 kilometers [283 miles] in the Southern hemisphere," period of 11 hours 54 minutes and 64.7° inclination. Tass said that the "main tasks of the 2 satellites . . . were to operate with previous Molniya-1 satellites in "the exploitation of the long distance telephone and telegraph radio communications system" and in relaying USSR Central Television programs to the 20 or more earth stations in the new "Orbita" network set up on the eve of the October Revolution's 50th anniversary in the USSR's "Extreme North, Siberia, Far East and Central Asia" districts.

58 Cosmos Satellites Orbited

The USSR orbited 58 Cosmos satellites in 1967. The purposes of most were unannounced.

Indications that *Cosmos 144*, launched Feb. 28, was a weather satellite, were confirmed Mar. 2 when, without warning, the U.S. National Environmental Satellite Center in Suitland, Md. began to receive weather data produced by *Cosmos 144*. Soviet scientists previously had used the U.S.-Soviet weather communications network to transmit data from the *Cosmos 122* weather satellite, which had gone dead before *Cosmos 144* was launched.

Pravda reported Apr. 12 that *Cosmos 149*, launched Mar. 21, which would have only a "short period of . . . active existence," also had a meteorological role: Its 2 "multi-channel photometers . . . scanned the earth" in "2 mutually perpendicular directions" and "determined the brightness of the planet in the narrow intervals of the spectrum"; its radiometer "measured the earth's own radiation in the . . . atmosphere transparency window" to provide data "on the composition of the atmosphere and the parameters of the earth's surface and the overcast." The satellite's "aerogyroscopic system of stabilization," in use for the first time, operated by means of "a special aerodynamic stabilizer and hydraulic damper."

Tass disclosed June 5 that *Cosmos 156*, launched Apr. 27, also was a weather satellite. *Cosmos 144* and *156*, "in combination with stations for receiving, processing and circulating meteorological information, form what is known as the experimental cosmic meteorological system 'Meteor,'" Tass said. Tass reported that Cosmos 156 carried TV equipment "for recording pictures of clouds, snow and ice zones on the earth's sunlit side; infra-red equipment for recording similar pictures on both the light and dark sides . . . , [and] actinometric equipment for recording the intensity of the radiation emitted and reflected by the earth-atmosphere system and for measuring the temperature of the clouds and underlying surface." (Tass had reported June 1 that single-stage weather-study rockets were being launched by Soviet scientists every Wednesday from an Arctic site in Heis Island [part of Franz Josef Land] on the shore of the Barents Sea. Phenomena studied were temperatures, air pressure and speed and direction of wind. The operation was part of of an international program, and the data were coordinated with findings of rockets launched at the same time in the U.S., France, Japan, Australia and India.)

According to Western sources, *Cosmos 146*, launched Mar. 10, might have been the heaviest man-made satellite launched so far. Its weight was estimated as possibly 40,000 to 65,000 pounds. (A similar unofficial estimate was given for *Cosmos 154*, launched Apr. 8.) The North American Air Defense Command, which routinely tracks all earth satellites and other space objects, reported Mar. 14 that a "2d payload for *Cosmos 146*" was circling the earth in an orbit slightly higher than the first orbit. According to some Western sources, a 3d object had accompanied the 2 *Cosmos 146* objects for awhile and then may have reentered the atmosphere, possibly in a heat-shield re-entry test.

As was customary in the Soviet space program, no hint was given that any of the shots were planned. The disclosure of each launching was made only after a successful orbit could be announced.

Dates and initial orbital parameters (where given) of Cosmos satellites sent into orbit by the USSR during 1967:

Cosmos 139–Jan. 25. Apogee–210 kilometers (130½ miles). Perigee–144 kilometers (89½ miles). Inclination to equatorial plane–50°.

Cosmos 140–Feb. 7. Apogee–150 miles. Perigee–140 miles. Inclination–51.7°. (Remained in orbit 2 days.)

Cosmos 141–Feb. 8.

Cosmos 142–Feb. 14. Apogee–1,362 kilometers (845.8 miles). Perigee–214 kilometers (132.9 miles). Period–100.3 minutes. Inclination–48.4°.

Cosmos 143–Feb. 27.

Cosmos 144–Feb. 28. North-south circular orbit about 387½ miles above earth's surface. Period–96.92 minutes.

Cosmos 145–Mar. 3. Apogee–1,325 miles. Perigee–136 miles.

Cosmos 146–Mar. 10. Apogee–310 kilometers (192½ miles). Perigee–

190 kilometers (118 miles). Period–89.2 minutes. Inclination–51½°. (Remained in orbit 8 days.)

Cosmos 147–Mar. 13.   Apogee–317 kilometers (196.9 miles).   Perigee–198 kilometers (123 miles). Period–89½ minutes. Inclination–65°.

Cosmos 148–Mar. 16.   Apogee–436 kilometers (270.8 miles).   Perigee–275 kilometers (170.8 miles). Period, 91.3 minutes. Inclination–71°.

Cosmos 149–Mar. 21.   Apogee–297 kilometers (184½ miles).   Perigee–248 kilometers (154.1 miles). Period–89.8 minutes. Inclination–48°21 minutes.

Cosmos 150–Mar. 22.   Apogee–373 kilometers (231.6 miles).   Perigee–206 kilometers (127.9 miles). Period–90.1 minutes. Inclination–65.7°.

Cosmos 151–Mar. 24.   Apogee–630 kilometers (391.2 miles). Period–97.1 minutes. Inclination–56°.

Cosmos 152–Mar. 25. Apogee–512 kilometers (318 miles). Perigee–283 kilometers (175½ miles). Period–92.2 minutes. Inclination–71°.

Cosmos 153–Apr. 4. Apogee–180 miles. Perigee–125 miles.

Cosmos 154–Apr. 8.   Apogee–232 kilometers (144.1 miles).   Perigee–186 kilometers (115½ miles). Period–88½ minutes. Inclination–51.6°. (Remained in orbit 8 days.)

Cosmos 155–Apr. 12.   Apogee–286 kilometers (177.6 miles).   Perigee–203 kilometers (126.1 miles). Period–89.2 minutes. Inclination–51.8°.

Cosmos 156–Apr. 27. Circular orbit about 390 miles high.

Cosmos 157–May 12.   Apogee–296 kilometers (183.8 miles).   Perigee–202 kilometers (125½ miles). Period–89.4 minutes. Inclination–51.3°.

Cosmos 158–May 15.

Cosmos 159–May 17.   Apogee–37,655 miles. Perigee–236.6 miles. Inclination–51°50 minutes.

Cosmos 160–May 17.

Cosmos 161–May 22.   Apogee–343 kilometers (213.1 miles).   Perigee–205 kilometers (127.4 miles). Period–89.3 minutes. Inclination–65.7°.

Cosmos 162–June 1. Apogee–280 kilometers (174 miles). Perigee–201 kilometers (125 miles). Period–89.2 minutes   Inclination–51.8°.

Cosmos 163–June 5.

Cosmos 164–June 8.   Apogee–198 miles. Perigee–125 miles. Period–89½ minutes.

Cosmos 165–June 12.   Apogee–1,542 kilometers (957½ miles).   Perigee–211 kilometers (131 miles). Period–102.1 minutes. Inclination–81.9°.

Cosmos 166–June 16. Apogee–578 kilometers (359 miles). Perigee–283 kilometers (176 miles). Period–92.9 minutes. Inclination–48.4°. (The satellite reportedly broke into 8 pieces and was destroyed on re-entering the atmosphere Oct. 25.)

Cosmos 167–June 17.

Cosmos 168–July 4.

Cosmos 169–July 17. Apogee–208 kilometers (129 miles). Perigee–144 kilometers (89½ miles). Inclination–50°.

Cosmos 170–July 31.

Cosmos 171–Aug. 8.

Cosmos 172–Aug. 9.

Cosmos 173–Aug. 24. Apogee, 528 kilometers (328 miles). Perigee, 280 kilometers (174 miles). Period 92.3 minutes. Inclination, 71°.

Cosmos 174–Aug. 31. Apogee, 39,750 kilometers (24,685 miles). Perigee, 500 kilometers (311 miles). Period, 11 hours 55 minutes. Inclination, 64.5°.

Cosmos 175–Sept. 11. Apogee, 386 kilometers (240 miles). Perigee, 210 kilometers (130 miles). Period, 92.2 minutes. Inclination, 72.9°.
Cosmos 176–Sept. 12. Apogee, 1,581 kilometers (982 miles). Perigee, 206 kilometers (128 miles). Period, 102½ minutes. Inclination, 81.9°.
Cosmos 177–Sept. 16.
Cosmos 178–Sept. 19. Apogee, 205 kilometers (127 miles). Perigee, 145 kilometers (90 miles). Inclination, 50°.
Cosmos 179–Sept. 22.
Cosmos 180–Sept. 26. Apogee, 370 kilometers (230 miles). Perigee, 212 kilometers (132 miles). Period, 90.1 minutes. Inclination, 72.9°.
Cosmos 181–Oct. 11.
Cosmos 182–Oct. 16.
Cosmos 183–Oct. 18.
Cosmos 184–Oct. 25. Circular orbit with altitude of about 635 kilometers (394 miles). Period, 97.14 minutes. Inclination, 81.2°.
Cosmos 185–Oct. 25. Apogee, 888 kilometers (551 miles). Perigee, 522 kilometers (324 miles). Period, 98.7 minutes. Inclination, 64.1°.
Cosmos 187–Oct. 28. Apogee, 210 kilometers (130 miles). Perigee, 145 kilometers (90 miles). Inclination, 50°.
Cosmos 189–Oct. 30. Apogee, 600 kilometers (373 miles). Perigee, 535 kilometers (332 miles). Period, 95.7 minutes. Inclination, 74°.
Cosmos 190–Nov. 3. Apogee, 347 kilometers (215 miles). Perigee, 201 kilometers (125 miles). Period, 89.9 minutes. Inclination, 65.7°.
Cosmos 191–Nov. 21. Apogee, 518 kilometers (322 miles). Perigee, 281 kilometers (174½ miles). Period, 92.2 minutes. Inclination, 71°.
Cosmos 192–Nov. 23. Circular orbit about 760 kilometers (472 miles) above the earth's surface. Period, 99.9 minutes. Inclination, 74°.
Cosmos 193–Nov. 25. Apogee, 354 kilometers (220 miles). Perigee, 203 kilometers (126 miles). Period, 89.9 minutes. Inclination, 65.7°.
Cosmos 194–Dec. 3. Apogee, 206 miles. Perigee, 127 miles. Period, 90 minutes. Inclination, 65.7°.
Cosmos 195–Dec. 16. Apogee, 375 kilometers (233 miles). Perigee, 211 kilometers (131 miles). Period, 90.1 minutes. Inclination, 65.7°.
Cosmos 196–Dec. 19. Apogee, 887 kilometers (551 miles). Perigee, 225 kilometers (140 miles). Period, 95½ minutes. Inclination, 49°.
Cosmos 197–Dec. 26. Apogee, 313 miles. Perigee, 136 miles. Period, 91½ minutes.

U.S. Scientific & Military Launchings

NASA and the Air Force orbited many scientific and military space vehicles during 1967. The launchings, in chronological order:

Jan. 14, May 22, Aug. 7, Dec. 5–"Secret satellites" were launched into near polar orbits by Air Force-industry teams at Vandenburg Air Force Base, Calif. by means of a Thor-Agena rocket Jan. 14, an Atlas-Agena May 22, a Thor-Agena Aug. 7 and a Titan 3-Agena D Dec. 5.

Jan. 18–8 U.S. military communications satellites were put into 21,000-mile-high orbits by means of a single 3-stage Air Force Titan-3C booster rocket launched from Cape Kennedy. The 100-pound satellites, built by the Philco Corp. at a cost of about $1 million each, joined 7 similar radio relay stations that had been orbited June 16, 1966. The first satellites had already been used for emergency communications with South Vietnam, where the U.S. had set up 2 ground stations (at Nahtrang and Baqueo).

The 36-inch satellites were orbited in a 6-hour operation during which (1) the 700-ton Titan-3C put its 3d stage (the transtage) into a circular parking orbit 103 miles high, (2) the transtage ignited and achieved an elliptical transfer orbit with an apogee of about 21,000 miles, and (3) the transtage ignited again at apogee to turn the orbit into a 21,000-mile-high circular one before releasing the 8 satellites one by one.

Jan. 26—NASA crews at Vandenburg Base launched the weather satellite *Essa 4* by means of a 3-stage TAID (thrust augmented improved Delta) rocket. The 290-pound satellite, launched at 9:32 a.m., achieved an orbit with an apogee of 894 miles, perigee of 822 miles, period of 113 minutes and 78° angle of inclination. NASA Feb. 8 turned over the operation of *Essa 4* to the Commerce Department's ESSA (Environmental Science Services Administration).

Mar. 8—The 627-pound *Oso 3* (for orbiting solar observatory) was launched from Cape Kennedy at 11:12 a.m. to provide scientific data for the study of the sun and its influence on the earth's atmosphere. Carried aloft by a 3-stage Thor-Delta, it achieved an orbit with a 377-mile apogee, 354-mile perigee, 96-minute period and 33° angle of inclination. NASA reported Mar. 14 that all 9 of *Oso 3*'s experiments had been put into operation and were working well.

Apr. 5—The 815-pound *ATS 2* (for applications technology satellite) was launched from Cape Kennedy at 10:23 p.m. by means of an Atlas-Agena-D. Its primary objective was to test a passive gravity gradient control system that used 4 123-foot-long booms to keep the spacecraft stabilized on 3 axes and pointed toward the earth. As *ATS 2* rose toward its 6,900-mile apogee, however, the Agena stage failed to execute a 2d ignition and put the satellite, as planned, into a 6,900-mile-high circular orbit. As a result, a NASA spokesman predicted Apr. 6, the booms, which had transformed *ATS 2* into an X-shaped satellite a record 252 feet long, would eventually "buckle or bend" and fail to carry out their stabilization assignment. The orbit remained elliptical with a 6,900-mile apogee and 29° angle of inclination. *ATS 2* carried equipment for communications, meteorological and other scientific studies, most of which require adequate stabilization.

Apr. 20—A NASA crew at Vandenburg Base launched the weather satellite *Essa 5* with a 3-stage TAID rocket. The 320-pound satellite was put into an orbit with an 833-mile apogee, 840-mile perigee and $113\frac{1}{2}$-minute period. NASA turned over operation of the satellite to ESSA May 8.

Apr. 28—5 satellites were put into orbit by means of a single Titan-3C rocket launched by the Air Force at Cape Kennedy at 5:01 a.m. 2 of the satellites were 731-pound Vela vehicles, whose mission was to detect possible nuclear tests in space. They were put into initial orbits with apogees of 69,000 miles and perigees of 5,400 miles. By means of radio command from ground stations, the orbits of the 2 Velas were made circular at altitudes of 69,000 miles, the first Vela went into its circular orbit Apr. 29, and the 2d did so on the opposite side of the earth from the first Vela May 1. 3 pairs of Vela nuclear-test monitors had been launched previously, but the new ones were considered much more advanced; they carried more sophisticated detection devices and rocket engines to adjust their orbits. The other 3 satellites, weighing 20 pounds each, were designed to (1) monitor radiation in the Van Allen belts circling the earth, (2) measure X-rays and other radiation from the sun and (3) provide date on the effects of friction on 16 metals in space.

May 24—The 163-pound *Explorer 34* (also labeled IMP-F for interplanetary monitoring platform) was launched from Vandenberg Base at 1:06 p.m. by means of a TAID rocket. The most complex of the spacecraft in the IMP series, its mission was to measure solar and galactic cosmic rays within and at the boundary of the earth's magnetosphere (the envelope formed by the earth's magnetic fields, which protect the earth from radiation) and in interplanetary space. The satellite's orbit had an apogee of 131,187 miles, perigee of 154 miles, period of 4 days 7 hours 51 minutes and angle of inclination of 67°.

July 1—6 military satellites were placed in independent 20,750-mile-high orbits by means of a 3-stage Titan-3C launched by the Air Force at Cape Kennedy at 9:15 a.m. EDT. 4 of the satellites were military communications devices that supplemented 15 similar satellites put into orbit previously to give the Pentagon almost constant contact with overseas units; this network was used largely for priority communications with 2 ground stations in South Vietnam, but it had also been employed for communications between Washington and U.S. officials in the Middle East during the Arab-Israeli war in June. The 5th of the satellites was an experimental device built by MIT's Lincoln Laboratory, to test a new system for communications with front-line troops. The 6th, a 437-pound satellite dubbed *Dodge* (for Department of Defense Gravity Experiment), was a TV-equipped experiment designed to grow silver "whiskers" up to about 150 feet long to see whether the pull of gravity on the silver booms would stabilize the satellite.

July 28—A 1,240-pound scientific satellite dubbed *Ogo 4* (for orbiting geophysical observatory) was launched by NASA personnel at Vandenberg Base into a near-polar orbit. The spacecraft, sent up in the nose of a 2-stage thrust-augmented Thor-Agena-D rocket, achieved an orbit with an apogee of 564 miles, perigee of 256 miles, period of 98 minutes and 86° angle of inclination. *Ogo 4*'s mission was to study the relationship between the sun and the earth's environment during a period of increased solar activity. The major emphasis of the study was on the interrelationships among particle activity, aurora and airglow, the geomagnetic field, the neutral and ionized composition of the atmosphere and the electromagnetic energy sources contributing to ionization and atmospheric heating. *Ogo 4* was the 4th spacecraft in a 6-satellite Ogo series devised to study the earth's environment in space.

Nov. 5—The multi-purpose *ATS 3*, an applications technology satellite, was sent into space by means of a 2-stage Atlas-Agena-D launched from Cape Kennedy. The Atlas stage and the Agena-D stage first put the satellite into a near-circular orbit about 114 miles above the earth. As *ATS 3* with the attached Agena-D crossed the equator, a 2d ignition of the Agena-D put the satellite into an elliptical orbit with its apogee just below the 22,300-mile altitude ultimately achieved. The Agena-D was then jettisoned. The satellite's apogee motor was fired over Brazil as *ATS 3* reached apogee a 2d time at a point above the equator at Longitude 52.7° W. *ATS 3* was then maneuvered by Nov. 16 into its planned synchronous orbit in which it appeared to hang motionless 22,300 miles above the equator at Longitude 47° W. over the Atlantic.

The $18 million *ATS 3* plotted its own course by releasing 3 small steel balls the size of baseballs and tracking them against the background of stars. It carried a camera to photograph the earth in full color, and carried equipment for navigation and communications experiments. The satellite was essentially a 6-foot-long cylinder, 5 feet in diameter, with projecting antennae that gave it a spider-like appearance. It had weighed 1,574 pounds on separation from the Agena-D but was down to 805 pounds after the expenditure of its fuel.

Nov. 10—The weather satellite *Essa 6* was launched from Vandenberg Base by means of a 3-stage Delta. The 290-pound, hat-box-shaped satellite achieved a near-polar orbit with an apogee of 922 miles, perigee of 874 miles and period of 114 minutes. *Essa 6* was equipped with 2 wide-angle cameras that took photos showing areas of $4\frac{1}{2}$ million square miles each. These cloud-cover photos were transmitted to 15 tracking stations, which made them available to users in the U.S. and 45 foreign countries. *Essa 6* began transmitting photos Nov. 12. The satellite was launched by NASA on behalf of ESSA.

Dec. 4—The 222-pound *OV3-6* ionosphere-study satellite was sent into polar orbit by means of a carrier rocket launched by a U.S. Air Force team at Vandenberg Base. Its equipment, designed by Air Force scientists at Cambridge, Mass., included 2 mass spectrometers, 3 ion density gauges and other devices to study the structure of the atmosphere at satellite altitudes and to measure electro-density and temperature irregularities.

Dec. 13–Pioneer 8, a 145-pound spacecraft, was shot into orbit around the sun by means of a 3-stage Delta booster rocket launched from Cape Kennedy at 9:08 a.m. On its way into space, the Delta ejected into orbit around the earth a 40-pound communications satellite dubbed *TTS 1* (test and training satellite). *Pioneer 8*'s principal tasks were to determine the exact shape of the earth's $3\frac{1}{2}$ million-mile-long magnetosphere and to monitor solar events (sun "storms") as the 11-year solar cycle reached a peak in 1969. The probe's orbit around the sun had an aphelion (farthest point from the sun) of 101,130,940 miles, a periphelion (nearest point) of 92,131,930 miles and a period of $387\frac{1}{2}$ days. *TTS 1* achieved an earth orbit with an apogee of 303 miles, perigee of 187 miles, period of 92 minutes and 33° angle of inclination to the equatorial plane. It was sent up to test the 18 communications stations in the Apollo tracking network.

Errant Satellites

U.S. space scientists reported that the 280-pound capsule of *Biosatellite 1*, launched Dec. 14, 1966, had apparently come out of orbit Feb. 15 and gone down perhaps 300 miles northeast of Perth, Australia, according to tracking data. A hunt for the capsule proved unsuccessful. The capsule, carrying a payload of insects and other organisms, had been supposed to come out of orbit 3 days after launching but had remained in orbit due to retro-rocket failure.

Scientists disclosed Sept. 26 that the Navy satellite *Transit 4B*, launched Nov. 21, 1961 but "silenced" as a result of damage due to the U.S. high-altitude thermonuclear test of July 9, 1962, had mysteriously begun transmitting again Mar. 23, 1967. It took until May 11 for U.S. space scientists to identify the source of the mysterious signals and to silence the satellite again so that the signals should not interfere with those of "working" satellites.

Launchings by Other Nations

6 other nations made space-exploration test firings. They were France, Japan, Italy, Great Britain, India and Australia. Details of the

missions:

France—2 French-built satellites, *Diadème 1* and *Diadème 2*, were launched by French scientists at France's Hammaguir base in the Algerian Sahara by means of French-built 3-stage Diamant rockets Feb. 8 and 15. *Diadème 1*, launched at 3:39 a.m. GMT, achieved an orbit with an apogee of 833 miles, perigee of 363 miles, period of 104 minutes and 40° angle of inclination. The orbit was too low for success in Diadème's principal mission, the use of ground-based laser beams directed at the satellite to calculate, by triangulation, exact distances between different points on the earth's surface and, particularly, of the Mediterranean basin. The cylindrical satellites weighed 50 pounds each; they were about 20 inches in diameter and 8 inches high. The 2 Diadèmes were France's 4th and 5th satellites, of which all but one had also been launched by France (one was launched by the U.S.).

2 small monkeys were rocketed almost 150 miles into space Mar. 7 and 13 by French-made Vesta rockets fired by French scientists at Hammaguir. The monkeys, used to test reaction to space environment, were landed safely by parachute.

Details of the French plans and equipment for each of the shots were made public long before the actual launchings.

(French and West German representatives agreed in Bonn Apr. 28 on joint French-West German development and construction of an experimental synchronous communications satellite to be launched in 1970 by the 6-nation European Launcher Development Organization from France's new launching site in French Guiana. The agreement was signed in Paris June 6. The cost was estimated unofficially at $40 million. The satellite, to be named *Symphonie*, was to be placed in a 22,790-mile-high orbit.)

2 Soviet-built rockets carrying French meteorological equipment were launched from the USSR's Heis Island rocket base in the Arctic Oct. 9 in the first joint Franco-Soviet space project under an agreement signed by the 2 nations June 30, 1966. The purpose of the shots was to measure temperatures in the upper atmosphere.

Japan—A 4-stage Lambda rocket was launched from Tokyo University's space center at Uchinoura Apr. 13 in the 3d successive unsuccessful attempt to put a Lambda's 4th stage into orbit. In this attempt, made before "live" TV cameras, the rocket's 3d stage failed to ignite. (Tokyo University's space scientists had developed a series of solid-fueled Lambda and Mu rockets and hoped to use the latter to orbit full-scale scientific satellites. The Japanese government's Science & Technology Agency had developed a rival liquid-fueled rocket with which it planned to launch communications satellites.)

Italy—The Italian-built satellite *San Marco 2* was launched by an Italian crew Apr. 26 from the San Marco floating platform in Formosa Bay about 3 miles off the coast of Kenya.

The 285-pound satellite, the first ever launched from such a platform, was sent into orbit by means of a 4-stage U.S.-built Scout rocket. It achieved an orbit with an apogee of about 425 miles, perigee of some 135 miles, period of about 94 minutes and $2\frac{1}{2}°$ angle of inclination. Its primary objective was to provide continuous air density measurements at satellite height. A 2d objective was to investigate ionospheric characteristics causing interference with long-range radio transmissions. The 26-inch sphere had 2 retractable 100-inch monopole antennas for the latter experiment.

The San Marco launching platform was stationed $2\frac{1}{2}°$ south of the equator. A smaller floating platform nearby, the Santa Rita, carried the control room and most of the instrumentation.

Italy's plans for the shot had been made public long before the launching.

Great Britain—Ariel 3 (UK 3), the first satellite designed and built entirely in Britain, was launched by a U.S. NASA crew at Vandenberg Air Force Base, Calif. May 5 in the nose of a 4-stage Scout rocket. The 198-pound scientific spacecraft achieved an orbit with a 373-mile apogee, 306-mile perigee, 95.6-minute period and 80° angle of inclination. *Ariel 3*'s mission was to supplement and extend atmospheric and ionospheric studies conducted by *Ariel 1* and *Ariel 2*, its Anglo-U.S. predecessors.

India—The first Indian-developed sounding rocket, the Rohini RH-75, was launched by an Indian team Nov. 20 at the Thumba Equatorial Rocket Launching Station (TERLS) near Trivandrum in Kerala. The rocket, launched for meteorological purposes, was developed by Rohini engineers at the Science & Technology Center at Thumba; its propellant was supplied by the Indian Defense Establishment. TERLS, operated by India's Department of Atomic Energy, was created in 1964 in response to a recommendation of the UN Committee on the Peaceful Uses of Outer Space. It was located 32 miles south of the magnetic equator. Major contributors of TERLS equipment were the U.S., the USSR and France.

Australia—Australia's first satellite, *Wresat 1* (for Weapons Research Establishment satellite), was sent into orbit Nov. 29 by means of a U.S. Redstone booster rocket launched at the Woomera rocket center in the South Australian desert at 4:19 a.m. The Australian satellite, whose mission was listed as "upper atmosphere and space research," achieved an orbit with an apogee of 1,248.91 kilometers (775.57 miles), perigee of 170.14 kilometers (105.67 miles), 83.306° angle of inclination and period of 98.9744 minutes.

OTHER DEVELOPMENTS

More Astronauts Selected

The U.S. Air Force June 30 announced the selection of 4 more Air Force pilots to man the Defense Department's Mol (manned orbiting laboratory), which was then scheduled to be put into orbit by 1970 and in which 2-man crews were to serve for periods of up to 30 days. Those selected included Maj. Robert H. Lawrence Jr., 31, of Chicago, the first Negro to be chosen as a U.S. astronaut. (Lawrence, who had earned a doctorate in physical chemistry from Ohio State University, died in a plane crash Dec. 8.) The other 3 named June 30 were Maj. James A. Abrahamson, 34, of Portland, Ore., Lt. Col. Robert T. Herres, 34, of Denver and Maj. Donald H. Peterson, 33, of Winona, Miss. 12 pilots—8 from the Air Force, 3 from the Navy and one from the Marine Corps—had been chosen previously.

NASA Aug. 4 announced the selection of 11 more civilian scientists as scientist-astronauts. Each held at least one doctorate, but none were pilots and, therefore, all were scheduled to be trained by the Air Force to qualify as jet pilots. Their appointment raised the number of NASA astronauts to 56.

2 of the scientists named Aug. 4 were the U.S.' first naturalized astronauts—Australian-born Philip K. Chapman, 32, a staff physicist at MIT's Experimental Astronomy Laboratory, and Welsh-born John A. Llewellyn, 34, an associate professor of chemistry at Florida State University. The other 9: Joseph P. Allen, 30, an associate research physicist at the University of Washington; Anthony W. England, at 25 the youngest man to be named an astronaut, an MIT graduate fellow in geophysics; Karl G. Henize, 40, astronomy professor at Northwestern University and an experimenter in the Gemini program; Donald L. Holmquest, 28, holder of an MD degree, who was completing internship requirements for a doctorate in physiology at the Baylor College of Medicine and was not to report for astronaut training until 1968; William B. Lenoir, 28, assistant electrical engineering professor at MIT; Franklin Story Musgrave, 31, holder of doctorates in medicine and physiology, post-doctoral fellow at the University of Kentucky; Brian T. O'Leary, 27, holder of an astronomy PhD and currently a NASA trainee at the Space Sciences Laboratory of the University of Cali-

fornia's astronomy department; Robert A. Parker, 30, assistant astronomy professor at the University of Wisconsin; William E. Thornton, 38, a physician who recently completed 2 years of duty with the Areospace Medical Division at Brooks Air Force Base, Tex.

NASA announced Aug. 3 that Navy Cmndr. M. Scott Carpenter, 42, one of the original 7 Mercury astronauts and simultaneously an "aquanaut" who had participated underwater in the Navy's Sealab program, was leaving the space program to work in the Navy's underwater research projects. In April, after surgery had failed to restore fully an arm broken in a 1965 motorbike accident, Carpenter had given up hope of ever getting another space-flight assignment.

Radiation Hazard Seen on Long Flights

The Space Science Board of the National Research Council reported July 17 that astronauts making space flights of long duration ("up to 3 years") might be exposed to radiation well above the levels considered acceptable under existing standards for industrial workers. The 12-man board, headed by Dr. Wright H. Langham of the University of California's Los Alamos (N.M.) Scientific Laboratory, made its findings known in a report entitled "Radiobiological Factors in Manned Space Flight." They said that manned space flights, as "high-risk endeavors," would have to be conducted under criteria different from those established for industry, in which standards anticipated occupational life-times of up to 40 years and were designed to provide maximum protection for each worker. A preface pointed out that the report "is focused more on the success of missions and the amounts of radiation that man can withstand rather than on maximum protection of the individual." The report said that "the application of existing standards ... would unduly limit the ability of this small group of specialists [astronauts] to achieve their objectives."

(Maj. Gen. James W. Humphreys Jr. of the Air Force became director of space medicine in NASA's Headquarters Office of Manned Space Flight June 1.)

Civilian Use of Navy Satellites Approved

Vice Pres. Hubert H. Humphrey disclosed in a speech at Bowdoin College in Brunswick, Me. July 29 that civilians would be permitted to use the previously secret all-weather Transit satellite navigation system that the Navy had been using since 1964. Pres. Johnson had authorized commercial manufacture, on an unclassified basis, of the Transit shipboard receivers. The system, developed by the applied physics laboratory of Johns Hopkins University, had made it possible for Polaris missile submarines to determine their exact locations within 1/10 mile; non-satellite navigation tolerated errors of 2 to 3 miles in clear weather and of up to 50 miles in cloudy weather. Currently, 3 Transit satellites were operating in polar orbits about 600 miles above the earth. The system had already cost the U.S. more than $100 million, and expenditures of $18 million were budgeted for it for fiscal 1968.

Communications Policy Review Set

In a message on the nation's foreign and domestic communications policy, Pres. Johnson informed Congress Aug. 14 that he had created a committee to undertake a broad review of communications policy.

The committee, headed by State Undersecy. Eugene V. Rostow, was to study: (a) the development of a domestic satellite system capable of carrying TV, phone messages, newspaper photos, telegraph and computer signals or combinations of these; (b) the feasibility of a merger of one or more of the international communications systems within the U.S., or a merger of the international operations of the U.S. carriers with the Communications Satellite Corp. (Comsat); (c) when a domestic satellite system would be economically feasible; (d) whether a domestic satellite system should be "general purpose or specialized" and whether there should be one or more such systems.

In the message the President repeated an invitation (first made in 1963) to the Soviet Union and East European nations to join the International Communications Satellite Consortium (Intelsat) linking 58 nations for international communications by satellite. "Nothing could better symbolize the truth that space belongs to all men," Johnson declared, "than an international undertaking that permits the free flow of communications."

The U.S.' commitment to a global system for commercial communications was also renewed in the message, and the President stated his opposition to any regional satellite systems competing against Intelsat. Any U.S. domestic satellite system should be compatible with the global system, he said. In regard to the existence of several international communications agencies within the U.S. (*e.g.*, International Telephone & Telegraph Co., Radio Corp. of America, Western Union International, American Telephone Co., Comsat), Johnson said there was a "legitimate question as to whether the present division of ownership continues to be in the public interest." He noted arguments against the division of ownership, such as duplication of facilities, disputes over ownership of ground stations, "a relatively poor bargaining position on communications matters with foreign counterparts" and a possible problem for defense communications.

U.S. Lag in Rocket Power Reported

NASA Director James E. Webb told a House Appropriations subcommittee Aug. 15 (in testimony made public Aug. 18) that "for a number of years to come" the USSR would probably continue to have more-powerful rocket boosters than the U.S. Webb noted that the U.S.' liquid-fueled Saturn-5 rocket booster, still awaiting final testing, had "the rough equivalent power in the first stage of 6,000 Boeing 707 airplanes." But the USSR "is building a larger booster," he said, "and will shortly, I believe, in calendar year 1968, be flying a booster larger than the Saturn-5." "There are very real grounds for saying" that the Soviets "wish to maintain the position of having bigger boosters than we have," Webb declared, "and they are developing the capability to make that wish come true."

The U.S.' most powerful rocket engine, the solid-fueled SL-3, had developed a record 5.4 million pounds of thrust as it was fired June 17 in a static test at the Aerojet-General Corp.'s plant near Homestead, Fla. The rocket motor, 260 inches in diameter and nearly 80 feet long, burned 1,645,000 pounds of propellant in 75 seconds. The test was probably the last firing of a U.S. large-diameter solid-fueled rocket under the current program. NASA had made no budget requests for continued development of such engines on the ground that work done so far had produced enough data on large-diameter solid-fueled rockets should it be decided later that such rockets would be useful in the space program. Each of the large-diameter rockets tested was a short-length version. A full-length rocket would be 120 to 200 feet long, depending on its mission, and would develop 5 to 8 million pounds of thrust.

The Air Force's Space Systems Division had started the large-diameter solid-fueled rocket program in 1963, but in mid-1964 most of the program was transferred to NASA, and the Air Force retained only the part of the program involving 156-inch engines, which was considered the largest practical size for the land mobility required in defense use. Under the NASA program, a 120-inch short-length engine had produced about 600,000 pounds of thrust in a test at the Homestead center in Sept. 1964, and the Thiokol Chemical Corp. produced 3 million pounds of thrust when it fired a 156-inch short-length engine in Brunswick, Ga. Feb. 27, 1965. Aerojet-General fired 2 short-length 260-inch engines in Sept. 1965 and Feb. 1966 and produced about $3\frac{1}{2}$

million pounds of thrust each time. The motor casing used June 17 was the same one used in the Sept. 1965 test. A larger nozzle and new catalysts made the faster burning possible; a portion of the nozzle exploded at the end of the June 17 test, but Aerojet-General Vice Pres. R. F. Cottrell described the test as "at least a 99% success." About $65 million in Air Force and NASA funds and nearly $50 million in company money had been spent on the large-diameter solid-fueled rocket program. Aerojet-General said that its part of the program had cost NASA $37 million and the company $23 million.

Red Chinese Progress Reported

Rep. James G. Fulton (R., Pa.) told the U.S. House of Representatives Oct. 23 that Communist China "is now proceeding with the development of a rocket with nuclear propulsion." He reported that the Chinese "program is based upon a nuclear reactor through which passes liquid hydrogen." The Chinese were also developing "a sounding rocket called Caditi for weather determination purposes," he said. "Likewise, work is being carried on . . . [in Kwangsi] on telecommunications satellites. The University of Nanking . . . is conducting training courses in . . . rocket programs. The Chinese space research center of Sinkiang now has programs in the study of materials necessary for space experiments. Intensive training is being given at the rocket institute in this center—Balon Roditi." Fulton predicted "that the Chinese will orbit the first Chinese satellite by Jan. 1968."

Red Banner, the organ of the Red Guard at the Peking Aeronautical Institute, had reported Jan. 21 that, according to "reliable sources," Communist China "will carry out this year [1967] the launching of a space vessel and a new experiment involving a missile with a nuclear warhead." The source, it was disclosed Jan. 25, was Yang Cheng-wu, then interim chief of staff of the armed forces.

USSR Nuclear Missile Development Hinted

U.S. Defense Secy. Robert S. McNamara told reporters Nov. 3 that the USSR might be developing a method of nuclear attack from orbit. McNamara said in a press conference statement that "a fractional orbital bombardment system, or FOBS," apparently was the method being perfected in Russia. In such a system, he reported, a missile "is fired into a very low orbit about 100 miles above the earth" and is brought out of orbit—"generally before the first orbit is complete"—by the firing of a retro-rocket.

McNamara based his announcement on "certain intelligence information we have collected on a series of space system flight tests being conducted by the Soviet Union." He estimated that a Soviet FOBS might be operational by as early as 1968. But he conceded that the tests might be for "some re-entry program" other than a FOBS. McNamara said the U.S. had "examined the desirability of the FOBS" several years previously but had decided not to develop it because "it would not improve our strategic offensive power."

McNamara said: The FOBS had both advantages and disadvantages when compared with "the traditional intercontinental ballistic missile," or ICBM. An ICBM "normally does not go into orbit" but "reaches a peak altitude of perhaps 800 miles," or 700 miles more than the FOBS. "Because of the low altitude of their orbits, some trajectories of a FOBS would avoid detection by some early warning radars, including our BMEWS [Ballistic Missile Early Warning System]"; "the impact point cannot be determined until [retro-rocket] ignition," and "the flight path can be as much as 10 minutes shorter than an ICBM." But a FOBS would be "significantly less" accurate than an ICBM, and its payload "would be but a fraction" of the ICBM's. A FOBS payload could be one to 3 megatons (equal in explosive force to one to 3 million tons of TNT), or about the same as a Polaris missile warhead. (The U.S. had deployed 650 of the latter submarine-launched weapons.) The FOBS "would not be accurate enough for a satisfactory attack upon United States Minutemen missiles, protected in their silos," but it might "provide a surprise nuclear strike against . . . soft land targets such as bomber bases." "Several years ago," however, the U.S. "initiated the deployment of equipment to deny" the Soviets the latter capability. "Already we are beginning to use operationally over-the-horizon [OTH] radars which possess a greater capability of detecting FOBS than do the BMEWS." OTH radars, the first of which had gone into service 2 months previously, "will give us more warning time [about 15 minutes] against a full-scale attack using FOBS than BMEWS does against the ICBM launch." "With 3-minute warning, 15-minute warning or no warning at all, we could still absorb a surprise attack and strike back with sufficient power to destroy the attacker." (McNamara announced that the limited anti-ballistic missile defense system the U.S. was deploying had been named the "Sentinel System" and that Lt. Gen. Alfred D. Starbird of the Army, director of the Defense Communications Agency, would be system manager.)

It was reported that 2 unannounced Soviet satellite launchings of Sept. 17 and Nov. 2, 1966, and perhaps 9 of the Cosmos satellites launched by the USSR on short-duration flights (one orbit or less) beginning Jan. 25, 1967 were believed to be FOBS development shots. The suspected Cosmos satellites were *Cosmos 139, 160, 169, 170, 171, 178, 179, 183 and 187*. The FOBS, in which nuclear warheads were not stationed in orbit, was not considered a violation of the treaty banning nuclear weapons in space.

Sen. Henry M. Jackson (D., Wash.) disclosed Nov. 4 that he had heard previously about the USSR's "new device" and that his Military Applications Subcommittee of the Joint Congressional Committee on Atomic Energy had scheduled hearings on it.

MIRV Development Revealed

John S. Foster Jr., director of defense research and engineering for the Defense Department, revealed in a speech in Dallas Dec. 13 that the U.S. was developing a spacecraft—which he described as a "space bus"—that could carry several thermonuclear warheads and drop them off one at a time as it passed over enemy cities. The bus, which Foster called "a major breakthrough in missile technology," was the U.S.' answer to "the Soviet deployment of additional ballistic missiles and defense against our ballistic missiles," Foster said.

The new weapon, also known as the MIRV (for multiple, independently targeted re-entry vehicles), could be sent into space by either the land-based Minuteman-3 intercontinental ballistic missile or by the submarine-mounted Poseidon missile. "After the main booster has cut off," Foster explained, "the bus keeps making minute adjustments to its speed and direction, and after each adjustment it ejects another [individual re-entry vehicle with a thermonuclear] warhead."

Foster asserted that the space bus "will assure penetration of Soviet antimissile defenses and can deliver unacceptable damage to the Soviet Union even after we have suffered an all-out nuclear attack."

Canadian Space Agency Urged

A 4-man panel of scientists urged the Canadian government Mar. 7 to create a centralized space agency that would coordinate planning and development of Canada's space program. The panel's report presented to the Science Council of Canada, warned that if Canada did not move quickly to "rent or buy" its own satellite launching facilities and to embark on its own communication satellite program within 5 years, it would be forced to rent U.S. satellites at high costs.

The panel, chaired by Dr. John H. Chapman, deputy superintendent of Canadian Defense Research Telecommunications, recommended that

Canada's yearly expenditures on space research be increased from the current $14 million to $60 million in 5 years. (Satellite communications were generally believed to be especially applicable in a large, lightly-populated country such as Canada. But the report noted that the growing U.S. space program might soon monopolize the limited space available for communication satellites.)

The report recommended that the joint U.S.-Canadian weather and high-altitude rocket program at Churchill, Manitoba be renegotiated when it expired in 1970 to ensure "complete Canadian control."

French Developments

France's shut-down of its space centers in the Algerian Sahara was completed June 30 with the transfer of the center at Colomb-Béchar to the Algerian government. The centers at Hammaguir and Reggan had been turned over to Algeria previously.

The French government announced July 1 that it was financing the development of a Super Diamant (Diamant B) satellite-launching rocket. The new rocket, a more powerful version of the current Diamant, was to have a first stage developing 28 tons of thrust, a 2d stage capable of 15 tons of thrust and a 15-ton-thrust 3d stage. Several Super Diamants were to be built, under current plans, and the first was to be launched when France's new space center in French Guiana was opened early in 1969.

Soviet Developments

Tass reported Aug. 22 that Italian-born physicist Bruno Pontecorvo, who had defected from Britain to the USSR in 1950, had become the chief of a new school of space physics established in Siberia on the shore of Lake Baikal.

The Soviet Defense Ministry newspaper *Krasnaya Zvezda (Red Star)* reported Aug. 24 that 10 Soviet cosmonauts were practicing parachuting into the sea in apparent preparation for future space missions. All Soviet manned space flights heretofore had ended with landings on dry ground.

Izvestia reported Nov. 30, in an article on "Stellar City, USSR," that Soviet Air Force Col. Gen. Nikolai Petrovich Kamanin, known to the cosmonauts as "Makarych," was "the political mentor of the cosmonauts." "Stellar City," whose location was not disclosed, was the Soviet space "capital" in which cosmonauts were trained and in which space missions were planned.

Col. Gen. Vladimir F. Tolubko, first deputy commander-in-chief of the USSR's strategic rocket forces, disclosed in an interview in *Trud* that his military group had launched all Soviet manned space shots, all 13 Luna shots, 4 Venus probes, all 7 Molniya-1 communications satellites, the first sputnik and presumably other space missions. He made no mention of any civilian-launched Soviet space shots.

Soviet scientist Boris Nikolayevich Petrov revealed in an interview in *Pravda* Dec. 8 that the USSR had concluded treaties for the establishment of space tracking stations in the United Arab Republic, Mali, and other nations in Africa and Asia. Some of the stations were said to be in operation already.

It was reported in Moscow Nov. 11 that Soviet Air Force Gen. Anatoly Stolyerov had been assigned to head a new commission to study unidentified flying objects (UFO's or "flying saucers").

European Rocket Cooperation

The West German aircraft company Bölkow announced in Munich Dec. 6 that 11 major European aircraft companies had formed a holding company named Société Européene pour l'Etude et l'Intégration des Systèmes Spatiaux (Setis) to develop space rockets. Its first assignment was to increase the power of the 3-stage Eldo (European Launcher Development Organization) Europa space booster to enable the Eldo rocket to put a 375-pound communications satellite into a synchronous orbit. Setis was to develop a thrust package to be mounted on the Europa's 3d stage and an extra motor for the satellite.

1968

3 U.S. astronauts became the first men to fly to the vicinity of the moon. They circled the moon 10 times in orbit and returned safely to earth in December. Earlier, 3 other astronauts had made the first manned flight in an Apollo spaceship—the vehicle designed for manned landings on the moon—when they orbited the earth 163 times in October. The USSR flew its first manned orbital mission in $1\frac{1}{2}$ years shortly after the October Apollo trial. Both the U.S. and USSR continued unmanned probes of the moon—and 3 Soviet probes returned to earth after having orbited the moon.

MAN IN SPACE

U.S. Moon Module Orbited

Prior to the manned lunar orbiting mission, NASA Jan. 22 had sent into earth orbit an unmanned working model of the landing craft that the U.S. planned to use for the Project Apollo manned landing on the moon. NASA officials reported Jan. 23 that despite a premature engine shutdown, all "essential objectives" had been accomplished on this first test flight of a lunar module (or LM). They indicated that the tested LM, designated LM-1, had proved itself sufficiently so that an LM would not have to be test flown again before one was used in a manned Apollo orbital flight.

The LM, a 2-stage vehicle built by the Grumman Aircraft Engineering Corp., weighed 31,700 pounds (with propellants). Its lower, or descent, stage, whose job was to power the manned LM from lunar orbit to the moon's surface, had a rocket engine whose thrust could be throttled down from 9,700 pounds to 1,050 pounds. It was planned that after a landing on the moon, the descent stage would serve as a launching pad when the ascent stage returned the astronauts to the command and service modules, which would have been left in lunar orbit. The ascent stage, requiring less power, had a rocket capable of delivering 3,500 pounds of thrust. Since the LM-1 was being used for testing in orbit only and not in a landing, it did not have the 4 landing legs or windows that were to be used in a manned lunar landing.

The LM-1 was sent into space in the nose of the 2-stage Saturn-1B (uprated Saturn) rocket atop which 3 astronauts had died in a fire in an Apollo capsule Jan. 27, 1967. The 181-foot Saturn-1B (designated AS-204) rose from Cape Kennedy's Launch Complex 37 at 5:48 p.m. EST Jan. 22 and put the LM-1 into an orbit with an apogee of 137 miles and perigee of 101 miles.

The test program had called for an initial 26-second burn of the descent rocket starting at 10% of the rated 10,500 pounds of thrust and building up to $92\frac{1}{2}$%. Instead, it was shut off after 4 seconds by an on-board computer because the rocket, although operating properly, did not build up power fast enough to indicate proper operation under the computer's rigid program. NASA officials pointed out later that if astronauts had been aboard, they would have corrected this computer misjudgment. The engines of both stages were ignited, however, and the test showed that the LM's stages could separate and fire properly.

After the 8-hour test, the 2 LM-1 stages were left in their separate orbits with their systems no longer functioning.

NASA announced Mar. 16 that a 2d unmanned test flight of the

Apollo lunar module was "not considered necessary" because of the module's successful operation in ground tests and in its single flight test Jan. 22.

Apollo Test Fails

An unmanned Apollo spaceship was shot into space Apr. 4 and then brought back to earth in a flight that was labeled a "failure" because of 3 engine failures in the 363-foot Saturn-5 launching rocket. The primary purpose of the mission, designated *Apollo 6*, was to qualify the Saturn-5 for manned flights. The failure meant that NASA would have to launch at least one more Saturn-5 (the 3d), at a cost of some $200 million, before entrusting it with a manned spaceship. NASA officials said that the failure might not have endangered astronauts if the flight had been a manned one.

The Saturn-5 with the *Apollo 6* spacecraft in its nose was launched from Cape Kennedy at 7 a.m. All 5 engines of the 138-foot first stage, generating $7\frac{1}{2}$ million pounds of thrust, operated properly and were shut down after $2\frac{1}{2}$ minutes.

The first stage was then detached and the 5 engines of the 81-foot 2d stage ignited. 2 of these engines, however, shut down almost 2 minutes before schedule. The 2d stage then dropped away, and the single engine of the 59-foot 3d stage ignited. Because of the 2d-stage failure, however, it put the space vehicle into an orbit with an apogee of 216 miles and perigee of 112 miles, although the program had called for a circular orbit 115 miles above the earth.

The 3d-stage engine then failed to reignite for a 2d burn that was supposed to bring the assembly to an altitude of 320,000 miles (further than the distance from the earth to the moon). As a result, when the 3d stage was dropped, *Apollo 6* was able to achieve an altitude of only 13,800 miles on its own power. Then, coming back to earth, it did not have enough power left to increase its re-entry speed to the 25,000 mph. that had been planned in order to simulate a re-entry after a lunar mission. *Apollo 6*'s re-entry speed was only 20,000 mph.

The command module was then eased down safely by parachute, and it splashed down in the planned landing area in the Pacific 580 miles northwest of Hawaii at 4:56 p.m. EST. It was recovered by a Navy helicopter carrier.

Apollo Spaceship Tested in Earth Orbit

The manned lunar orbital mission was preceded by the flight of *Apollo 7*, the first manned flight test of the Apollo series. 3 U.S.

astronauts circled the earth 163 times aboard *Apollo 7* Oct. 11-22. The 260-hour, $4\frac{1}{2}$ million-mile journey was the 2d longest manned space flight made so far and the first trip by Americans into orbit since Nov. 1966.

The commander of *Apollo 7* was Navy Capt. Walter Marty (Wally) Schirra Jr., 45, the oldest man to make a space flight and the first man to go into orbit for a 3d time. (He was the first man to pilot all 3 U.S. manned spacecraft models—Mercury, Gemini and Apollo.) The other 2 crew members, each making his first space flight, were Air Force Maj. Donn Fulton Eisele, 38, the navigator, and civilian R(onnie) Walter Cunningham, 36, who monitored the controls and handled communications. (Eisele was promoted to lieutenant colonel Oct. 23.)

The flight, from lift-off at Cape Kennedy to safe splashdown in the Atlantic, was described as virtually perfect. The most notable difficulty, which in no way marred the success of the flight, was the fact that all of the 3 astronauts suffered colds during their 11 days in space. But the astronauts and the ground personnel declared themselves completely delighted with the performance of the spaceship.

Air Force Lt. Gen. Samuel C. Phillips, the Apollo program director, said after the flight was completed Oct. 22 that it was "the first space operation that has accomplished more than 100% of its preplanned objectives. Our official count is that we have accomplished 101% of our intended objectives." The additional 1% consisted of tests added after the flight had started.

The launching took place Oct. 11 at 11:03 a.m. EDT when a 1.3 million-pound, 224-foot assembly consisting of the spacecraft and a 2-stage Saturn-1B booster rocket rose from Cape Kennedy's Launch Complex 34.

The astronauts went into orbit in a 69,000-pound 100-foot space vehicle consisting of the booster rocket's 2d (S4B) stage and the 34-foot 9-inch *Apollo 7* spacecraft. Flying in an orbit with an apogee (initially) of 176 miles and perigee of 141 miles, Schirra fired the S4B stage's engines 6 times and practiced steering the spaceship while it was still attached to the S4B stage.

Schirra then detached the S4B and thus reduced the manned space vehicle to a 41,358-pound 2-module assembly consisting of (1) the conical command module, $10\frac{1}{2}$ feet high and 12 feet 10 inches in diameter at its base, in which the 3 astronauts rode, and (2) the $22\frac{1}{2}$-foot-long cylindrical service module, housing the engines, fuel tanks, fuel cells and some communications and control equipment. Thus reduced in size, the vehicle was believed to be still the biggest manned spaceship flown so far and perhaps some 4,000 pounds heavier than the largest manned Soviet spaceship.

Using the service module's 20,500-pound-thrust rocket—which they fired a total of 8 times during the 11-day flight—the astronauts put the spaceship through a variety of maneuvers while flying in "formation" with the S4B Oct. 11. Then they dropped the spaceship into a slightly lower orbit and drew away from the S4B. They also made themselves more comfortable by removing their bulky space suits and wearing 2-piece Teflon coveralls for most of the flight.

In a key exercise Oct. 12, the astronauts located the abandoned S4B and executed a successful "rendezvous," coming to within 70 feet of the S4B but deciding to approach no closer because it was tumbling dangerously. The exercise was considered important because it duplicated the maneuvers an Apollo spacecraft would have to execute to rescue astronauts in a lunar landing module should the landing craft become disabled in lunar orbit.

Schirra's cold was in full bloom by Oct. 12, and he doctored himself with decongestants and aspirin. Cunningham was showing cold symptoms by Oct. 13, and Eisele came down with a cold by Oct. 15. Aside from the colds, the astronauts appeared to be in good health. They took turns sleeping in 2 hammocklike sleeping bags, they ate normally, and they exercised regularly with an elastic-cord device. (All 3, however, were later found to have suffered a temporary diminution in strength as a result of 11 days of weightlessness.) Forbidden to shave during the flight, the 3 astronauts came back to earth with untidy, 11-day growths of beard.

The 5th of the 8 firings of *Apollo 7*'s 20,500-pound-thrust engine took place Oct. 18 when the engine was ignited for 66 seconds (this was the longest firing of the flight) to raise the orbit's apogee to 282 miles. (The perigee: 102 miles.)

Schirra had refused, with irritation, to transmit a telecast of himself and his crew members in space Oct. 12. He complained that they were too busy. But the first of a series of 7 live TV broadcasts of the *Apollo 7* crew in flight was transmitted to earth Oct. 14. The astronauts complained freely and frequently about being told to perform new experiments that they considered useless and "idiotic." In one of the tests, *Apollo 7* was allowed to tumble in the sun's rays to see whether the heat changes made changes in anything inside the spaceship. Schirra finally snapped to the ground personnel Oct. 10: "I have had it up [to] here today and from now on I am going to be an onboard flight director for these updates. We are not going to accept any new games like doing some crazy tests we never heard of before. Each test is going to be thoroughly reviewed before we act on it."

The re-entry procedure started at 6:42 a.m. EDT Oct. 22 over the Pacific Ocean. With the 20,500-pound-thrust engine facing in the direction of flight, it was fired for 10 seconds to reduce the spacecraft's speed and bring it out of orbit. About 2 minutes later the service module was detached automatically from the manned command module. The service module then dropped into the atmosphere separately and burned up in the heat created by friction with the air.

The re-entry flight line of the command module was adjusted by the use of the command module's small thrusters as the module made its gradual approach to the upper atmosphere.

The command module, with its heat shield-protected base facing in the direction of flight, entered the atmosphere at about 6:56 a.m. at a speed of about 16,675 mph. and at an altitude of about 400,000 feet approximately over New Orleans. Although the temperature of the blunt heat shield rose to about 3,200°F., the temperature inside the command module remained below 80°.

The astronauts had put on their space suits for the re-entry. They did not wear their space helmets, however, because their colds caused fear that their ear drums might rupture or their sinuses become clogged as pressure built up in the spaceship during the descent. In preparation for descent, therefore, they all took decongestants. Then when the cabin vents opened automatically at an altitude of 27,000 feet and the cabin air pressure started to rise, the astronauts closed their mouths, held their noses and forced air into their ear and nasal passages. (Physical examination later showed that this had apparently worked and that the descent had produced no ill effects.)

At an altitude of about 24,000 feet, while the spacecraft was falling at a speed of 316 mph., 2 nylon parachutes were unfurled automatically to slow the descent to about 139 mph. The 3 main 'chutes inflated at an altitude of about 10,000 feet, and they slowed the rate of descent to about 22 mph.

The command module splashed into the Atlantic at 7:11 a.m. EDT Oct. 22 about 325 miles south of Bermuda, $7\frac{1}{2}$ miles from the waiting recovery ship, the carrier *Essex,* and, it was reported later, only $\frac{1}{3}$ mile from the aiming point. The capsule capsized in the water but was quickly righted by inflated balloons. Recovery helicopters from the *Essex* took 20 minutes to locate the capsule. The astronauts were then flown by helicopter to the *Essex,* and the *Apollo 7* was hauled aboard later. (It was reported that Schirra, the only Navy man among the 3 occupants of *Apollo 7,* suffered seasickness while waiting to be taken from the floating spaceship.)

Soviets Disclaim Race to Moon

Prof. Leonid I. Sedov, Soviet space scientist and head of the Soviet delegation to the 19th Congress of the International Astronautics Federation, told newsmen in New York Oct. 14 that the USSR was not racing the U.S. to get a man on the moon. "The question of sending cosmonauts to the moon at this time is not ... on our agenda," he declared. "The exploration of the moon ... is not a priority." He said that Soviet lunar exploration depended on the success of a series of unmanned flights then in process by Soviet probes designated as Zond. (Sedov and the rest of his 40-member Soviet delegation were invited by the U.S. to inspect selected parts of the Cape Kennedy space center, but Sedov declined because, he noted, "quite frankly, we cannot reciprocate.")

The Mexico City newspaper *Ultimas Noticias* Oct. 23 quoted Soviet cosmonaut Gherman S. Titov as saying in an interview in Mexico City that the USSR would get to the moon before the U.S. did and that he (Titov) would be a member of the Soviet crew.

Tass asserted Nov. 29 that the successful unmanned Zond flights around the moon and back to earth had opened "the space route earth-moon-earth" for Soviet cosmonauts. "Automatic space probes always precede manned flights," Tass said. It reported that the Zond missions had helped to solve such problems as protection of cosmonauts from solar flares and recovery of spaceships after interplanetary flights.

U.S. Astronauts Fly Around Moon

3 U.S. astronauts, aboard the U.S. Spaceship *Apollo 8,* became the first men to fly to the vicinity of the moon. They circled the earth's natural satellite 10 times in orbit Dec. 24–25—on Christmas Eve and Christmas. They were not equipped to land on the moon, however. After completing the 10 revolutions around the moon, they brought *Apollo 8* back to earth and a safe splashdown in the Pacific Ocean Dec. 27.

The historic feat was accomplished by Air Force Col. Frank Borman, 40, the spaceship commander, Navy Capt. James A. Lovell, Jr., command module pilot (and navigator), and Air Force Maj. William A. Anders, 35, who was largely in charge of the trip's photographic mission. (Borman and Lovell had flown in space together in *Gemini 7,* and Lovell had flown in orbit again as command pilot of *Gemini 12.* Of the 3, only Anders had not made a space flight.)

The 6-day flight, from launching at 7:51 a.m. EST Dec. 21 at Cape Kennedy, Fla. to the watery landing at 10:51 EST (4:51 a.m. local time) Dec. 27 in the Pacific, established many records or "firsts" for space flight:

- This was the first time men had flown so far (about 233,000 miles) from the earth.

- The trip was man's longest space flight (about 550,000 miles round-trip).
- This was the first time men had gone into orbit around the moon or had even reached its vicinity.
- The 3 astronauts were the first men to reach a place where the gravitation of an astronomical body other than the earth was dominant.
- They were the first men to see the side of the moon that is always turned away from the earth.
- On returning, the astronauts achieved a speed record of 24,530 mph. as they re-entered the atmosphere.
- A record weight of 63,433 pounds (the spacecraft and the booster rocket's 3d stage) was put into earth orbit as a single assembly in the launching Dec. 21.
- Lovell, who had held the record for time spent in space flight (425 hours 10 minutes) even before the *Apollo 8* flight, increased his record to 572 hours 10 minutes.

The 3 astronauts received congratulatory messages from U.S. and other world leaders, including Pres. Johnson, Pope Paul VI, Soviet Pres. Nikolai V. Podgorny and Un Secy. Gen. U Thant. Press coverage and comment in the USSR was highly laudatory.

A major purpose of the flight was to get closeup photos of the moon and especially of areas that might be suitable for a manned landing. The photos the astronauts took, of the earth as well as of the moon, were described as of exceptionally high quality. The astronauts also transmitted to earth excellent views of the moon and of the earth during 6 "live" telecasts.

All phases of the flight and of the planning that preceded it were covered extensively by all news media. The schedule of possible launching times and optional missions (should the lunar orbital mission be canceled) were published almost as soon as the details were decided on.

The flight was said to have taken place from start to finish with "textbook" perfection. It began and ended almost precisely to the minute scheduled, and the trajectories to and from the moon were almost exactly as planned.

Apollo 8 and its 3 astronauts were sent into space in the nose of a 3-stage Saturn-5 rocket launched from Cape Kennedy's Launch Complex 39A at 7:51 a.m. Dec. 21. The 363-foot rocket-spaceship launching vehicle, weighing 6,218,558 pounds at ignition, was carried to an altitude of about 40 miles by the Saturn-5's first (S1C) stage. The S1C, its 5 engines developing a total of $7\frac{1}{2}$ million pounds of thrust, accelerated the assembly to a speed of 6,068 mph. and dropped away.

The 2d (S2) stage carried the remainder of the assembly to an altitude of about 115 miles, speeded it to a velocity of about 15,250 mph. and also separated. The 3d (S4B) stage then put itself and the attached spaceship into 1 115-mile-high orbit with an 88.2-minute period and 32.5° angle of inclination in which it traveled at a speed of 17,435 mph.

An hour and a half of checking by the 3 orbiting astronauts and various ground personnel showed that all equipment was operating satisfactorily. The Manned Space Flight Center in Houston, Tex. then radioed the *Apollo 8* crew members to start on the 234,100-mile curved trajectory toward the moon.

As *Apollo 8* passed over the Western Pacific Dec. 21 on its 2d revolution around the earth, Borman ignited the S4B stage for a 2d time, and the spaceship was accelerated to an "earth escape" velocity of 24,200 mph. At 11:12 a.m. EST, about 20 minutes after completing its 2d ignition, the S4B separated, and the *Apollo 8* continued toward the moon on its own.

The 363-foot structure that had risen from the launch pad had been divested piecemeal of its 138-foot first stage, its $81\frac{1}{2}$-foot 2d stage and $58\frac{1}{2}$-foot 3d stage. The slim 33-foot launch escape system that had stood atop the launching assembly had also been jettisoned. The remaining 3-section 52.9-foot spaceship speeding toward the moon consisted of: (1) The command module, a conical capsule 12 feet high, 12 feet 10 inches in diameter at the base, 13,392 pounds in weight (launching weight), encased in heat shields; the command module, containing the crew compartment, was the only part of the spaceship to complete the trip and to return to earth. (2) The service module, a cylinder 22 feet long, 12 feet 10 inches in diameter, 51,258 pounds in weight (launching weight), containing the service propulsion system (a rocket engine). (3) The spacecraft-LM (lunar module) adapter, a truncated cone, 28 feet long, tapering from a diameter of 21 feet 8 inches to a diameter of 12 feet 10 inches, carrying a dummy lunar landing module, 4,150 pounds.

As *Apollo 8* raced toward the moon, it rotated gently at a rate of about 6 turns an hour to keep the sunlit side from getting too hot and the dark side from getting too cold.

Apollo 8 used its own service propulsion system for the first time at 6:51 p.m. EST Dec. 21 when Borman fired it for 2.4 seconds to make the first scheduled mid-course correction. This ignition increased the speed of the spaceship by 16 mph. As anticipated, *Apollo 8* had been losing speed, due to the pull of the earth's gravity, ever since the S4B ignition had ended. It continued to lose speed until 3:30 p.m. Dec. 23, when, at a velocity of 2,216 mph. (its slowest speed of the mission), it entered the "sphere of gravitational influence" of the moon—the area in which the pull of the moon's gravity was greater than the pull of the earth's. At this point *Apollo 8* was about 214,000 miles from the earth and 38,900 miles from the moon. It was then speeded up by the pull of the moon's gravity.

Borman Dec. 22 began suffering symptoms (vomiting and diarrhea) of viral gastroenteritis, a 24-hour sickness commonly (but mistakenly) called intestinal 'flu. All 3 of the astronauts reported feelings of nausea, especially when they removed their spacesuits shortly after they entered the lunar trajectory (they completed the flight in comfortable coveralls) and when they moved around the capsule. Dr. Charles A. Berry, chief flight surgeon for the mission, said at the Manned Space Center in Houston that the nausea was probably due to "a form of motion sickness." But all 3 were sufficiently improved by 3:01 p.m. Dec. 22 to transmit to earth the first of their 6 "live" telecasts from space. And all appeared to be in excellent health by the time they went into lunar orbit.

Another minor course correction was made as scheduled shortly after 9 p.m. Dec. 23 at a distance of less than 27,000 miles from the moon. The pre-correction trajectory, however, was so close to the one desired that if the course correction had not been made, the spaceship would have swung around the moon at a distance of only 80 miles instead of the 71.6 miles originally planned.

Apollo 8 was pulled around the moon by lunar gravity at 4:49 a.m. EST Dec. 24. At 4:59 a.m., while behind the moon and hidden from earth, the spaceship's rocket was fired in the direction of the line of flight. 4 minutes of this retrofire slowed *Apollo 8* from its 5,758-mph. speed to a velocity of 3,643 mph. This put the spaceship, at 5:03 a.m., into a lunar orbit with an apocynthion (maximum altitude from the moon's surface) of $194\frac{1}{2}$ miles and pericynthion (closest distance to the lunar surface) of 69.6 miles.

Lovell's first close-up description of the moon, radioed to ground controllers shortly thereafter, was: "The moon is essentially gray. No color. Looks like plaster of Paris or sort of grayish beach sand." The astronauts reported being able to see much detail. They easily and quickly recognized many of the craters, faults, peaks and maria already charted by telescope and unmanned lunar probes.

After circling the moon twice in this preliminary orbit, the astronauts fired their engine briefly again to put the spaceship into a circular orbit 69.8 miles above the surface. *Apollo 8*'s speed in the circular orbit was 3,551 mph.

The astronauts transmitted 2 live telecasts to earth Dec. 24 while in lunar orbit. At the close of the 2d Christmas Eve telecast, Borman announced that the spaceship crew had "a message that we would like to send to you." He then began reading the first verses of Genesis: "In the beginning God created the heaven and the earth...." The other astronauts took up the reading, and Borman concluded: "... and God saw that it was good."

The astronauts Dec. 24 unofficially bestowed their own names and the names of other astronauts, NASA colleagues and friends on previously unnamed lunar craters, peaks and other landmarks. Most of the names had been picked before the flight by the 3 astronauts and scientist-astronaut Harrison A. Schmitt, a geologist. But NASA officials quickly disclosed that the names had been given only for NASA and astronaut identification and that there was no plan to submit them officially to the International Astronomical Union.

After circling the moon 10 times in 20 hours, the astronauts fired their engine again at 1:10 a.m. Dec. 25 to bring the spaceship out of lunar orbit and into a trajectory for returning to the earth. *Apollo 8* passed into the earth's "sphere of gravitational influence" at 12:38 p.m. Dec. 25, and a minor mid-course correction was made at 3:51 p.m. Dec. 25.

A final scheduled course correction was dropped Dec. 27 because the spaceship already was on a virtually perfect course that would bring it through a 36-mile-wide re-entry "corridor" to its Pacific target area.

The service module was jettisoned at 10:23 a.m. EST. An on-board computer in the 11,000-pound command module then fired 6 thruster rockets to turn the conical capsule's blunt base in the direction of flight.

The manned capsule plunged into the atmosphere at 10:37 a.m. EST at a searing speed of 24,530 mph., and the heat shield was heated by friction with the air to temperatures of up to 5,000°F. The 3 astronauts inside the capsule, however, were protected from the heat. For about 5 minutes communications were "blacked out," as expected, by the envelope of charged particles that formed around the spaceship as it hurtled through the resisting atmosphere.

2 14-foot drogue parachutes were deployed automatically at an altitude of 23,300 feet, when the atmosphere had already slowed the spaceship to some 300 mph. The 3 small chutes pulled out the 3 83-foot main 'chutes. The removal of a cover over the parachutes uncovered a flashing beacon, which was seen in the darkness by men aboard the waiting carrier *Yorktown*.

The spaceship splashed into the water at 10:51 a.m. EST (4:51 a.m. local time) Dec. 27 at a speed of 20 mph. It came down only 7,100 yards from the *Yorktown*, about 1,000 miles southwest of Hawaii.

The astronauts remained in the floating capsule for about 90 minutes until it was light enough for a helicopter to lower 3 frogmen into the water. The swimmers secured the capsule with a flotation collar, and the astronauts then left the capsule and were brought by helicopter to the *Yorktown*. The spaceship was hoisted aboard the ship an hour later.

Preliminary medical examinations aboard the *Yorktown* indicated that all 3 astronauts were in good condition after their 6 days in space.

(Sir Bernard Lovell, director of Britain's Jodrell Bank radiotelescope observatory, had said Nov. 20 that there was no scientific justification for the "undue risks to human life" inherent in the *Apollo 8* flight.)

Cosmonaut Orbits Earth 64 Times

Soviet Air Force Col. Georgi Timofeyevich Beregovoi circled the earth 64 times in the spaceship *Soyuz 3* Oct. 26-30 and then landed safely after a 4-day test of the USSR's Soyuz-class spacecraft. Beregovoi, 47, was the oldest of the 32 people who had flown in orbit so far. He was the first Soviet cosmonaut to go into space since Col. Vladimir M. Komarov was killed in a re-entry accident after an Apr. 1967 space flight.

Soyuz 3, with Beregovoi the lone crew member, was sent into space by means of a large booster rocket launched from the Baikonur cosmodrome at Tyuratam, northeast of the Aral Sea, in Kazakhstan. Before entering his spaceship, Beregovoi dedicated the mission to the 51st anniversary (Nov. 7) of the Bolshevik revolution. The launching took place at 11:34 a.m. (Moscow time) Oct. 26, and the spaceship achieved an orbit with an initial apogee of 225 kilometers (140 miles), perigee of 205 kilometers (127 miles), period of 88.6 minutes and inclination of $51°\,40'$ to the equatorial plane. The launching was not disclosed by the USSR until *Soyuz 3* was successfully in orbit.

The *Soyuz 3* shot was preceded by the secret launching Oct. 25 of the unmanned spaceship *Soyuz 2*, which achieved an orbit with an initial apogee of 224 kilometers (139 miles), perigee of 185 kilometers (115 miles), period of $88\frac{1}{2}$ minutes and inclination of $51.7°$.

The orbiting of *Soyuz 2* was disclosed by the USSR Oct. 26 after Beregovoi in *Soyuz 3* had located *Soyuz 2* and completed the first of 2 successful rendezvous maneuvers with the target spacecraft. According to Tass, the first rendezvous took place Oct. 26 during *Soyuz 3*'s first orbit. The initial approach, "up to a distance of 200 meters [656 feet], was carried out by an automatic system," Tass reported. "The subsequent operations were performed by Pilot-Cosmonaut

Beregovoi, using the manual control system." The 2d rendezvous, a 90-minute experiment, took place Oct. 27. Tass did not report whether "docking" was attempted during either of the 2 rendezvous exercises. (It was reported later that docking was not attempted.)

Following the 2d rendezvous, the 2 spaceships were in slightly altered orbits with these parameters: *Soyuz 3*—252-kilometer ($156\frac{1}{2}$-mile) apogee, 179-kilometer (111-mile) perigee, 88.6-minute period and 51.7° inclination. *Soyuz 2*—231-kilometer ($143\frac{1}{2}$-mile) apogee, 181-kilometer ($112\frac{1}{2}$-mile) perigee, 88.4-minute period and 51.7° inclination.

Beregovoi's health and efficiency were reported to be good throughout the flight and after his landing. He slept in a special compartment "adjoining the cosmonaut's cabin and intended for scientific research and the cosmonaut's rest." He exercised regularly and ate with good appetite. He reported back to earth both by radio and TV, and he sent greetings to Soviet party and government leaders and to the people of various countries, including (Oct. 27) "the courageous Vietnamese people who are waging a heroic struggle, against the American aggressors, for their freedom and independence."

Soviet officials reported that Beregovoi had completed successfully all experiments assigned to the mission. His experiments included observation of luminescent particles and photographing of the overcast and twilight horizons of the earth.

Soyuz 2 was brought out of orbit by the firing of its retro-rocket Oct. 28, and it made a parachute-slowed landing "in the pre-set area of the Soviet Union."

Soyuz 3 was brought down in the same manner. As described by Tass: "*Soyuz 3* made a controlled descent with the use of aerodynamics [the braking and lifting characteristics of the atmosphere]." The spaceship was first turned so that its thruster rocket faced in the direction of flight. A 145-second blast of the retro-rocket slowed the spacecraft and brought it out of orbit. "Then the capsule was separated from the spaceship and was operated with the engines for guided descent to orient it for entrance into the dense layers of the atmosphere.... The guidance system issued commands to orient the spacecraft and insured its landing precisely in the preset [target] area...." The parachutes then continued to slow the spacecraft, "and close to the earth the engines for soft landing were switched on."

Soyuz 3 made a soft landing in snow "in the area of Karaganda" (the province in which the Baikonur cosmodrome is located) at 10:25 a.m. (Moscow time) Oct. 30. Beregovoi reported in Moscow Nov. 1 that the spaceship came down so close to the target that "after the landing and before the unfastening of the hatch I saw through the portholes my comrades coming to meet me. More than that, the craft was detected in the air by planes and, for the first time, a parachute descent of the spacecraft and its landing ... was filmed."

The Soviet newspaper *Pravda* reported Nov. 17 that the Soyuz spaceships consisted of these 3 "main compartments": "an orbital compartment, which is a scientific laboratory where the cosmonaut conducts research and rests; a module-descending canister designated for the orbiting of the crew and the return of the latter to earth; and an instrument compartment containing the spaceship's main systems and engines." The orbital compartment, in the front of the spaceship, "communicates with the module-descending canister with the help of a pressurized hatch," *Pravda* reported. The size of the "working compartments" was given as "up to 9 cubic meters [318 feet]."

Pravda disclosed that *Soyuz 3* had been equipped with "an automatic docking system." *Pravda* reported that Soyuz spacecraft had engines that made it possible to maneuver "at an altitude of up to 1,300 kilometers [808 miles]."

Pravda said that because of the use of aerodynamics in Soyuz descents, the spaceship's crew was subjected to forces only 3 to 4 times the pull of gravity as compared with 8 to 10 Gs experienced in the ballistic descents made in previous spacecraft. In a descent, *Pravda* reported, the retro-rockets were fired, the module-descending canister (carrying the crew) was detached from the rest of the spacecraft, "the decelerating parachute opens at an altitude of some 9 kilometers [$5\frac{1}{2}$ miles]"; "the main cupola of the parachute system" then opens; at "about one meter [3.28 feet] from the ground, the soft-landing power retro-rockets come into play," and "landing speed does not exceed 2 to 3 meters a second."

Pravda said that the Soyuz program, designed to investigate "near-earth space," provided for "large-scale scientific and technical investigations and ... the establishment of manned orbital stations in the future."

UNMANNED LUNAR ORBITAL MISSIONS

Soviet Probes Return After Moon Orbits

The USSR launched 3 Zond (Probe) unmanned spacecraft during 1968. 2 of them orbited the moon and returned to earth.

The first probe, *Zond 4*, was launched from an undisclosed site in the USSR Mar. 2 on a space flight that apparently took it 240,000 miles into space by Mar. 5-6 and then back to earth, where, according to some Western observers, it might have crashed Mar. 9.

The announcement of the *Zond 4* launching was made by Tass, which disclosed that the probe had been placed in a 170-mile-high "parking orbit" around the earth before being sent further into space. Soviet silence on the fate of *Zond 4* thereafter led to Western speculation that the mission was a failure. The initial announcement said that the probe's aim "is to study outlying regions of near-earth space and improve new systems and units" aboard the spacecraft.

According to Western sources, *Zond 4* reached a peak altitude "comparable to lunar altitude" before radio signals from the ground activated its braking rocket to send it back to earth. Calculations indicated that the return to earth took place Mar. 9, the day the USSR withdrew from the Pacific and Indian Oceans the tracking ships it had deployed there.

Zond 5 was more successful. Launched by the USSR Sept. 15, it flew around the moon Sept. 18 and then returned to earth. It splashed down in the Indian Ocean Sept. 21 and was recovered by a Soviet ship the next day. The space probe was the first ever recovered after a flight to the vicinity of the moon.

Zond 5 launched at 00:42 Sept. 15 (Moscow time), had been put into a parking orbit around the earth in preparation for its lunar trip. The parking orbit had an apogee of 219 kilometers (136 miles) and perigee of 187 kilometers (116 miles). It was inclined at an angle of 51.5° to the equatorial plane. The USSR announced the launching only after the probe had left the parking orbit and had achieved what appeared to be a successful lunar trajectory. This first announcement said nothing about the lunar mission.

Some controversy was provoked Sept. 18 when Sir Bernard Lovell, director of Britain's Jodrell Bank radio-telescope observatory, reported that the probe had passed within 1,000 miles of the moon that day. He speculated that the probe had circled the moon and was en route back to earth for an attempted recovery. A Soviet Foreign Ministry spokesman, denouncing Lovell's report as a "canard," asserted that it "does not correspond with reality."

Tass reported Sept. 20 that *Zond 5* had flown around the moon Sept. 18 and had come within 1,950 kilometers (1,212 miles) of the lunar surface. It said that during the passage, the probe's automatic equipment had studied "the physical characteristics of outer space in the area of the moon." Tass did not disclose that a return of the probe to the earth would be attempted. Instead, Tass said that *Zond 5*'s research mission had been completed but that the probe would continue to send scientific information to earth stations.

Dr. Heinz Kaminski, director of West Germany's Bochum Satellite & Space Research Center, reported Sept. 20 that voices (probably recorded in advance or radioed from a Soviet earth station) had been transmitted by the probe in what appeared to be a voice-transmission test in preparation for a manned flight to the area of the moon.

Zond 5 re-entered the earth's atmosphere at 6:54 p.m. Moscow time Sept. 21. Smashing into the atmosphere at a speed of 24,500 mph., it was slowed aerodynamically before its parachute was deployed. It then made a parachute-slowed descent to the Indian Ocean, hit the surface of the water at 7:08 p.m. Sept. 21 and was picked up by a recovery ship Sept. 22. This was the first water landing of a Soviet probe returning from space.

Pravda reported Nov. 15 that *Zond 5* had carried 2 live turtles, wine [fruit] flies, yellow mealworms, spiderworts with buds, cells in culture, seeds of wheat, pine and barley, chlorella on nutritive mediums, lyzogen bacteria and other specimens. The turtles were alive and well on landing, but they had lost 10% of their weight, and changes had taken place in their liver cells and spleen structure.

The flight of *Zond 6* was also a success. The probe, launched Nov. 10, circled the moon Nov. 14 and then returned to earth for a successful soft landing in the USSR Nov. 17. A unique feature of *Zond 6*'s return was its use of aerodynamic drag and lift in 2 "plunges" into the atmosphere to decelerate before it made its parachute-aided landing: Smashing into the atmosphere for the first "plunge," *Zond 6* was slowed from a velocity of more than 11 kilometers (perhaps 7 miles) a second to a speed of some 7.6 kilometers (4.7 miles) a second. The deceleration was caused by air resistance and by aerodynamic lift, which soon thrust the spaceship back up and thus enabled it to re-enter for a 2d time and to benefit from a 2d aerodynamic deceleration. Soviet spokesmen said the use of aerodynamics made possible a "controlled descent" and a landing close to the target point.

Zond 6 had been launched Nov. 10 in the nose of a multi-stage carrier rocket sent up from the Baikonur cosmodrome at $10:11\frac{1}{2}$ p.m. (Moscow time). The probe, with the carrier rocket's final stage still attached, achieved a parking orbit around the earth with an apogee of 210 kilometers ($130\frac{1}{2}$ miles), perigee of 185 kilometers (115 miles) and 51.4° angle of inclination. The final stage's rocket engine was ignited for a 2d burn by radio signal from the earth at $11:18\frac{1}{2}$ p.m. to send *Zond 6* into a lunar trajectory. The probe then separated from the carrier rocket stage.

After a course correction was made by the use of *Zond 6*'s onboard engine Nov. 12, the spaceship flew around the moon Nov. 14 and then headed back towards the earth. Its nearest approach to the moon was given as 2,420 kilometers (1,504 miles).

Approaching the earth, *Zond 6* executed 2 course corrections Nov. 16 and 17 before the final "plunge" into the atmosphere Nov. 17 at 4:58 p.m. At an altitude of $7\frac{1}{2}$ kilometers ($4\frac{2}{3}$ miles), with the speed reduced to about 200 meters (656 feet) a second, the parachutes were deployed, and *Zond 6* made its soft landing in an unspecified "predetermined" area presumed to be in Kazakhstan.

Tass and the Soviet press reported that the flights of *Zond 6* and of its predecessors *Zond 4* and *5* had paved the way for a manned Soviet flight to the moon. Tass said Nov. 19 that the *Zond 6* flight had shown that the Russians could protect cosmonauts from cosmic radiation during a lunar flight and recover them when they returned to earth. Tass added Nov. 23 that the spaceship had carried "biological objects" to check on the effects of radiation. It said Nov. 29 that "the total amount of radiation *Zond 6* underwent in the course of its flight does not exceed the permissible level." *Izvestia* had reported Nov. 25 that *Zond 6* had found the radiation level "almost 100 times less than the permissible level."

The launching of *Zond 6* had not been announced by the USSR until the probe had been injected successfully into its lunar trajectory. The initial announcement did not reveal that the mission called for a turn around the moon and return to earth, and the USSR then remained silent about its space probe for 4 days. *Zond 6*'s rounding of the moon was first made known by Britain's Jodrell Bank observatory Nov. 14 and was confirmed by Tass several hours later. The return to earth was reported by Jodrell Bank Nov. 17 before the USSR announced it. Jodrell Bank had reported Nov. 13 the radio transmission of 2 voices (presumably recorded in advance for test purposes) from the unmanned spaceship.

Pravda asserted Nov. 24 that the Zond flights were part of a consistent Soviet "scientifically grounded program of moon exploration," which the USSR "regards as an important object of research."

Luna 14 Orbits Moon

The USSR put another Luna spaceship in orbit around the moon during 1968. The unmanned *Luna 14* was launched at 1:09 p.m. Moscow time Apr. 7 and was put into orbit around the moon Apr. 10 by a radio signal from the USSR. The probe achieved a selenocentric orbit whose apocynthion (maximum distance from the moon's surface) was given as 870 kilometers (540 miles), pericynthion (minimum distance) as 160 kilometers (99 miles) and period as 2 hours 40 minutes.

Tass reported Apr. 11 that *Luna 14*'s mission was to "study the relation between the masses of the earth and the moon and the moon's gravitational field," to study "the propagation and stability of radio signals" between the earth and probe, "to measure the stream of charged particles given off by the sun" and "to collect more data essential to the formulation of a precise theory of the moon's movement."

The announcement of the launching was made about 6 hours after it took place—when *Luna 14* appeared to be on a successful lunar trajectory. The announcement of the lunar orbit was made almost a day after it was achieved. (Britain's Jodrell Bank observatory had monitored the probe and had disclosed the lunar orbit nearly a day before the Russians did.)

Last Surveyor Lands on Moon

The unmanned U.S. lunar probe *Surveyor 7* was launched from Cape Kennedy early Jan. 7 on a 244,360-mile space flight that brought it at 8:05 p.m. EST Jan. 9 to a successful soft landing on the moon. It came down almost exactly at its aiming point north of the crater Tycho in the moon's Southern hemisphere. The 10-foot-high craft, the 7th and last probe in NASA's Surveyor program, began transmitting TV pictures back to earth 45 minutes after landing.

Surveyor 7, the most fully equipped scientifically of all Surveyor probes, carried a TV camera, alpha-particle scattering equipment to analyze the chemical composition of the lunar soil, a "surface sampler" to dig into the moon's soil, magnets on 2 of the footpads and on the surface sampler scoop and mirrors to give views under the probe, to provide stereo views and to detect dust. It was equipped to produce data on the radar reflectivity, magnetic properties and thermal conditions of the moon's surface.

The probe weighed 3,288 pounds when it rose from Cape Kennedy in the nose of the 2-stage Atlas-Centaur carrier rocket. On landing, after jettisoning its retro-rocket and using up its liquid propellants and attitude control gas, its weight (earth weight) was down to about 637 pounds.

Surveyor 7's original target had been the crater Hipparchus, and the navigational instructions built into its electronic guidance system were for a landing in Hipparchus. The change in target was made partly because earlier Surveyors had provided sufficient data to accomplish the Surveyor program's primary mission—to supply information on surface conditions at sites considered suitable for manned landings. NASA officials, therefore, decided to use *Surveyor 7* more intensively for scientific purposes and to send it to Tycho, a younger crater, where, it was believed, the soil would yield more information to the probe's sampling equipment. The change in course was accomplished during the single mid-course correction made in *Surveyor 7*'s 65-hour flight.

OTHER U.S. DEVELOPMENTS

U.S. Space Program 'Gearing Down'

John Noble Wilford reported in the *N.Y. Times* Apr. 16 that the U.S. space program was "gearing down to a slower pace." Civilian space spending had dropped from the record $5.9 billion in fiscal 1966 to $4.7 billion in the current year and was expected to decline further. Employment in civilian space work dropped from 420,000 in 1966 to fewer than 300,000 currently and was still dropping at the rate of 4,000 a month. Military space operations, however, were steadily increasing. The military space expenditure rate had reached $1.9 billion in the current year and was expected to rise to $2.2 billion in fiscal 1969.

Although "Apollo Project planners still expect to land men on the moon by the end of next year," the U.S. might then find itself with 5 or 6 Saturn-5 rocket boosters but with no missions for them, Wilford said. Members of the 55-man astronaut team were reported to be victims of a "sense of frustration" and a fear that they might be too old for space flight by the time missions might finally be assigned to them.

NASA Administrator James E. Webb had told the House Science & Astronautics Committee Feb. 7 that the USSR would probably outdo the U.S. in space during the next few years because "we are reducing our effort by $\frac{1}{3}$" at a time when "the USSR is still increasing its effort." Webb said: "The Soviet space program has consistently utilized larger boosters than were currently available to the U.S. . . . During 1968, or shortly thereafter, they will have available a booster with over 10 million pounds of thrust. . . . We believe they have the capability to do a flyby of the moon with some form of life, which some believe could be man. . . . New test and launch facilities are steadily added to expand their resource base, and a number of space-flight systems more advanced than any heretofore used are nearing completion." As for the U.S. program, Apollo had been cut back from the proposed 13 Saturn-5 flights to "no more than 9 ... by the end of 1969. ... It has been necessary to cancel the Pilgrim Project and to delay Sunblazer. ... The Voyager program was eliminated." Many planned flights were dropped.

Pres. Lyndon B. Johnson's budget for fiscal year 1969 allotted the the smallest amount for the space program in 5 years—about $220 million less in new obligational authority than the fiscal 1968 amount. The reduction in expenditures—$230 million below 1968, $850 million below 1967 and over $1.3 billion less than in 1966," Johnson said, "reflects our progress beyond the costly research and development phases of the manned lunar mission, as well as the immediate need to postpone spending for new projects wherever possible." But "we will resume manned flight tests of the Apollo spacecraft this year and proceed toward the manned lunar expedition," and "we will not abandon ... planetary exploration." Development of a new spacecraft was planned "for launch in 1973 to orbit and land on Mars." Production of the large Saturn-class space boosters was to be continued, "but at a reduced rate." Development of a nuclear rocket engine (the Nerva) was also to be continued, "but at a smaller size and thrust [75,000-pound thrust] than originally planned [200,000-pound thrust]."

NASA Sets Up New Units

NASA announced Jan. 4 that an Apollo Lunar Exploration Office was being established in NASA headquarters in Washington to unify the U.S.' unmanned and manned programs for scientific exploration of the moon. Retired Navy Capt. Lee R. Scherer, who had been active in NASA lunar programs, was named director of the new office. Retired Navy Capt. William T. O'Bryant was chosen to head the new unit's flight systems development program, and Dr. Richard J. Allenby Jr. was named head of its lunar science activities.

NASA announced Jan. 9 that it had set up a Research & Technology Advisory Council, headed by Dr. Raymond L. Bisplinghoff of MIT's Aeronautics & Astronautics Department, to help NASA plan and evaluate research and technology for aeronautics and space. The council was to "assess and render judgments on the relative importance of ongoing research, suggest additional work that should be undertaken and advise on the methods for further developing the nation's resources."

Adm. W. F. Boone had retired, effective Jan. 1, as NASA assistant administrator for defense affairs. His functions were transferred to a new NASA Office of Department of Defense & Interagency Affairs, which was to handle NASA relations with other government units as well as the Defense Department. Gen. Jacob E. Smart was named head of the new office, and Dr. Alfred J. Eggers was named to succeed Smart as NASA assistant administrator for policy.

(Pres. Johnson Mar. 5 presented the Goddard Trophy to ex-Deputy NASA Administrator Robert C. Seamans for advancing U.S. leadership in astronautics.)

U.S. Tests Hybrid Rocket

The first successful flight test of a hybrid rocket—using a combination of solid propellant and liquid nitric oxidizer—was announced Jan. 7 by the U.S. Air Force and the rocket's manufacturer, United Aircraft Corp.'s United Technology Center of Sunnyvale, Calif. The test had been conducted secretly Dec. 12, 1967 at Elgin Air Force Base, Fla.

The rocket tested was designed for the Sandpiper missile and for possible later use in space missions. (The Sandpiper, with a planned speed of about 2,000 mph., was being evaluated for use as a target in anti-missile defense training.) A unique feature of the rocket was a device enabling thrust levels to be changed from 60 pounds to 300. The tested rocket, 114 inches long and 13 inches in diameter, was dropped from an F-4 Phantom II jet plane at an altitude of 49,000 feet; it achieved a speed of about 1,320 mph. during its 5-minute flight.

The test "proved for the first time that hybrids [hybrid rockets] ... can fly," a company official declared.

UFO Study Proposed

A new study of UFOs (unidentified flying objects) was urged at a House Science & Astronautics Committee hearing July 29 by Dr. J. Allen Hynek of Northwestern University and 5 other scientists. Hynek proposed that the U.S. ask the UN to cooperate in creating "an international clearing house" for UFO data. The scientists complained that UFO reports were largely treated with derision rather than investigated. Besides Hynek, the scientists urging a UFO study were Dr. Carl Sagan of Cornell, Dr. James E. McDonald of the University of Arizona, Dr. Robert L. Hall of the University of Illinois, Dr. Robert M. L. Baker Jr. of Computer Sciences Corp. and Dr. James A. Harder of the University of California.

Unmanned Interplanetary Flights Urged

An increase in unmanned flights to the planets was called for Aug. 14 by the Space Science Board of the National Academy of Sciences. But it simultaneously opposed any early program for sending men on such flights. "While at some time in the future it may be in the national interest to undertake manned missions to the planets," the 23-member board said in a report to NASA, "we do not believe man is essential for scientific planetary investigation at this stage." It strongly opposed proposals that a manned flyby of Mars should be made the U.S.' next major space goal.

Specific projects recommended by the board included: (a) orbital flights around Venus and Mars every 2 years for 8 years; (b) the landing of an instrumented capsule on Mars in 1973 "to search for signs of life"; (c) the landing of instrumented capsules on Mars and Venus in 1975; (d) flights past Jupiter in 1972 and 1973; (e) a flight past Mercury in 1973. "Jupiter is probably the most interesting planet in the solar system," the report said, "and ... we recommend that Jupiter missions be given high priority."

The board suggested that a single probe could be used in a flyby of Venus and Mercury when the 2 planets would be lined up in 1972 or 1973. In 1977, in a situation existing only once in about 100 years, the planets Jupiter, Saturn, Uranus and Neptune would also be lined up for possible inspection in a "grand tour" by a single passing probe.

The report was based on studies by a committee of scientists headed by Dr. Gordon J. F. MacDonald of the Institute for Defense Analyses of Arlington, Va.

2 Biosatellites Canceled, X-15 Program Ends

NASA announced Dec. 16 that it had canceled work on 2 of the 4 biosatellites it had planned to orbit for studies of the effects of space conditions on living organisms. The 2 canceled biosatellites had been designed for 21-day missions beginning in 1971. The 2 retained were designed for 30-day missions, the first scheduled for 1969's 2d quarter. The cancellation was caused by budget cuts, but NASA said also that "success with smaller spacecraft and studies of non-recoverable satellites and bioscience experiments on manned flights have indicated attractive possibilities for experiments with greater flexibility in the early 1970 time period." The biosatellites were being built under contract with the General Electric Co.'s Reentry Systems division in Philadelphia.

The flight program of the X-15 experimental rocket plane, of which 3 were built and flown, was ended Dec. 20 after the scheduled 200th flight of an X-15 was canceled because of bad weather. The 199th and final X-15 flight had been made Oct. 24 by NASA research pilot William H. Dana, who piloted his X-15 to an altitude of 255,000 feet. Since the first X-15 flight, made June 8, 1959, the experimental planes had set records of 354,200 feet for altitude and 4,520 mph.

OTHER U.S. DEVELOPMENTS

(6.7 Mach) for speed in total flight time of slightly more than 30 hours. In addition to Dana, X-15s were piloted by A. Scott Crossfield, Joseph A. Walker, Col. Robert M. White, Capt. Forrest S. Petersen, John B. McKay, Col. Robert A. Rushworth, Neil A. Armstrong, Maj. Joe H. Engle, Milton O. Thompson, Maj. William J. Knight and Maj. Michael J. Adams. The X-15 program was a joint enterprise of NASA, the Air Force and the Navy.

U.S. Launchings

In addition to those recorded above, a number of other NASA and military satellites and space probes were sent up during 1968. The launchings, in chronological order:

Jan. 11—The 468-pound *Geos 2* (the U.S.' 2d geodetic earth orbiting satellite) was shot into near-polar orbit by means of a Delta-DS-V3E rocket launched from Vandenberg Air Force Base, Calif. at 8:16 a.m. The 4-foot scientific spacecraft, known also as *Explorer 36*, achieved an orbit with an apogee of 977 miles, perigee of 670 miles, period of 112 minutes and 74° angle of inclination. The satellite, built by the Applied Physics Laboratory of Johns Hopkins University, was launched by NASA as part of the U.S. National Geodetic Satellite Program, a project whose participants included the Coast & Geodetic Survey (of the Commerce Department) and the Defense Department. Its mission was to transmit data to help develop more precise knowledge of the earth's gravitational field and to provide more accurate information on the size and shape of the earth. The satellite used an extendable 60-foot gravity-gradient boom for stabilization.

Jan. 17, Apr. 17, Oct. 5—Secret satellites were sent into polar orbit from Vandenberg Air Force Base by means of a thrust-augmented Thor-Agena-D rocket launched by the Air Force Jan. 17, a Titan-3B-Agena launched Apr. 17 and a Thor-Agena launched by the Air Force Oct. 5.

Mar. 4—The 1347-pound unmanned satellite *Ogo 5*, the U.S. 5th "orbiting geophysical observatory," was shot into space to seek additional data on the effects of the sun on the earth's environment. The heaviest satellite in the Ogo series, *Ogo 5* was sent up in the nose of a modified Atlas-Agena booster rocket launched from Cape Kennedy's Launch Complex 13. It achieved an initial orbit with an apogee of 91,195 miles, perigee of 181 miles and period of almost 63 hours. Initially, all of the satellite's record 24 experiments were reported to be operating satisfactorily, but a low-energy-particle detector that had been turned on Mar. 7 stopped functioning Mar. 11.

Mar. 5—The 198-pound *Explorer 37*, equipped to monitor solar radiation, was sent into orbit by NASA and the Naval Research Laboratory (NRL) by means of a 4-stage Scout rocket launched from Wallops Island, Va. at 1:28 p.m. *Explorer 37* achieved an orbit with an apogee of 545 miles, perigee of 324 miles, period of 98.77 minutes and inclination of 59.4° to the equatorial plane. The satellite was one of a series of NRL solar radiation (SOLRAD) spacecraft designed to develop data on solar X-radiation throughout the solar cycle.

Apr. 6—2 satellites—one weighing 220 pounds and the other 236 pounds—were sent into orbit by means of a single F2 booster rocket launched from Vandenberg Air Force Base. Their mission was to measure cosmic radiation in space.

May 18—A 1,260-pound spacecraft designed to serve as a weather satellite was launched by NASA from Vandenberg Base but was destroyed by the Western Test Range safety officer after only 120 seconds of flight because its Thorad Agena-D rocket booster malfunctioned and threw the experimental craft dangerously off-course. The spacecraft, designated Nimbus-B, carried 7 meteorological experiments. The total cost of the Nimbus-B shot (including the design of the spacecraft) was estimated at $61.9 million. Nimbus-B carried 2 nuclear-power generators fueled with $\frac{1}{2}$ pound of plutonium each. The rocket also carried, as a "hitchhiker," the Army's 10th Secor (Sequential Collation of Range) satellite.

June 13—A single 3-stage Titan-3 space booster was launched by the Air Force from Cape Kennedy to put 8 military communications satellites into orbit about 22,000 miles above the surface of the earth. The satellites, costing about $1 million each, amplified a Defense Department satellite communications network that already had 17 satellites in orbit. The network was used to relay communications between Washington and Vietnam and between the Pentagon and other military posts around the world.

July 4—A radio astronomy satellite dubbed *Explorer 38* (Radio Astronomy Explorer-A) was launched by NASA from the Western Test Range at Point Mogu, Calif. at 1:27 p.m. EDT into a transfer orbit with an initial apogee of about 3,725 miles and perigee of some 400 miles. It was launched by means of a 3-stage improved Delta rocket. An onboard rocket fired July 7 by radio signal from earth then put the 417-pound satellite into an almost perfectly round orbit with an apogee of 3,641 miles, perigee of 3,636 miles, period of 3 hours 44 minutes and inclination of 59.2°. The spin rate was also reduced from about 92 revolutions a minute to almost zero, and 4 antennas were to be slowly extended—eventually to their full lengths of 750 feet each—so that at full extension the satellite would form a 1,500-foot X. The mission of *Ex-*

OTHER U.S. DEVELOPMENTS

plorer 38 was to monitor low-frequency radio signals from the Milky Way, other galaxies, the sun, Jupiter and near-earth sources. The ionosphere normally blocks such signals from earth-bound antennas. *Explorer 38* was the first of 2 satellites in a $24 million radio-astronomy satellite program.

July 11—The Air Force launched a 600-pound sphere into polar orbit by means of an Atlas booster fired from Vandenberg Base. The 23-inch satellite, named *Loads* (for low altitude density satellite) but nicknamed "Cannonball," achieved an orbit with an apogee of 110 miles and a record-low perigee of 90 miles. Its mission was to record atmospheric densities at these altitudes to learn how the atmosphere decelerates satellites and missiles. *Loads,* encased in a one-inch brass shell, was the densest satellite ever sent up.

Aug. 6—A secret satellite presumed to be a spy satellite was launched by an Air Force-industry team at Cape Kennedy's Launch Complex 13 at 7:08 a.m. EDT by means of an Atlas-Agena rocket assembly. The AP cited sources as indicating that the launching was the the first of 2 authorized in a program labeled "17" and that the AF had the option to order 2 more. The AP said that the satellite, nicknamed "Spook Bird," was "seeking intelligence data high over Russia, Red China, Southeast Asia and other potential trouble spots." The *N.Y. Times* reported Aug. 7 that "the vehicle may be able to hover over a target area 8 to 9 hours." The orbit was reported to be east-west and "very high," perhaps 18,000–19,000 miles up.

Aug. 8—2 satellites were sent into near-polar orbit by means of a Scout booster rocket launched by NASA at the Western Test Range in a program to continue the study of density and radiation characteristics of the upper atmosphere during high solar activity. The 2 payloads, weighing a total of 178 pounds, consisted of *Explorer 39* (Air Density Explorer) and *Explorer 40* (Injun Explorer). The former, a 12-foot polka-dotted sphere, was to acquire data on density and temperature variations in the polar regions and at intermediate latitudes. The latter, a 6-sided cylinder, carried 12 separate detectors to measure the bombardment of the atmosphere by energetic particles from space and the intensity of very-low-frequency radio emissions. Their orbits had apogees of about 1,550 miles, perigees of about 435 miles, periods of about 118.2 minutes and 82° inclinations.

Aug. 10—An Atlas-Centaur rocket was launched by NASA at Cape Kennedy at 6:33 p.m. EDT in an unsuccessful attempt to put the 864-pound *ATS 4* (applications technology satellite) into a synchronous orbit from which it could check on hurricanes and other storms. The booster put the satellite into a successful transfer orbit with an apogee of 460 miles and perigee of 115 miles, but the Centaur stage then failed to ignite a 2d time. The satellite's goal had been a "stationary" orbit 22,300 miles above the equator about 400 miles west of Quito, Ecuador. *ATS 4* carried an experimental gravity gradient stabilization system and 2 experiments—in microwave communications and ion-engine use—designed for use aboard a gravity gradient satellite.

Aug. 16—The 320-pound U.S. weather satellite *Essa 7* was launched by NASA from Vandenberg Base into a near-polar orbit. Its 2 cameras were designed to take cloud-cover photos that were stored in on-board tape recorders and then transmitted to ground stations. The satellite's orbit gave the cameras a view of every part of the earth once every 24 hours.

A record 12 satellites were sent into polar orbit by means of a single Atlas rocket launched by the Air Force at Vandenberg Base. The launching was the 4th in the Defense Department's space experiments support program.

Nov. 8—The 148-pound interplanetary probe *Pioneer 9* was sent into orbit around the sun by means of a 3-stage Delta rocket launched from Cape Kennedy. The scientific spacecraft achieved an orbit with an aphelion of 93 million miles, perihelion of 70 million miles and period of $297\frac{1}{2}$ days. It carried 8 experiments to gather data on solar plasma and energetic particles and magnetic fields propagated by the sun towards the earth.

The Delta also carried aloft a 40-pound "hitchhiker" satellite dubbed *TETR 2* (for test and training satellite), which was ejected into earth orbit with an apogee of about 500 miles, perigee of some 200 miles and 91-minute period. *TETR 2*'s mission was to serve as an orbiting target to test tracking and data-acquisition equipment and to train personnel under conditions similar to those provided by an orbiting Apollo spaceship.

Dec. 7—The 4,400-pound satellite *OAO 2* (for orbiting astronomical observatory), also known as *Stargazer,* was sent into orbit to gather data on extremely hot young stars and other phenomena from above the atmosphere, which distorts and filters out much of the ultraviolet and other radiation emitted by celestial bodies. A 2-stage Atlas-Centaur rocket, launched from Cape Kennedy at 3:40 a.m., put the $75 million satellite—the heaviest, most automated and most expensive unmanned spacecraft developed by the U.S.—into an orbit with an apogee of 485 miles, perigee of 279 miles, period of about 100 minutes and 35° angle of inclination.

OAO 2, 10 feet high and 21 feet wide after its solar panels were unfolded in space, carried 11 telescopes. 7 of the telescopes were designed by University of Wisconsin scientists to examine about 15 stars a day in great detail and to report on their temperature, structure and composition. This set of telescopes was turned on successfully Dec. 11. A 2d set of telescopes, successfully turned on Dec. 14, was a battery of 4 designed by the Smithsonian Astrophysical Observatory. In conjunction with a Westinghouse Electric Corp. TV camera sensitive only to ultraviolet rays, the Smithsonian instrument was assigned to map broad areas of the sky—about 700 stars a day—and to search for previously undetected stars.

OAO 2 was the 2d of 4 satellites in a program expected to cost a total of about $378 million. The first of the OAO satellites had been orbited Apr. 8, 1966 but had failed to function because of electrical troubles. *OAO 3* was tentatively scheduled for launching late in 1969; *OAO 4* was to go up late in 1970. The Grumman Aircraft Engineering Corp. of Bethpage, N.Y. was the prime contractor in charge of building the satellites.

According to NASA, *OAO 2* was capable of collecting in one day twice as much ultra-violet data as NASA had collected in 15 years of activity with about 40 sounding rockets.

Dec. 15–Essa 8, a weather satellite, was sent into near polar orbit by means of a Delta rocket launched by NASA from Vandenberg Air Base. The satellite's mission was to photograph the earth's cloud cover from an altitude of 900 miles.

INTERNATIONAL DEVELOPMENTS

USSR Orbits Satellites

The Soviet Union launched its 8th and 9th Molniya-1 communications satellites during 1968. The first was orbited Apr. 22; Tass reported that it had achieved "a high elliptical orbit with an apogee of 39,700 kilometers [24,654 miles] in the Northern hemisphere and a perigee of 460 kilometers [286 miles] in the Southern hemisphere." The period was given as 11 hours 53 minutes, the inclination as 65°. The satellite's "main tasks" were reported to be "exploitation of the system of long-range telephone-telegraph-radio communications and also transmission of the program of the Orbit network in the extreme North, Siberia, the Far East and Central Asia."

The 2d Molniya-1 of 1968 was launched Oct. 5 into an elliptical orbit with an apogee of 39,600 kilometers (24,600 miles) in the Northern hemisphere, perigee of 490 kilometers (304 miles) in the Southern hemisphere, period of 11 hours 52 minutes and 65° angle of inclination. The satellite's function was the same as that of the previous Molniya-1s.

Proton 4, a satellite described by the USSR as "the world's biggest automatic [un-manned] space station," was launched by Soviet space scientists Nov. 16 into an orbit with an apogee of 495 kilometers (307 miles), perigee of 255 kilometers (158 miles), period of 91.75 minutes and 51° 30′ angle of inclination. Tass reported that *Proton 4* weighed "about" 17 metric tons (37,478 pounds) and carried about $12\frac{1}{2}$ metric tons (27,458 pounds) of scientific apparatus. (By comparison, the U.S.' manned Apollo spaceships—including the command and service modules—weighed about $20\frac{1}{2}$ tons.) *Proton 4* was reported to be continuing the cosmic-ray research started by its 3 predecessor Proton satellites.

2 unmanned Soviet satellites—*Cosmos 212* and *Cosmos 213*—had linked up automatically in space Apr. 15 and then uncoupled after flying together for 3 hours and 50 minutes. This was the 2d time that the docking of unmanned satellites in space had been accomplished—in both cases by Soviet satellites.

Cosmos 212 had been launched at 1 p.m. Moscow time Apr. 14. It achieved an orbit with an apogee given as 239 kilometers ($148\frac{1}{2}$ miles), perigee of 210 kilometers ($130\frac{1}{2}$ miles), period of 88.75 minutes and 51.7° angle of inclination. Western observers pointed out that the orbit's parameters were almost identical to the parameters of the orbit of *Soyuz 1*, the spaceship in which Soviet cosmonaut Vladimir M. Komarov had been killed in a crash after re-entry. The USSR did not disclose that docking would be attempted.

Cosmos 213 was launched at 12:34 p.m. Moscow time Apr. 15 as the orbiting *Cosmos 212* was approaching overhead. *Cosmos 213* achieved an orbit with an apogee of 291 kilometers (183.8 miles), perigee of 205 kilometers ($127\frac{1}{4}$ miles), period of 89.16 minutes and 51.4° angle of inclination.

The 2 spacecraft then began the maneuver of locating each other, and the docking took place at 1:21 p.m. Moscow time over the Pacific Ocean. Tass reported that the satellites were "fitted out with special closing-in systems, radio detecting and computing devices," and that they "carried out an automatic mutual search, closing, docking and rigid coupling to each other." *Cosmos 212*, however, was the "active" satellite, equipped with 2 propulsion systems, and had to alter its orbit to match that of *Cosmos 213*.

The 2 satellites disengaged from each other at 5:11 p.m. in response to radio command from earth. Tass said that they then "were switched over to different orbits" and "are continuing their flight and studies of outer space."

Tass reported Apr. 20 that *Cosmos 212* had been brought back to earth Apr. 19 and that *Cosmos 213* had been recovered Apr. 20 after the 2 spaceships had docked automatically in space and then had completed other maneuvering, orientation, communication and engine tests.

The USSR announced the launching of 53 other satellites in its Cosmos series during 1968. Some were apparently reconnaisance (or spy) satellites. Others were for astronomical and meteorological studies, and some were thought to be tests of weapons. Tass disclosed Jan. 3 that *Cosmos 184*, launched Oct. 25, 1967, was a weather satellite assigned to check on Arctic conditions. Its data was shared with the U.S.

Evart Clark reported in the *N.Y. Times* Apr. 3 that the *Cosmos 185* (launched Oct. 27, 1967), *Cosmos 198* (launched Dec. 27, 1967) and *Cosmos 209* shots apparently were flight tests of "a maneuverable rocket stage that could be used to guide bombs down from orbit or to send instruments to the moon." All 3 were launched from Tyuratam, Kazakhstan (the Baikonur "cosmodrome") into low orbits from which they "climbed to near-circular orbits about 500 miles above the earth," Clark said. The angle of inclination in each case was about 65°. Clark reported fears that the tests indicated that the USSR had developed an MOBS (multiple-orbit bombardment system), in which nuclear warheads could be stationed in orbit, perhaps for months, or could be "sent up during a time of crisis, for psychological purposes or for use," and then "recalled safely to earth... if not used during the crisis." Clark reported in the *Times* Apr. 26 that, according to "authoritative sources," *Cosmos 218* apparently was launched as a further test of an orbital bombing system.

Pravda reported June 9 that *Cosmos 215*, which had been recovered "quite recently," was an astronomical satellite that had used 8 telescopes to gather data using wavelengths of light that are absorbed by the atmosphere and therefore invisible to earth-based telescopes. The telescopes were used to observe "hot" (or "young") stars. The satellite also carried an X-ray telescope and 2 photometers to measure solar radiation dispersed in the atmosphere's upper layers.

George C. Wilson asserted in the *Washington Post* Aug. 6 that *Cosmos 234* was the USSR's 100th reconnaisance satellite. Other spy satellites, he said, included *Cosmos 208*, which went up Mar. 21 and landed Apr. 2; and *Cosmos 228*, which was launched June 21 and landed July 3. He said these 2 were the first Soviet spy satellites to stay up 12 days instead of the usual 7 or 8. *Cosmos 231* was also a spy satellite, he added.

Wilson reported in the *Post* Oct. 8 that the *Cosmos 244* launching was the 13th flight test of a Soviet FOBS (fractional orbital bombardment system), a system for dropping nuclear bombs from orbit. Wilson reported Dec. 4 that *Cosmos 248, 249* and *252*, launched Oct. 19, Oct. 20 and Nov. 1, respectively, from the Tyuratam space center, appeared to have been test versions of "a new space vehicle that could inspect—and possibly destroy—American satellites...."

Launching dates and orbital data (where available) given for Soviet Cosmos satellites launched during 1968:

Cosmos 199—Jan. 16.

Cosmos 200—Jan. 20. Circular orbit 536 kilometers (333 miles) high. Period, 95.2 minutes. Inclination to equatorial plane, 74°.

Cosmos 201—Feb. 6. Apogee, 355 kilometers (220½ miles). Perigee, 210 kilometers (130½ miles). Period, 89.9 minutes. Inclination, 65°.

Cosmos 202—Feb. 20.

Cosmos 203—Feb. 20.

Cosmos 204—Mar. 5.

Cosmos 205—Mar. 5.

Cosmos 206—Mar. 14. Circular orbit 630 kilometers (391¼ miles) high. Inclination, 81°. (Tass announced Mar. 18 that *Cosmos 206* was part of the USSR's Meteor weather-satellite system.)

Cosmos 207—Mar. 16.

Cosmos 208—Mar. 21. Apogee, 305 kilometers (189½ miles). Perigee, 207 kilometers (128½ miles). Period, 89.4 minutes. Inclination, 65°.

Cosmos 209—Mar. 22. Apogee, 282 kilometers (175 miles). Perigee, 250 kilometers (155¼ miles). Period, 89.6 minutes. Inclination, 65.1°.

Cosmos 210—Apr. 3.

Cosmos 211—Apr. 9. Apogee, 1,574 kilometers (977½ miles). Perigee, 210 kilometers (130½ miles). Period, 102.5 minutes. Inclination, 81.9°.

Cosmos 214—Apr. 18. Apogee, 403 kilometers (250¼ miles). Perigee, 211 kilometers (131 miles). Period, 90.3 minutes. Inclination, 81.4°.

Cosmos 215—Apr. 19. Apogee, 264 miles. Perigee, 161 miles. Inclination, 48½°.

Cosmos 216—Apr. 20. Apogee, 172 miles. Perigee, 123 miles. Inclination, 51.8°.

Cosmos 217—Apr. 24. Apogee, 520 kilometers (323 miles). Perigee, 396 kilometers (246 miles). Period, 93.4 minutes. Inclination, 62.2°.

Cosmos 218—Apr. 25. Apogee, 210 kilometers (130½ miles). Perigee, 144 kilometers (89½ miles). Inclination, 50°.

Cosmos 219—Apr. 26. Apogee, 1,770 kilometers (1,099 miles). Perigee, 222 kilometers (137.9 miles). Period, 104.7 minutes. Inclination, 48.4°.

Cosmos 220—May 7. Apogee, 760 kilometers (472 miles). Perigee, 670 kilometers (416 miles). Period, 99.2 minutes. Inclination, 74°.

Cosmos 221—May 24. Apogee, 2,108 kilometers (1,309 miles). Perigee, 220 kilometers (137 miles). Period, 108.3 minutes. Inclination, 48.4°.

Cosmos 222—May 30. Apogee, 327 miles. Perigee, 172 miles. Period, 92.3 minutes. Inclination, 71°.

Cosmos 223—June 1.

Cosmos 224—June 4. Apogee, 270 miles. Perigee, 124 miles. Period, 89 minutes. Inclination, 51.8°.

Cosmos 225—June 12. Apogee, 530 kilometers (328 miles). Perigee, 257 kilometers (159½ miles). Period 92.2 minutes. Inclination, 48.4°.

INTERNATIONAL DEVELOPMENTS 161

Cosmos 226—June 12. Apogee, 650 kilometers (404 miles). Perigee, 603 kilometers (374½ miles). Period, 96.9 minutes. Inclination, 81.2°.
Cosmos 227—June 18. Apogee, 281 kilometers (174½ miles). Perigee, 194 kilometers (120½ miles). Period, 89.1 minutes. Inclination, 51.8°.
Cosmos 228—June 21. Apogee, 160 miles. Perigee, 128 miles. Period, 89 minutes. Inclination, 51.6°.
Cosmos 229—June 26. Apogee, 220 miles. Perigee, 130½ miles. Inclination, 72.8°.
Cosmos 230—July 5.
Cosmos 231—July 10. Apogee, 330 kilometers (205 miles). Perigee, 211 kilometers (131 miles). Period, 89.7 minutes. Inclination, 65°.
Cosmos 232—July 16. Apogee, 352 kilometers (219 miles). Perigee, 202 kilometers (125½ miles). Period 89.8 minutes. Inclination, 65°.
Cosmos 233—July 18.
Cosmos 234—July 30. Apogee, 310 kilometers (193 miles). Perigee, 210 kilometers (130½ miles). Period, 89½ minutes. Inclination, 51.8°. (West Germany's Bochum space observatory reported Aug. 5 that *Cosmos 234* had made a soft landing near the Baikonur cosmodrome. According to the Bochum observers, orbital data indicated that the satellite was a Soyuz-class spacecraft.)
Cosmos 235—Aug. 9. Apogee, 303 kilometers (188¼ miles). Perigee, 207 kilometers (128½ miles). Period, 89.4 minutes. Inclination, 51.8°.
Cosmos 236—Aug. 27. Apogee, 655 kilometers (407 miles). Perigee, 600 kilometers (373 miles). Period, 96.9 minutes. Inclination, 65°.
Cosmos 237—Aug. 27. Apogee, 343 kilometers (213 miles). Perigee, 201 kilometers (125 miles). Period, 89.7 minutes. Inclination, 65.4°.
Cosmos 238—Aug. 28.
Cosmos 239—Sept. 5. Apogee, 175 miles. Perigee, 125 miles. Period, 89.2 minutes.
Cosmos 240—Sept. 14.
Cosmos 241—Sept. 16.
Cosmos 242—Sept. 20. Apogee, 440 kilometers (273½ miles). Perigee, 280 kilometers (174 miles). Period, 91.3 minutes. Inclination, 71°.
Cosmos 243—Sept. 23. Apogee, 319 kilometers (198¼ miles). Perigee, 210 kilometers (130½ miles). Period, 89.6 minutes. Inclination, 71.3°.
Cosmos 244—Oct. 2. Apogee, 205 kilometers (127⅓ miles). Perigee, 135 kilometers (84 miles). Inclination, 50°.
Cosmos 245—Oct. 3.
Cosmos 246—Oct. 7. Apogee, 348 kilometers (216¼ miles). Perigee, 147 kilometers (91⅓ miles). Period, 89.4 minutes. Inclination, 65.4°.
Cosmos 247—Oct. 11. Apogee, 362 kilometers (225 miles). Perigee, 205 kilometers (127⅓ miles). Period, 89.9 minutes. Inclination, 65.4°.
Cosmos 248—Oct. 19. Apogee, 551 kilometers (342⅓ miles). Perigee, 328 kilometers (204 miles). Inclination, 62.3°.
Cosmos 249—Oct. 20. Apogee, 2,177 kilometers (1,353 miles). Perigee, 514 kilometers (319⅓ miles). Period, 112.2 minutes. Inclination, 62.4°.
Cosmos 250—Oct. 31.
Cosmos 251—Oct. 31. Apogee, 270 kilometers (167¾ miles). Perigee, 194 kilometers (120½ miles). Period, 89.1 minutes. Inclination, 65°.
Cosmos 252—Nov. 1. Apogee, 2,172 kilometers (1,349⅔ miles). Perigee, 538 kilometers (334¼ miles). Period, 112½ minutes. Inclination, 61.8°.
Cosmos 253—Nov. 13. Apogee, 220 miles. Perigee, 128 miles. Period, 89.9 minutes. Inclination, 65.4°.

Cosmos 254—Nov. 20. Apogee, 217 miles. Perigee, 126 miles. Period, 89.8 minutes. Inclination, 65.4°.
Cosmos 255—Nov. 29. Apogee, 336 kilometers (208½ miles). Perigee, 201 kilometers (124.8 miles). Period, 89.7 minutes. Inclination, 65.4°.
Cosmos 256—Nov. 30. Apogee, 1,234 kilometers (766⅓ miles). Perigee, 1,169 kilometers (725⅓ miles). Period, 109.3 minutes. Inclination, 74.06°.
Cosmos 257—Dec. 3. Apogee, 470 kilometers (292 miles). Perigee, 282 kilometers (175 miles). Period, 91.7 minutes. Inclination, 71°.
Cosmos 258—Dec. 10. Apogee, 325 kilometers (201.8 miles). Perigee, 210 kilometers (130½ miles). Period, 89.6 minutes. Inclination, 65°.
Cosmos 259—Dec. 14. Apogee, 707 miles. Perigee, 137 miles.
Cosmos 260—Dec. 16. Apogee, 39,000 kilometers (24,235 miles). Perigee, 500 kilometers (311 miles). Period, 11 hours 52 minutes. Inclination, 65°.
Cosmos 261—Dec. 20. Apogee, 309 miles. Perigee, 139 miles.
Cosmos 262—Dec. 26. Apogee, 818 kilometers (508¼ miles). Perigee, 263 kilometers (163½ miles). Period, 95.2 minutes. Inclination, 48.5°.

Canadian Plans

Canadian Industry Min. Charles M. Drury announced Apr. 1 that the Canadian government planned to set up a $100 million public and private corporation that would provide Canada with an operational domestic satellite communications system by 1971-2. The system, outlined in a white paper, envisioned 2 satellites in synchronous orbit. It would relay phone calls, TV broadcasts and other services and would bring TV economically to people living in remote northern areas. Drury said that Dr. R. M. MacIntosh, Bank of Nova Scotia general manager, had been selected to prepare a plan for the satellite corporation's financial, corporate and senior management structure and to determine the amount of private participation in the organization. 2 groups of Canadian companies—one involving the Power Corp. of Canada (Montreal) and the other including the Bell Telephone Co. of Canada and communications units of the Canadian Pacific Railway and the Canadian National Railway—had expressed interest in participating in the organization.

Quebec Premier Daniel Johnson said June 8 that there could be no equality for French Canadians unless Quebec maintained the right "to freely establish necessary communications with the outside [foreign countries] for the full exercise of its internal jurisdictions." Speaking at the University of Sherbrooke commencement, Johnson said work was continuing on an eventual communications satellite linkup between France and Quebec. He said, "several Quebec engineers" were currently working at the French national space research center, and "a close cooperation has been established" between Radio-Quebec and the French government broadcasting system. Johnson added that Radio-Quebec, only recently established, would make a final report to the provincial government by August. "We can be assured," he said, "that in a few years Quebec will have a telecommunications system with virtually unlimited possibilities."

Astronaut Rescue Pact

In parallel ceremonies in Washington, London and Moscow Apr. 22, the U.S., Britain and USSR signed a treaty pledging international cooperation in the rescue and return of astronauts in danger on the earth, in space or on the moon. 41 other nations joined in the signing ceremonies. France and Communist China were among the countries that neither signed nor indicated an intention to sign the pact.

The treaty committed signatories to "take all possible steps" to rescue and aid astronauts landing outside of their own nation's jurisdiction. Astronauts were to be "safely and promptly returned to representatives of the launching authority"; expenses were to be borne by the launching powers. Recovery and return of spacecraft were similarly covered in the treaty.

The treaty was to enter into force on ratification by the 3 depository governments and any 2 other governments. The U.S. Senate ratified the treaty Oct. 8 by 66–0 vote.

European Disagreements & Cooperation

Disagreements within the European space organizations promoted efforts to reorganize and unify them.

Esro (European Space Research Organization) announced Apr. 25 that it was canceling its 2 biggest satellites—TD-1, scheduled for launching in 1970 to study relationships between the earth and sun; and TD-2, scheduled for launching in 1971 to study ultraviolet radiation and electromagnetic waves. Esro had negotiated a 100 million-franc ($20 million) contract for the construction of the 2 satellites by a consortium including Hawker Siddeley Dynamics of Britain, Matra of France, ERNO of West Germany and Saab of Sweden. The cancellation was caused by Italy's withdrawal of financial support from the project; Italy had complained that it was not getting a fair share of the contracts for the satellites.

Britain's Technology Ministry had announced Apr. 16 that Britain was withdrawing from Eldo (European Launcher Development Organization) after its current commitment expired in 1971 and that it would not contribute extra funds to defray pre-1971 costs that exceeded the sums it had agreed to pay. It simultaneously disclosed that Britain would not participate in plans for an Esro communications satellite designed for direct broadcast of TV programs to domestic receivers (instead of to ground relay stations). The British held that Eldo costs had exceeded estimates too greatly and that developing duplicate European facilities was uneconomic since U.S. rocket launchers and communications satellites already in existence were available at considerably lower cost.

In an effort to resolve the differences that were splitting Europe's space organizations, Belgian Scientific Policy Min. Theo Lefèvre proposed at an Eldo ministerial meeting in Paris Oct. 1 that the European nations (a) develop a common policy covering all aspects of technical and scientific cooperation, (b) concentrate on projects taking greater account of the European peoples' needs and capacities, (c) make sure that each country gets its fair share of contracts under the programs and (d) reorganize the agencies in charge of the programs.

The problem was not resolved at the Paris meeting, but it was taken up again in Bad Godesberg, West Germany Nov. 12 at the 3d European Space Conference, where British Technology Min. Anthony Wedgwood Benn and his delegation indicated that Britain was willing to participate in a restructured European space organization combining Eldo and Esro in a body similar to the U.S.' NASA. The British proposal was accepted Nov. 14, and a committee was formed to draft a convention for the new space agency.

3 Esro satellites—*Esro 2, Esro 1* and *Heos 1* (for highly eccentric orbit satellite)—were shot into orbit by NASA rockets May 16, Oct. 3 and Dec. 5, respectively. The 2 *Esro* satellites, both sent into near-polar orbits, were launched (*Esro 2* before *Esro 1*) from the U.S.' Western Test Range in California by means of 4-stage U.S. Scout rockets under a U.S.-Esro cooperative program providing for the U.S. to bear the launching costs. *Heos 1* was launched from Cape Kennedy's Launch Pad 17B at 1:55 p.m. Dec. 5 by means of a 3-stage Thor-Delta rocket under an Esro-NASA agreement for reimbursable launchings, and Esro paid the U.S. $$3\frac{3}{4}$$ million for the rocket and the launching costs.

The 164-pound *Esro 2* carried 7 experiments to study solar and cosmic radiation. Its sun-synchronous orbit had an apogee of about 680 miles, perigee of some 215 miles, period of 98 minutes and $98°$ angle of inclination.

The 185-pound *Esro 1* carried 8 experiments to measure the energies and pitch angles of particles impinging on the polar ionosphere during both magnetic storms and quiet periods. Its orbit had an apogee of about 932 miles, perigee of some 171 miles, period of 103 minutes and inclination of $94°$ retrograde to the equator.

The 238-pound *Heos 1* carried 8 experiments to study interplanetary physics (particularly magnetic fields), cosmic radiation and solar wind outside the magnetosphere and the earth's shock-wave. For this purpose the European scientists selected a highly elliptical orbit with an apogee of about 138,000 miles, equivalent to 2/3 the distance to the moon. The chosen perigee was about 274 miles, the period about $4\frac{1}{2}$ days and the angle of inclination 28.3°.

All 3 satellites were prepared under the technical supervision of Esro's European Space Technology Center (ESTEC) in Noordwijk, the Netherlands.

Eldo launched the 104-foot F7 Europa-1 rocket from the Woomera base in Australia Nov. 30 in an unsuccessful attempt to put a 550-pound Italian scientific satellite into polar orbit. The British-made Blue Streak first stage and France's Coralie 2d stage apparently functioned satisfactorily, but the West German 3d stage of the rocket reportedly burned out prematurely, and the payload was lost.

Red Network Planned

The USSR and 7 other Communist countries submitted to the UN Committee for the Peaceful Uses of Outer Space Aug. 5 a proposal to create a worldwide satellite communications network. They planned to call the system Intersputnik and said membership would be open to all countries. The countries joining the USSR in the plan were Poland, Czechoslovakia, Bulgaria, Rumania, Hungary, Cuba and Mongolia. At the request of the sponsors, their draft treaty was circulated Aug. 13 as a document of the UN committee.

A fresh announcement of the plan was made by Soviet Premier Aleksei N. Kosygin in a message read for him in Vienna Aug. 14 at the opening session of the UN Conference on Exploration & Peaceful Uses of Outer Space. The Soviets proposed a single vote for each Intersputnik member. This would contrast with the U.S.-sponsored Intelsat (International Telecommunications Satellite Consortium), already in existence with 62 member nations but with weighted voting giving control to the U.S. and its allies.

Intelsat Activities

An Intelsat-3A, the 5th operational communications satellite owned by the multination International Telecommunications Consortium (Intelsat), was shot into a preliminary orbit Dec. 18 by means of a 3-stage Thor-Delta rocket launched from Cape Kennedy at 7:32 p.m. Comsat (Communications Satellite Corp.), the government-chartered private U.S. group that managed the Intelsat program and was Intelsat's U.S. member, had agreed to pay NASA $4.7 million to launch the 642-pound Intelsat-3A. NASA transferred control of the Intelsat-3A to Comsat Dec. 19, and Comsat officials in Washington Dec. 20 sent a radio signal that ignited the satellite's on-board engine for 27 seconds to move the commercial communications satellite to its permanent synchronous orbit, in which it would appear to hang motionless 22,300 miles above the equator off Brazil's Atlantic coast.

An attempt to put a communications satellite (also an Intelsat-3A) into a synchronous orbit failed Sept. 18 when the Thor-Delta booster rocket veered off-course and was destroyed by the range safety officer at 8:11 p.m. EDT, less than 2 minutes after the 8:09 p.m. launching at Cape Kennedy.

3 more satellites were planned to complete the Intelsat-3 program. The 4 Intelsat-3s would be expected to provide global communications coverage. The drum-shaped satellites, each 66 inches high and 56 inches in diameter, were being built by TRW, Inc. of Redondo Beach, Calif. under a $32 million contract with Comsat. Each was to be capable of relaying the equivalent of 1,200 2-way voice circuits for phone use or 4 TV programs. (The 4 Intelsat satellites previously in orbit had a combined capacity of only 960 2-way voice circuits or one TV program.)

Life Forms Reported in Meteorites

Dr. Boris Vasilyevich Timofeyev reported in an interview in the Soviet newspaper *Vecherny Leningrad* Feb. 21 that he had discovered "shells of algae" in carbon-containing meteorites. The presumably extraterrestrial microorganisms bore no resemblance to any life forms found on earth, he said. Timofeyev, head of the biostratigraphy laboratory of the Pre-Cambrian Geology & Geochronology Institute of the Soviet Academy of Sciences, noted that previous "sensational" claims about biological discoveries in meteorites "were later refuted." His findings were disputed by many Soviet and non-Soviet scientists. But, *Vecherny Leningrad* reported, Soviet scientist A. S. Lopukhin and some non-Soviet scientists had also conducted research "confirming that there exist in outer space forms of life differing from terrestrial" life.

Gagarin Dies in Plane Crash

Soviet Col. Yuri Alexeyevich Gagarin, 34, the first man known to have made an orbital flight, was killed Mar. 27 in the crash of a jet plane about 40 miles northeast of Moscow during what was described by the

Soviets as a training flight. Engineer Col. Vladimir Sergeyevich Seryogin, 45, described as a test pilot and commander of a Red Air Force unit, died with Gagarin in the crash. Gagarin and Seryogin received a full state funeral Mar. 30, and their ashes were interned in the Kremlin wall. The cosmonaut, who had made a single-orbit flight Apr. 12, 1961, was the first human being known to have gone into orbit.

1969

U.S. astronauts Neil A. Armstrong and Edwin E. Aldrin Jr. became the first men to set foot on an astronomical body other than earth when they walked on the moon July 20. Thanks to TV, their landing, activities on the moon and safe return to earth were watched "live" by most of the civilized world. 2 more U.S. astronauts landed on the moon in November, spent 8 hours on the lunar surface outside their spaceship and returned safely to earth. The moon landings climaxed an 8-year space effort by the U.S. Both the U.S. and USSR sent other manned flights into space during 1969 and continued unmanned exploratory probes.

REHEARSING FOR MOON LANDING

Moon Lander Flown in Test

2 manned Apollo missions had been flown in the months preceding the July moon landing. The first, *Apollo 9*, tested the complete Apollo spaceship in earth orbit; the 2d, *Apollo 10*, was the 2d manned spaceship to circle the moon.

3 astronauts flew *Apollo 9* for 10 days in earth orbit Mar. 3–13. The successful flight, which cost a record $340 million, proved the spaceworthiness of the fragile Apollo lunar module (LM). During the flight, 2 astronauts flew the LM on its own for $6\frac{1}{2}$ hours and then brought it back to link up again with the *Apollo 9* command module. The LM flight and docking simulated the maneuvers required of an LM in a lunar landing and in the return to an Apollo in lunar orbit.

The 3.7 million-mile *Apollo 9* mission took 241 hours from blast-off at Cape Kennedy, Fla. Mar. 3 to splashdown in the Atlantic northwest of Puerto Rico Mar. 13. The launching, originally scheduled for Feb. 28, had been delayed for 3 days after all 3 astronauts developed colds. The astronauts were: Air Force Col. James Alton McDivitt, 39, the spacecraft commander; Air Force Col. David Randolph Scott, 36, command module pilot (2d in command), and Russell Louis (Rusty) Schweickart, 33, LM pilot, a civilian.

Apollo 9 rose from Cape Kennedy's Launch Pad 39A at 11 a.m. EST Mar. 3 in the nose of a 3-stage Saturn-5 booster rocket. The 363-foot, 6,483,320-pound rocket-and-spacecraft assembly was the heaviest object sent aloft so far. The first and 2d stages of the booster rocket dropped away into the Atlantic after exhausting their fuel, but the 3d (S4B) stage, its fuel only partly expended, was still attached to *Apollo 9* when it achieved its initial orbit about 11 minutes after blast-off. The orbit had an apogee of 119 miles and perigee of 118 miles. The 297,000-pound orbiting assembly, consisting of the command module, service module and S4B (in which the LM was carried), was the heaviest payload sent into space so far.

The critical task of mating the conical command module (nicknamed *Gumdrop*) to the LM (nicknamed *Spider*) began at 1:43 p.m. EST Mar. 3 when explosive bolts were set off to separate the command module-service module assembly from the S4B. The separation simultaneously threw aside the shroud that had covered the LM. By firing *Apollo 9*'s small maneuvering jets, Scott moved the spaceship about 50 feet forward of the S4B and turned it completely around to face the S4B. With *Apollo 9*'s nose pointed toward the S4B and LM, Scott brought the spaceship back toward the S4B and eased the *Apollo*'s docking probe into an opening in the LM. *Apollo 9* and the LM were then locked tightly together by 12 clamps. The connection formed an airtight tunnel with one airtight hatch at the LM end and another at the command-module end.

After the docking, Scott fired explosive springs to separate the LM (attached to the command module) from the S4B. *Apollo 9*, now a 60-foot assembly with the LM attached to its nose, moved away from the S4B, and the S4B's engine was ignited twice by ground controllers. The 2d ignition sent the S4B away into orbit around the sun. The astronauts also ignited their service propulsion system (SPS) engine and shifted *Apollo 9* into an altered orbit with an apogee of 143 miles and perigee of 124 miles.

The astronauts tested the strength of the *Apollo 9* assembly Mar. 4 by swiveling the SPS engine as they fired it. This maneuver, tried twice with increasing roughness, was done to make the orbiting spaceship whip in space and to put strain on the link between the command module and the LM. The spaceship passed the test without bending or breaking. A final firing of the SPS engine Mar. 4 moved *Apollo 9* into an orbit 10° eastward to put it in better range of ground trackers. After the day's maneuvers were over, *Apollo 9*'s orbit had an apogee of 314 miles and perigee of 125 miles.

Schweickart and McDivitt entered the LM Mar. 5 after Scott had carefully removed all the hatch and latching devices from the 38-inch-long tunnel connecting the command module and the LM. Schweickart, despite nausea that made him vomit once before leaving the command module and once in the LM, was the first to go through the 32-inch-wide tunnel. Pulling himself into the LM shortly after 6:30 a.m. EST, he attached a set of restraints to his waist to anchor himself in place in the LM, and he then began turning on the LM's electrical equipment. McDivitt followed him about 50 minutes later, and Scott remained alone in the command module.

McDivitt and Schweickart spent about 9 hours in the LM Mar. 5 before returning to the command module. While in the LM, they transmitted a live 7-minute telecast to earth; their activities inside the LM appeared clearly on TV screens, but communications troubles blanked out most of their conversation. They tested the LM's descent rocket successfully with a 6-minute ignition at full throttle, and the firing moved the entire spaceship assembly more than 400 miles eastward.

Because of Schweickart's nausea, ground controllers Mar. 5 canceled a 2-hour EVA (extra-vehicular activity) that had called for him to leave the LM Mar. 6 and to move in open space to the command module and back. Schweickart felt well again Mar. 6, however, so McDivitt reinstated part of the EVA. First, all 3 astronauts donned their space suits, and Schweickart and McDivitt again used the tunnel to enter the LM. The outer hatches of the LM and command module were opened, and Schweickart went out through the LM's outside hatch. He wore on his back a self-contained life-support system containing oxygen, electrical supply, cooling system and communications equipment. The back-pack was designed to enable astronauts to work on the moon or elsewhere in space without being attached to their spaceships by the "umbilicals" through which these needs had previously been supplied. The suit and back-pack cost more than a quarter-million dollars. Schweickart spent about 40 minutes outside. After he returned to the LM, the LM and command module assembly were repressurized so that the astronauts could remove their space helmets for the mission's 2d and last live telecast.

The LM, with McDivitt and Schweickart aboard for the 3d time, was detached from the command module Mar. 7 for a successful independent flight of $6\frac{1}{2}$ hours. The LM dropped as much as 114 miles behind the command module before catching up and docking with it. A critical aspect of the flight was the fact that the LM could not return to earth—it would burn up on re-entry into the atmosphere. The LM, therefore, had to rendezvous and dock with the command module if the 2 astronauts were to get back to earth alive.

McDivitt and Schweickart had crawled through the tunnel into the LM at about 3:30 a.m. EST Mar. 7 to prepare the LM for its trip. They were ready shortly after 7 a.m. Scott tripped a switch in the command module at 7:39 a.m. to release the latches holding the LM to the command module, but the spaceships did not separate until he tried a 2d time. After separation, McDivitt used the LM's small (100-pound-thrust) maneuvering jets to move the LM about 50 feet away from the command module and then to move it completely around the command module so that Scott could examine and photograph it.

The 2 astronauts in the LM fired the throttleable descent rocket at 8:47 a.m. and tested it at 10% to 40% of its 9,870-pound-thrust capacity. (The function of the descent rocket was to lower a manned LM gently to the moon.) The firing put the LM into an orbit ranging from $13\frac{1}{2}$ miles above to $13\frac{1}{2}$ miles below that of the command module, and the spaceships were separated by a distance of about 57 miles. A 2d firing of the descent rocket put the LM into a circular orbit 13 miles above the command module's. Since it was in a lower orbit, the command module continued to increase its distance ahead of the LM. The descent stage was jettisoned by the astronauts at 11:19 a.m. by the firing of explosive bolts and was left in its own separate orbit.

REHEARSING FOR MOON LANDING

Using only the small reaction jets (maneuvering engines) to adjust their orbit, the astronauts in the LM continued to drop back until they eventually were about 114 miles behind the command module. At 11:58 a.m. they fired the LM's 3,500-pound-thrust ascent engine (which had been designed to take an LM off the moon and up to an orbiting command module). The firing put the LM into a new orbit $11\frac{1}{2}$ miles below the command module's. Since it was in a lower orbit, the LM began to catch up to the higher command module. The ascent engine was fired again when the spaceships were 23 miles apart. This ignition raised the LM's altitude, and it approached the command module more slowly. The LM's small reaction jets were then used to make final adjustments, and McDivitt sighted through a gun-sight device to bring the LM up to the command module's nose. The command module's docking probe was slowly eased into the opening in the LM, and the docking was completed at 2 p.m., about 8 minutes ahead of schedule.

After McDivitt and Schweickart returned to the command module and fastened the hatch, the LM was separated and sent off into orbit by itself. The 2 spaceships had flown docked together for a record 54 hours 47 minutes.

Most of the remainder of the mission, after the eventful first 5 days, was devoted to more leisurely tasks that included 3 more firings of the SPS, navigation experiments and photography. A 25-second ignition Mar. 10 put *Apollo 9* into an elliptical orbit with an apogee of 288 miles and perigee of 113 miles.

The astronauts Mar. 11 saw and tracked the unmanned satellite *Pegasus 3* twice—the first time as it passed at a distance of 1,100 miles, the 2d time at a distance of 850 miles. *Pegasus 3*, with a wingspan of 96 feet, had been launched July 30, 1965 to record meteorite density. They also tracked the LM's upper stage Mar. 12 as it passed 345 miles above them and 750 miles in front of them.

Because of stormy weather in the primary recovery area, a patch of the Atlantic 195 miles southwest of Bermuda, ground controllers Mar. 12 added an additional revolution around the earth to *Apollo 9*'s mission and shifted the landing to a spot about 480 miles further south.

The astronauts brought *Apollo 9* out of orbit by an 11.8-second firing of the SPS in the direction of flight at 11:30 a.m. EST Mar. 13 during the spaceship's 191st revolution around the earth. This took place over the Pacific. The spaceship was then turned around by means of the maneuvering jets, and the service module was jettisoned by the firing of explosive bolts.

Re-entering the atmosphere at a speed of 17,500 m.p.h., the service module was slowed first by friction with the air and finally by parachutes. Although the air friction heated up the outside of the capsule to temperatures of 2,700°F., the astronauts inside the air-conditioned spaceship remained relatively comfortable.

Apollo 9 splashed into the Atlantic at 12:01 p.m. EST Mar. 13 only one mile from its target point about 360 miles northwest of San Juan, P.R. A helicopter from the recovery ship *Guadalcanal*, which had been waiting only $4\frac{1}{2}$ miles from the splashdown point, then took the 3 astronauts to the ship. Initial physical checks showed them to be in good health. The *Apollo 9* command module, its exterior charred from the heat of re-entry, was hauled aboard later.

Apollo 10 Orbits Moon

Air Force Col. Thomas Patten Stafford, 38, and Navy Cmndr. Eugene Andrew Cernan, 35, brought a U.S. lunar landing craft nicknamed *Snoopy* to less than 9 miles from the moon's surface May 22 in the closest approach to the moon so far made by any human beings. (A landing on the moon was neither planned nor attempted on this mission.) After circling the moon twice in their lunar module (LM), Stafford and Cernan flew it back up as planned and linked up again with their *Apollo 10*'s command and service module (CSM). The CSM, nicknamed *Charlie Brown*, had been waiting for them in a 69-mile-high lunar orbit with Navy Cmndr. John Watts Young, 38, at the controls. The 3 astronauts were the 2d crew to fly around the moon.

The 8-day exercise, a "dress rehearsal" for the first manned landing on the moon, had started with a launching from Cape Kennedy May 18. Duplicating, as closely as possible, the conditions expected on the first lunar landing flight, the *Apollo 10* astronauts flew their spaceship a quarter of a million miles through space and brought it into orbit around the moon May 21. While CSM pilot Young remained aboard *Charlie Brown*, *Apollo 10* commander Stafford and LM pilot Cernan entered *Snoopy* and swooped down in this lunar vehicle for their close approach to the moon. After the LM occupants' return to the CSM, *Apollo 10* continued on in lunar orbit while the astronauts continued their careful inspection of the moon and of the various sites selected for manned lunar landings. After circling the moon 31 times in $61\frac{1}{2}$ hours, *Apollo 10* headed back towards the earth May 24 and brought its 3 astronauts to a safe splashdown in the Pacific May 26. The total distance covered on the mission was 705,000 miles.

The $350 million *Apollo 10* lunar flight, with its spacecraft named for characters in Charles M. Schultz' newspaper cartoon *Peanuts*, was described as a near-perfect mission. (NASA spokesmen said that *Apollo 10*'s Saturn-5 booster rocket cost $185 million, the command

REHEARSING FOR MOON LANDING

and service modules $55 million, the lunar module $41 million, launching and mission operations [including the cost of deploying the Navy's recovery fleet] $69 million. Total: $350 million.)

Apollo 10, with its 3 astronauts aboard, had begun its journey atop a 363-foot, 6,493,800-pound Saturn-5 rocket launched from Cape Kennedy's Launch Pad 39B at 12:49 p.m. EDT May 18. The first and 2d stages of the 3-stage Saturn-5 dropped away as they exhausted their fuel. The remaining 3d (or S4B) stage—with the LM (lunar module, or *Snoopy*) encased in its forward section and the CSM (command and service module, or *Charlie Brown*) attached to its nose—then carried *Apollo 10* into a parking orbit around the earth. In this orbit, with an apogee of about 118 miles and perigee of 115 miles, the spacecraft circled the earth $1\frac{1}{2}$ times. The S4B rocket engine was then reignited. This 2d ignition increased the space vehicle's speed to about 24,200 mph., and it brought *Apollo 10* out of orbit and into a translunar trajectory.

Soon after *Apollo 10* had achieved its lunar trajectory, the astronauts detached the CSM from the S4B. Young then maneuvered it about 500 feet ahead of the S4B, turned it around so that the command module's nose faced the LM in the S4B's forward section, and brought the CSM back to dock (link up) with the LM. A color telecast of the docking was transmitted "live" to earth. The first of 19 telecasts made from *Apollo 10*, it was the first live color telecast made from space. The LM was then disengaged from the S4B and the remaining 59-foot-long assembly, consisting of joined command, service and lunar modules, continued its flight toward the moon. (The S4B, its engine ignited again by radio command from earth, also went on toward the moon, whose gravity whipped it on into a trajectory that put it into orbit around the sun.)

Apollo 10's lunar trajectory was achieved with such precision that an initial course correction scheduled for May 18 was dropped. A 2d scheduled mid-course correction was accomplished at 3:22 p.m. EDT May 19 by means of a 7-second firing of the service engine. 2 other programmed course corrections were found to be unnecessary.

Shortly before 2 a.m. May 21, when the spacecraft reached a point about 219,000 miles from the earth and 38,900 miles from the moon, *Apollo 10* passed out of the primary influence of the earth's gravity and into the primary domination of the moon's gravity. The spaceship's speed, which had been dropping ever since the initial rocket burst that brought it out of parking orbit, was down to 2,150 mph. at this point, but it then began to pick up as *Apollo 10* approached (in effect, fell toward) the moon.

Apollo 10 and its crew went into orbit around the moon at 4:45 p.m. EDT May 21 after a 6-minute firing of the spaceship's main service engine. The orbit was achieved as the spacecraft passed around the moon and, as anticipated, lost radio contact with the earth. Ground officials waited tensely for 35 minutes for *Apollo 10* to round the far side of moon and regain radio contact.

The initial 2-hour orbit had an apocynthion (farthest point from the moon's surface) of 196 miles and a pericynthion (closest approach to the surface) of 70 miles. The visibility of the moon from the spaceship was magnificent, and Young's first description of the crater-pocked moon was: "Boy, this is really a rugged planet."

As the spacecraft's 3d revolution around the moon was starting, a 14-second engine firing at 9:11 p.m. reduced the orbit's apocynthion to 70 miles and the pericynthion to 69 miles. Shortly thereafter the astronauts transmitted the first live color telecast of the moon from space. Later May 21 Cernan unsealed the hatches between the lunar module and the command module and crawled through the short connecting tunnel into the LM to inspect its apparatus. Aside from loose insulation inside the tunnel, all was found to be in good condition.

Preparations for the $7\frac{1}{2}$-hour flight aboard *Snoopy*, the 15-ton lunar module, began May 22 with the 3 astronauts donning their space suits and with first Cernan, then Stafford, crawling through the tunnel from the command module to the LM.

2 problems presented themselves almost immediately. The astronauts were unable at first to depressurize the tunnel, apparently because of a clogged vent. They solved this problem by using other vents. The 2d problem was a $3\frac{1}{2}°$ slippage (or twisting) of the connection between the LM and the command module. Just before the spaceship lost radio contact with earth as it passed around the far side of the moon, ground controllers warned the astronauts that "if it [the $3\frac{1}{2}°$ slippage] goes double, do not undock." But 35 minutes later, when the spacecraft reappeared around the moon and contact was reestablished, *Snoopy* and *Cnarlie Brown* were 30 to 40 feet apart and "station keeping" (flying in close formation). The separation took place at 3:36 p.m. EDT.

At 4:35 p.m. Stafford fired the LM's descent rocket. With Cernan and Stafford standing in their places in the landing craft, the LM began to drop from its altitude of 69 miles above the lunar surface. Young remained at the controls in *Charlie Brown* as the CSM continued in its 69-to-70-mile-high orbit to await the LM's return.

Snoopy made its closest approach to the moon—less than 9 miles (not quite 47,000 feet) from the surface—at 5:33 p.m. as it passed above the selected *Apollo 11* landing site in the moon's dry Sea of

Tranquility. The 2 astronauts aboard the LM reported that the site seemed suitable for the manned landing. Cernan said that "it looks like very wet clay, with lots of holes." Stafford reported that "the crater looks flat and smooth at the bottom; it should be real easy for A-11 [*Apollo 11*]." A brief firing of the engine then brought the LM up to a peak altitude of 220 miles as it completed its first independent orbit.

Coming down again, *Snoopy*'s astronauts jettisoned the LM's descent section at 7:34 p.m.; the ascent section, with Stafford and Cernan in it, then began to gyrate wildly. The gyrations were stopped by the repositioning of a guidance control switch, and the LM passed over the landing site again at an altitude of about 10 miles at 7:43 p.m. (NASA disclosed later that a ground-staff error—the inadvertant omission of instructions to reposition the switch—was the cause of the LM's gyrations.)

During the 2 descents toward the lunar surface the astronauts successfully tested the LM's landing radar in lunar conditions. The flight also provided additional successful tests of the LM descent and ascent rockets.

The LM flew little closer to the moon's surface than about 50,000 feet because (a) this was the altitude at which the landing radar became effective, (b) at this altitude the descent engine in an actual landing mission would be fired for the final approach, and (c) it was the lowest altitude to which the CSM could descend to rescue the LM's occupants if their engines failed.

Snoopy's maneuvers took it to a maximum of some 360 miles from the orbiting command module. After the 2d descent toward the moon's surface, the LM's ascent engine was ignited at 7:44 p.m. to bring the ascent portion of the landing vehicle back up toward the CSM. Rendezvous was achieved at 10:54 p.m., and the 2 spacecraft were linked together again at 11:11 p.m. After Cernan and Stafford crawled through the tunnel back into the CSM, the airtight hatches were sealed again, and the astronauts were able to remove their bulky space suits. The LM ascent stage was cast adrift into lunar orbit early May 23.

Apollo 10 was brought out of lunar orbit May 24 by means of a 164-second burning of its service engine. The "burn" began at 6:25 a.m. EDT as the spaceship passed behind the moon during its 31st lunar revolution. The ignition put the CSM on such a precise course toward the earth that ground controllers found course corrections scheduled for May 24 and 25 to be unnecessary.

Preparing for their return, the astronauts shaved May 25 with brushless shaving cream and a straight razor. They were the first men to shave in space. NASA had barred shaving during previous missions because of the danger of sparks from electric shavers and because of fear that bristles would float around the cabin and clog instruments. The *Apollo 10* astronauts, however, received permission to try wiping the razor on a paper towel after each stroke, and the experiment worked.

The only course correction of the 230,000-mile return trip was made the final day, May 26, by means of a 6.6-second firing of the maneuvering rockets. The service module was jettisoned at 12:22 p.m. EDT May 26 by means of explosive bolts. The conical command module was then turned so that its heat-shielded base faced in the direction of flight.

At an altitude of 400,000 feet over Asia, while still 1,400 miles from its landing point, *Apollo 10* smashed into the outer fringes of the earth's atmosphere. This took place at 12:38 p.m. at a velocity of 24,790 mph., a speed faster than any other men had flown. Friction with the air then began to slow the hurtling spacecraft and at the same time to heat its outer surface to temperatures as high as 5,000°F. The astronauts in the air-conditioned cabin, however, were unaffected by the heat. The module's descent was slowed further as the first 2 parachutes unfurled at 12:46 p.m. and the main 3 parachutes popped open a minute later.

Apollo 10, with its astronauts strapped to their couches inside, splashed down safely in the Pacific at 12:52 p.m. EDT May 26 about 440 miles east of Pago Pago, American Samoa and only about 7,000 yards from the waiting recovery ship, the aircraft carrier *Princeton*. The astronauts left the capsule with the aid of Navy swimmers. They were quickly brought by helicopter to the *Princeton*, and the capsule was hauled aboard shortly thereafter. Initial examination showed the astronauts to be in excellent physical condition after 8 days of space flight.

NASA Administrator Thomas O. Paine, jubilant at the success of the *Apollo 10* mission, said at a press conference at the Manned Spacecraft Center in Houston, Tex. May 26 that "today we see no obstacles on the path to the moon." "2 weeks from today, when we have carefully reviewed the flight data and debriefed the crew of *Apollo 10*, we will know whether we will be ready to set forth on July 16 [target date for *Apollo 11*'s launching]." But Paine insisted that "we will not hesitate to postpone the *Apollo 11* mission if we feel we are not ready in all respects. Once the voyage has begun, we have no commitment that would make us hesitate to bring home the crew immediately if we encounter problems." (Paine added that "while the moon has been the focus of our efforts, . . . the real goal is to develop and demonstrate the capability for interplanetary travel.")

Pres. Nixon and other world leaders congratulated the *Apollo 10* crew on their achievement. Soviet officials joined in the praise, and Soviet scientists told their countrymen May 27 that Americans would be the first to attempt a manned landing on the moon. (The Soviet press service Novosty, however, had asserted in a dispatch from New York May 22 that "the launching of rockets to the moon" had become "a new kind of entertainment for the American high society." It said that such members of American "nobility" as Vice Pres. Spiro T. Agnew and ex-Vice Pres. Hubert H. Humphrey and his wife had gathered at Cape Kennedy May 18 to "look for thrills" in the launching. Other notables who watched the launching included King Baudouin and Queen Fabiola of Belgium, U.S. Supreme Court Justices Byron White and Thurgood Marshall, and several members of Congress.)

MAN ON THE MOON

2 Americans Walk on the Moon

2 U.S. astronauts flew a fragile 4-legged spaceship to a safe landing on the moon at 40 seconds past 4:17 P.M. EDT July 20. With most of the civilized world watching by TV from a distance of 241,500 miles, the spacesuit-clad Americans climbed out of their lunar module $6\frac{1}{2}$ hours later. Then, for $2\frac{1}{4}$ hours, they walked, worked, performed experiments, collected samples of lunar rock and took photos in a barren, airless world to which no man had ever gone before them. This was the first landing of man on an astronomical body other than the earth, and the 2 astronauts stayed on it for 21 hours $36\frac{1}{4}$ minutes before blasting off in their spaceship July 21 for a safe journey to their home planet.

The manned landing on the moon—and the astronauts' safe return to earth 4 days later—climaxed an 8-year program set in motion by the late Pres. John F. Kennedy May 25, 1961 when he called on the U.S. to achieve the goal, "before this decade is out, of landing a man on the moon and returning him safely to the earth." At least 400,000 people and some 20,000 business organizations and universities were employed in the effort that finally set the 2 astronauts on the moon. Various estimates were given as to the money cost, but the most widely accepted figure was $24 billion.

The first man to put his foot on the moon was Neil Alden Armstrong 38, civilian commander of the *Apollo 11* spaceship, which had been launched from Cape Kennedy, Fla. July 16 on the flight that brought the astronauts to the moon. After gingerly backing out of the spaceship's hatch, Armstrong slowly climbed down a ladder and placed his clumsy left spaceboot on the dry lunar soil at 20 seconds after 10:56 p.m. EDT July 20. He was followed at 11:14 p.m. by Air Force Col. Edwin Eugene (Buzz) Aldrin Jr., 39, pilot of the lunar module (LM), which the astronauts had code-named *Eagle*.

The LM had descended to the moon from *Apollo 11*'s orbiting command and service modules (CSM), code-named *Columbia*. The 3d member of the *Apollo 11* crew, Air Force Lt. Col. Michael Collins, 38, the CSM pilot, remained behind aboard the *Columbia* as it continued in orbit around the moon.

Armstrong and Aldrin, moving easily in the slight gravity of the moon ($\frac{1}{6}$ the gravity of the earth), completed their EVA (extra-vehicular activity) on the moon early July 21. Armstrong, the last to quit the lunar surface, re-entered the LM at 1:11 a.m. They took with them a solar wind sampler, films of the scenes they had photographed on the moon and about 50 pounds (earth weight) of lunar rock and soil for scientific analysis on earth. They left an American flag, other mementos and several scientific experiments.

At 1:54 p.m. EDT July 21 the astronauts ignited the engine of the ascent portion of the LM. With the 4-legged descent stage remaining on the moon, Armstrong and Aldrin rose in the ascent section and linked up with the orbiting CSM. Later, after Armstrong and Aldrin had rejoined Collins in the CSM, the LM was cast adrift in space.

Apollo was blasted out of lunar orbit early July 22 and sent into a trans-earth trajectory that brought the 3 astronauts to a safe splashdown in the Pacific July 24.

Claims for these 6 world records in manned flight to the moon were filed by the U.S. with the International Aeronautic Federation in Paris July 28: (a) Duration of stay on lunar surface outside spacecraft—Armstrong, 2 hours 21 minutes 16 seconds. (b) Duration in lunar orbit Collins, 59 hours 27 minutes 55 seconds. (c) Duration of stay on lunar surface—Armstrong and Aldrin, 21 hours 36 minutes 16 seconds. (d) Duration of stay on lunar surface inside spacecraft—Aldrin, 19 hours 45 minutes 52 seconds. (e) Greatest mass landed on moon—Armstrong and Aldrin, 7,211 kilograms (15,897 pounds). (f) Greatest mass lifted into lunar orbit from lunar surface—Armstrong and Aldrin, 2,648 kilograms (5,838 pounds).

3 days before the *Apollo 11* launching the USSR had sent an unmanned spacecraft, *Luna 15*, toward the moon. *Luna 15* was launched July 13 with no advance announcement, and the Russians never explained the exact mission of their space probe. Western observers speculated that the Russians planned to have *Luna 15* land on the moon, scoop up some lunar surface samples and return them to earth before

Apollo 11 got back. *Luna 15*, however, went into orbit around the moon July 17 and then crashed into the moon July 21.

The day following the astronauts' lunar landing was observed in the U.S. as a holiday. Pres. Richard M. Nixon July 16 had issued a proclamation designating July 21 as a "National Day of Participation," ordering federal offices to be closed that day and urging governors, mayors, other government and school officials and employers to allow as many people as possible to share in the observances. Many state and local governments, and business enterprises, however, continued with business as usual.

Full information on all aspects of the *Apollo 11* mission—goals, target dates and times (down to the minute), equipment, personnel and even disputes over controversial aspects of the project—had been made public by the U.S. months and even years in advance of the flight. The launching, the flight, the activities on the moon and the splashdown that concluded the mission were televised to all parts of the world. It was reported that in addition to the worldwide TV audience, 750,000 to 1,000,000 people viewed the launching from beaches, boats, roads and other vantage points near the launching site, and it was estimated that a record audience of more than a half billion (possibly 600 million) people around the world saw the telecast of the first step on the moon.

Invitations to the launching were accepted by 234 members of the U.S. Congress (of whom 199 took along one member of his family each), 69 of the foreign ambassadors in Washington, members of the cabinet and other American and foreign dignitaries. To take these guests to Cape Kennedy, the Air Force used 2 of its own transport planes and 6 chartered commercial airliners (at a cost to the National Aeronautics & Space Administration of $54,300 for the commercial flights alone).

Ex-Pres. Lyndon B. Johnson and his wife attended the launching on Pres. Nixon's invitation, but Nixon watched by TV from the White House. (The President had planned to dine with the astronauts July 15, on the eve of the launching, but had canceled the plan on the advice of the astronauts' chief physician, Dr. Charles Alden Berry, who feared exposing his charges to any unnecessary infection. The men had been kept in isolation.)

Vice Pres. Spiro T. Agnew, who was among the launching spectators, said he had proposed to the President that the U.S. "articulate a simple, ambitious goal of manned flight to Mars by the end of the century." He asserted that "whether we say it or not, someone's going to do it."

The spectators at the launching included the Rev. Ralph David

Abernathy and a 25-family delegation representing the Poor People's Campaign. Hosea Williams, an official of the campaign, said: "We do not oppose the moon shot. We feel the effort is laudable. Our purpose for being here is to protest America's inability to choose human priorities."

The 3 astronauts had started their lunar voyage July 16 atop a 363-foot Saturn-5 booster rocket launched from Cape Kennedy's Launch Pad 39A precisely on schedule at 9:32 a.m. EDT. Strapped to their couches in the *Apollo 11* command module, Armstrong, Aldrin and Collins shot upward in an arc over the Atlantic.

The first and 2d stages of the 6,493,035-pound Saturn-5 dropped away as each of the 2 stages expended its fuel. The 3d (or S4B) stage then completed the job of carrying the spaceship up into a 118-mile-high parking orbit around the earth slightly less than 12 minutes after launching. The 297,000-pound payload—consisting of the S4B, the *Apollo 11* modules and the crew and equipment—was the heaviest man-made object ever put into orbit.

During the next $2\frac{1}{2}$ hours the astronauts checked their spacecraft and determined that it was in condition for the flight to the moon.

As *Apollo 11*, still attached to the S4B, was completing its 2d $88\frac{1}{2}$-minute orbit around the earth, the S4B was fired again at 12:16 p.m. This 2d S4B ignition increased the spaceship's speed by 10,435 feet a second to a velocity of about 35,575 feet a second (24,200 mph.) and headed *Apollo 11* toward the moon. The spaceship was put into a lunar trajectory so accurate than an initial course correction planned for July 16 was dropped.

Confident that they were in a successful trajectory toward the moon, the astronauts at 12:47 p.m. EDT July 16 began the one-hour process of removing the lunar landing craft (lunar module, or LM) from its shielded position inside the S4B. The first step was to separate the *Apollo 11* command and service modules (CSM) from the S4B by the firing of explosive bolts. Collins then moved the CSM about 100 feet ahead of the S4B and turned it completely around so that the docking probe in its nose was pointed toward the docking ring in the top of the 33,205-pound LM, which was still in the S4B. Slowly he eased the CSM back toward the S4B, and the probe entered the ring. After the LM was locked to the nose of the command module, the LM was ejected from the S4B by the release of springs, and *Apollo 11* moved off about 2,500 feet from the S4B. The S4B's engine was then fired by radio signal from earth to send the S4B out of *Apollo 11*'s path and into orbit around the sun.

As *Apollo 11* hurtled toward the moon at a decreasing velocity, the spaceship rotated on its long axis about 3 times an hour so that the sun warmed each side evenly. The astronauts practiced astral navigation, and at about 8 p.m. July 16 they transmitted the first of a series of color telecasts from their spaceship to earth. They went to sleep an hour later.

A course change was made at 12:17 p.m. EDT July 17 by means of a 2.91-second firing of the CSM engine. The ignition added 21 feet a second to *Apollo 11*'s velocity, which was then 3,430 mph. and decreasing as the earth's gravity continued to slow the spaceship. The course correction was so accurate that no further ones were necessary for the rest of the trip to lunar orbit.

Looking down at the earth July 17, Armstrong noted rain clouds over Houston (where it actually was raining) but informed the Manned Spacecraft Center personnel there that "it ought to clear up pretty soon" because "the western edge of the weather isn't very far from you." A half-hour later the rain stopped.

Armstrong and Aldrin inspected the LM July 18 and found it and its equipment in excellent shape. To enter the LM, they opened the hatches at each end of the "tunnel" separating the LM from the CSM. They then crawled through the tunnel. A one-hour 36-minute telecast began at 4:44 p.m. while the LM inspection was starting, and the TV audience was able to see the interior of the LM as well as of the CSM.

Apollo 11 entered the moon's sphere of gravitational influence at 11:32 p.m. July 18 at a distance of about 33,000 miles from the moon and 215,000 miles from the earth. Thereafter the spaceship's speed, which had been slowed constantly by the pull of the earth's gravity, began to increase as *Apollo 11*, in effect, fell toward the moon. Ultimately *Apollo 11* attained a speed of 5,700 mph.

Apollo 11 was pulled around the moon by lunar gravity at 1:13 p.m. EDT July 19 after a trip of 244,930 miles. At 1:22 p.m., while behind the moon and out of radio contact with the earth, the astronauts fired their main 20,500-pound-thrust engine for 6 minutes 2 seconds to slow the spaceship's velocity. They thus put *Apollo 11* into orbit around the moon.

In its first revolution around the moon the spaceship traveled in an orbit with an apocynthion (high point) of 195 miles and pericynthion (low point) of 70 miles. After 2 revolutions around the moon in the initial orbit, the engine was ignited again to put *Apollo 11* into an orbit with an apocynthion of 75 miles and pericynthion of 62 miles.

While still in the initial orbit, Armstrong had observed the glow that some astronomers had reported in the crater Aristarchus and which they thought indicated volcanic activity. Armstrong said at

2:45 p.m., as Apollo passed about 40 miles south of the crater: "... There's an area there that's considerably more illuminated than the surrounding area. It's got a slight amount of fluorescence to it. That area is definitely lighter than anything else I can see out of this window."

Preparations for the lunar landing began shortly after 9:30 a.m. EDT July 20 as *Apollo 11* was starting its 11th revolution around the moon. Aldrin first and then Armstrong, clad in their spacesuits, crawled through the tunnel from the CSM to the LM and closed the hatch. They fastened themselves loosely with harnesses, Armstrong standing at the left and Aldrin at the right.

All equipment was checked, and the LM's 4 landing legs were extended by the pushing of a button at 12:32 p.m. This stretching out of the legs increased the height of the $16\frac{1}{2}$-ton LM to 22 feet 11 inches and its width to 31 feet. 24 minutes later Mission Control Center in Houston radioed its approval for undocking, and *Apollo 11* then lost radio contact with the earth as it swung around the far side of the moon to start its 13th lunar revolution.

The LM was separated from the CSM at 1:47 p.m. EDT. The separation was made when Collins, who remained alone in the CSM, released the last of 15 spring latches that had held the 2 spaceships together. After radio contact with the ground was restored, Armstrong announced: "*Eagle* has wings."

They then flew in close formation while equipment checking continued. At 2:12 Collins, using the CSM's small maneuvering rockets, altered the CSM's orbit and brought his spacecraft about 7/10 mile away from the LM.

At 3:08 p.m. EDT July 20, with the LM again behind the moon and out of radio contact, its on-board guidance-and-navigation computer ignited the engine and brought the LM out of orbit from an altitude of $65\frac{1}{2}$ miles. The firing, in which Aerozine 50 burned for 29.8 seconds on contact with nitrogen oxide, headed the LM down toward the surface of the moon. The spacecraft was a scant 21 miles above the surface before radio contact was again established with Houston.

At an altitude of about 50,000 feet above the moon, Armstrong made the decision to continue with the attempt to land. The LM's 9,870-pound-thrust engine was switched on again, and the powered phase of the descent began as the LM sped toward the previously designated landing site about 250 miles ahead.

The guidance computer directed the descent until the LM had come down to an altitude of about 300 feet. At this point Armstrong took semi-manual control of guidance because, as he explained later, he saw the computer aiming the LM to a landing "right into a football

field-sized crater with a large number of big boulders and rocks."
(Armstrong would have taken control at this altitude in any case).

With the computer continuing to control the engine, Armstrong guided the LM down to a soft landing in a level and relatively clear area on the moon's dry Sea of Tranquility. The LM came down about 120 miles southwest of the crater Maskelyne in a position given as Latitude $0.6914°$ N., Longitude $23.461°$ E., although ground controllers in Houston failed in initial attempts to confirm the exact landing point.

"Houston," Armstrong announced to Mission Control immediately after the LM touched down at 4:17:40 p.m., "Tranquility Base here. The *Eagle* has landed!" (A check of the automatically radioed biological data showed that Armstrong's heartbeat, which normally had a rate of 77 a minute, had reached a rate of 156 a minute on landing.)

At the Manned Spacecraft Center in Houston, the safe landing was confirmed immediately by Dr. Thomas O. Paine, administrator of the NASA, who immediately phoned Pres. Nixon at the White House and announced: "Mr. President, it is my honor on behalf of the entire NASA team to report to you that the *Eagle* has landed on the Sea of Tranquility, and our astronauts are safe and looking forward to starting the exploration of the moon."

The first minutes after the lunar landing were spent in checking to make sure that no harm had been done to the LM or to any LM equipment and that it would not be necessary to try to send *Eagle*—already being referred to as Tranquility Base—immediately back to the orbiting *Columbia*.

Then, confident that it was safe to remain on the moon, Armstrong and Aldrin removed their helmets and gloves and ate the first meal ever eaten by human beings on the moon.

Throughout their stay on the moon the astronauts made repeated efforts to describe the close-up appearance of the moon to their spellbound audience on earth. Among their first reports on what they saw from the window of the LM after *Eagle* landed:

Armstrong—"It looks like a collection of just about every variety of shape, angularity, granularity, about every variety of rock you could find. The colors vary pretty much depending on how you are looking relative to the zero phase length. There doesn't appear to be too much of a general color at all...." The LM was on a "relatively level plain cratered with a fairly large number of craters of the 5- to 50-foot variety, and some ridges, small, 20 to 30 feet high, I would guess, and literally thousands of little one- and two-foot craters around the area.... There is a hill in view just about on the ground track ahead of us...."

Aldrin—I'd say the color of the local surface is very comparable to

that we observed from orbit at this sun angle—about $10°$ sun angle....
It's pretty much without color. It's gray and it's very white as you
look into the zero phase line. And it's considerably darker gray, more
like an ashen gray, as you look out $90°$ to the sun. Some of the surface
rocks in close here that have been fractured or disturbed by the rocket-
engine plume are coated with this light gray on the outside. But where
they've been broken, they display a dark, very dark, gray interior, and
it looks like it could be country basalt."

Armstrong said he and Aldrin had been unable to locate their precise landing point, and "I haven't been able to pick out the things on
the horizon as a reference yet." (Throughout their entire stay of more
than $21\frac{1}{2}$ hours on the moon, the astronauts—and Mission Control Center in Houston as well—remained unable to pinpoint the exact landing
site. And while the LM was on the surface of the moon, Collins was
never able to spot it from the orbiting *Columbia*.)

Armstrong reported that he and Aldrin had noticed no "difficulty
at all in adapting to $\frac{1}{6}$ G [the lunar gravity exerts a pull only $\frac{1}{6}$ that of
the force of the earth's gravity]." "It seems immediately natural to
move around in this environment," he declared.

The astronauts had been scheduled to sleep aboard the LM before
stepping out on the surface. The beginning of the walk on the moon
had been planned for about 2:12 a.m. EDT July 21. But it was decided
a few hours after the landing that an earlier moon-walk should be attempted. Among the reasons for this decision: (a) The mission had
gone extremely well so far, (b) the astronauts felt no need for rest and
(c) it probably would have been impossible for normal human beings to
sleep under these exciting circumstances (in anticipation of this excite-
ment, the astronauts had been provided with sleeping drugs).

At about 8 p.m. EDT July 20 Armstrong and Aldrin requested and
received permission to postpone the sleep period and to start the EVA
(extra-vehicular activity) on the surface of the moon early. (NASA officials had decided several days previously that they would give permission for an early EVA if conditions were right but only if the astronauts asked for permission. They withheld this decision from the astronauts in fear that knowledge of it might tempt them to request an early
EVA even if conditions were not favorable.)

Armed with authorization for the early moon-walk, the astronauts
donned their cumbersome PLSS (portable life-support system) backpacks, helmets and gloves and slowly depressurized the interior of the
LM. They then opened the exit hatch.

Armstrong was the first of the astronauts to leave the LM. He
backed carefully out through the narrow hatch, adjusting his clumsy

movements under Aldrin's guidance. At 10:51 p.m. EDT he stood on the platform just outside the hatch. He then started cautiously to back down the 9-step ladder attached to one of the LM's 4 legs. Aldrin leaned out of the LM hatch to take motion pictures of Armstrong's movements.

On reaching the 2d step, Armstrong pulled a cord to uncover a TV camera. The camera immediately began transmitting the first faltering human steps on the moon, and man's initial actions on the moon were seen by hundreds of millions of people watching TV sets in all parts of the earth.

Armstrong, the first man to step on the surface of a world other than his own, placed his space-booted left foot on the moon at 10:56:20 p.m.

"It's one small step for [a] man, one giant leap for mankind," he said. (His words were originally reported as " ... one small step for man ... " He disclosed later that he had said [or had meant to say] " ... for *a* man ... .")

Before stepping off the ladder, Armstrong had reported that "the LM footbeds are only depressed in the surface about one or 2 inches, although the surface appears to be very, very fine-grained as you get close to it. It's almost like a powder."

He then moved off the ladder and gave this first description of walking on the moon: "The surface is fine and powdery. I can kick it up loosely with my toes. It does adhere in fine layers like powdered charcoal to the sole and the sides of my boots. I only go in a small fraction of an inch, maybe an eighth of an inch, but I can see the footprints of my boots and the treads in the fine sandy particles. There seems to be no difficulty in moving around this, and we suspect that it's even perhaps easier than the simulations of $\frac{1}{6}$ G that we performed ... on the ground [on earth]. Actually no trouble to walk around."

Armstrong reported that the LM's "descent engine did not leave a crater of any size. It [the engine] has about one foot clearance on the ground. ... I can see some evidence of rays [ground marks] emanating from the descent engine, but a very insignificant amount."

One of Armstrong's first tasks on the lunar surface was to get a "contingency sample"—a pocketful of lunar surface material taken immediately in case it became necessary to make an emergency departure and forego the rest of the mission's experiments. Since his ability to bend was limited in his pressurized space suit, Armstrong had to scoop up his samples into a small canvas bag at the end of a long handle. (Bringing back samples of the moon's surface had frequently been described as the mission's "first scientific priority.")

"It's a very soft surface," Armstrong reported, "but here and there where I bore with the contingency sample collector, I run into very hard surface, but it appears to be very cohesive material of the same sort." He said that he had dug as far as "about 6 or 8 inches into the surface." "It's easy to push on it," he declared. "I'm sure I could push it in farther, but it's hard for me to bend down farther than that."

Armstrong said that the area of the moon in which they had landed "has a stark beauty all its own. It's like much of the high desert of the United States. It's different, but it's very pretty out here."

Aldrin followed Armstrong out of the LM at 11:11 p.m. At 11:14 p.m. July 20 he became the 2d man to walk on the surface of the moon.

"Beautiful," Aldrin murmured as he gazed at the lunar scene. "It's a magnificent sight," Armstrong agreed.

Under Armstrong's inspection, Aldrin began a series of experimental walking, loping and jumping movements on an airless, dust-covered planetoid where he weighed only $\frac{1}{6}$ of what he weighed on earth. Because of the backpack, he said, "there's a slight tendency ... to tip backwards." But "if I'm about to lose my balance in one direction," he continued, "recovery is quite natural and easy. You've just got to be careful leaning in the direction you want to go in." (The backpack weighed 183 pounds on earth, only about 30 pounds on the moon. It pressurized the spacesuit, provided radio communication and oxygen for breathing. The astronauts were encased in a portable, temperature-controlled atmosphere that was kept at about 68°F. although the surface temperature at the time and in the place of the EVA was estimated at 40° to 90°F. in the sun and -150°F. in the shade.)

The astronauts found that some of the lunar rocks felt slippery underfoot because of their coating of lunar dust.

Armstrong moved the TV camera to a spot about 30 feet from the LM and inserted a telephoto lens. Then he and Aldrin read to the TV audience the wording on a small stainless-steel plaque fastened to one of the LM's legs:

> HERE MEN FROM THE PLANET EARTH
> FIRST SET FOOT UPON THE MOON
> JULY 1969, A. D.
> WE CAME IN PEACE FOR ALL MANKIND

"It has the crew members' signatures and the signature of the President of the United States," Armstrong noted.

Armstrong then repositioned the TV camera about 40 feet from the LM to cover the rest of his and Aldrin's activities on the surface of the moon. Shortly after 11:30 p.m. the astronauts removed a plastic American flag and a flagstaff from one of the LM's legs, set the flag up on the lunar surface and saluted.

At 11:48 p.m. Armstrong and Aldrin were informed that the President would then speak to them. Talking into a phone at the White House, his face sharing the world's TV screens with the view of the two astronauts standing on the moon, Pres. Nixon said: " . . . This certainly has to be the most historic telephone call ever made. I just can't tell you how proud we all are of what you've done. For every American this has to be the proudest day of our lives. And for people all over the world, I am sure they, too, join with Americans in recognizing what an immense feat this is. Because of what you have done, the heavens become a part of man's world. And as you talk to us from the Sea of Tranquility, it inspires us to redouble our efforts to bring peace and tranquility to earth. For one priceless moment in the whole history of man, all the people on this earth are truly one—one in their pride in what you have done and in our prayers that you will return safely to earth."

Armstrong replied that "it's a great honor and privilege for us to be here representing not only the United States but men of all nations, men with interest and a curiosity and men with a vision for the future." At the conclusion of his remarks, the 2 astronauts saluted.

During their 2 hours $21\frac{1}{4}$ minutes on the surface of the moon outside the LM, the astronauts deployed a variety of scientific equipment. Aldrin unrolled and set up on a pole a one-by-4-foot sheet of aluminum foil to pick up particles of the solar wind—atomic and subatomic matter streaming at high speed from the sun. When they returned to the LM, the astronauts took the aluminum sheet back with them for testing on earth.

Aldrin set up a seismic detector consisting of 3 long-period seismometers and one short-period vertical seismometer for detecting and measuring meteroid impacts and possible "moonquakes." The detector was left on the moon, and solar panels supplied the power for radioing the seismic data to the earth. (The detector was so sensitive that it recorded Aldrin's hammering on tubes he used to get core samples of the lunar surface. After the astronauts returned to the LM, the detector recorded the thumps when their heavy backpacks were thrown out onto the moon.)

Armstrong set up a laser reflector at which laser beams were aimed (at first unsuccessfully) from the world's 3d largest telescope, the 107-inch telescope of the University of Texas in Fort Davis. (Through this experiment, it was said, scientists could determine the distance from the earth to the moon within 6 inches. They could thus measure very subtle changes in earth-moon distances. Various observatories planned to aim laser beams at the reflector over a period of 10 years to test a questioned part of Einstein's theory of relativity, to check on continental drift on the earth, to get data on the internal construction of the moon and to measure the periodic 14-month "Chandler wobble" of the rotating earth.)

The astronauts filled 2 airtight boxes with specimens of the moon's surface material. Among the items was what Aldrin jokingly referred to as "purple rock." The material included core samples Aldrin secured by pounding 2 tubes into the soil. (After returning to the LM, Aldrin described what had taken place: " ... I could get down about the first 2 inches without much of a problem, and then ... I would pound it in about as hard as I could do it. The 2d one took 2 hands on the hammer, and I was putting pretty good dents in the top of the extension rod. And it just wouldn't go much more than—I think the total depth might have been about 8 or 9 inches. But even there it—for some reason—didn't seem to want to stand up straight. So that I'd keep driving it in, and it would dig some sort of a hole, but it wouldn't just penetrate in a way that would support it and keep it from falling over.... And also I noticed when I took the bit off that the material was quite well-packed, a good bit darker; and ... the way it adhered to the core tube gave me the distinct impression of being moist.")

Among the other items left on the moon was an aluminum capsule containing a $1\frac{1}{2}$-inch silicon disc on which was inscribed, in microscopically small print, messages of goodwill from the heads of 73 nations (not including the USSR). The disc carried an excerpt from the U.S. National Aeronautics & Space Act of 1958, signed by Pres. Dwight D. Eisenhower, and parts of statements on space by Presidents Kennedy, Johnson and Nixon. It also bore the names of Vice Pres. Spiro T. Agnew and of 49 Representatives and 28 Senators who were members of Congressional committees involved in the U.S. space program.

The astronauts left on the moon 5 medals honoring dead U.S. and Soviet astronauts—Virgil I. Grissom, Edward H. White 2d and Roger B. Chaffee of the U.S. and Yuri A. Gagarin and Vladimir M. Komarov of the USSR. (While Col. Frank A. Borman, the U.S. astronaut, was visiting Russia in July, he had received the Russian medals from the widows of the Soviet cosmonauts.) Another memento left on the moon was a cloth shoulder patch of the design the 3 dead U.S. astronauts had planned to wear.

Warned to return to the LM soon because of the depletion of the backpack oxygen supply, Aldrin climbed up the ladder at 12:58 a.m.

EDT July 21. With Armstrong still on the moon's surface, the 2 astronauts used a rope-and-pulley arrangement to haul up the 2 lunar sample boxes. Armstrong then climbed inside at 1:11 a.m.

The astronauts left behind on the moon their cameras (but took their exposed film with them), various experiments and discarded tools. Repressurizing the LM cabin, the astronauts transferred from the backpack life-support system to the LM's life-support system. They then depressurized the LM and jettisoned their backpacks, lunar boots and other things not needed for the return trip or for study on earth. The discarded items, thrown through the hatch to the moon, were disposed of to make the LM's ascent stage as light as possible. It was estimated that the value of this litter exceeded $1 million. The astronauts closed the hatch for the last time, again repressurized the LM, removed their bulky space suits and slept fitfully in the crowded, uncomfortable cabin. When they blasted off from the moon later in the LM's ascent stage, the biggest man-made object they left behind was the LM's descent stage.

(Among items carried to the moon and back were 4-by-6-inch flags of the UN, the 136 nations with which the U.S. had diplomatic relations, the 50 U.S. states, the District of Columbia and 4 U.S. possessions—Guam, Puerto Rico, the Virgin Islands and American Samoa. 2 5-by-8-foot U.S. flags were carried aboard the command module and and brought back for presentation to the 2 houses of Congress. The astronauts also carried the torn halves of 4 $1 bills. On their return, the bill halves were compared with the remaining bill halves—which had been left on earth—to prove that the 3 men who had left Cape Kennedy July 16 were the ones who returned to earth July 24.)

At 1:54 p.m. EDT July 21 Armstrong fired the 3,500-pound-thrust rocket of the LM's ascent stage to lift this upper section of the LM off the 4-legged descent stage. The latter section, which remained on the moon, served as the launching pad for the ascent stage. Before the engine was ignited, the astronauts had fired the explosive bolts that had held the ascent and descent sections together.

Eagle, the LM, was the first man-made vehicle ever to leave an astronomical body other than the earth. The 7-minute 18-second firing of the rocket took *Eagle* upward and westward into a preliminary lunar orbit with a perilune (low point) of 10 miles and apolune (high point) of 54 miles. At this time *Columbia*, the CSM, with Collins aboard, was in its 25th revolution around the moon in a circular orbit about 69 miles high. Collins, above and behind the 2 lunar explorers, reported seeing *Eagle* 20 minutes after it had risen from the moon.

Using the LM's small thrusters, Armstrong boosted *Eagle* to higher altitudes until, at 5:35 p.m. EDT, while the 2 *Apollo 11* spacecraft were behind the moon and out of radio contact with the earth, the 2 spaceships docked.

After the link-up, the astronauts opened the hatches between the LM and the CSM and carried their photos, lunar samples and other mementos back into the command module. The CSM hatch was then fastened again, and the LM ascent stage was detached and cast adrift in orbit around the moon.

Apollo 11 was brought out of lunar orbit by a firing of the main engine at 12:56 a.m. EDT July 22 as the spaceship passed behind the moon on its 31st lunar revolution.

As *Apollo* hurtled toward the earth July 22, Aldrin revealed that he had walked to the edge of a crater about 200 feet away from the LM. He described the crater as "maybe 70 or 80 feet in diameter and 15 or 20 feet deep." He said that "it had rocks on the bottom that were bigger than anything on the surface, and I took quite a few pictures of them."

Armstrong and Aldrin reported having seen a white light with a yellow tinge that flashed from the earth while they were on the moon, and they speculated that it might have been a laser beam. But project scientists said that such a beam could not have come from any source connected with Apollo.

At 4 p.m. July 22 the astronauts executed a course correction by means of a $10\frac{1}{2}$-second blast of the maneuvering rockets. They thereby reduced speed by 4.8 feet a second, primarily to improve the flight path angle for re-entry into the earth's atmosphere. The correction put them into a trajectory so accurate that an additional course correction scheduled for July 23 was cancelled.

In *Apollo 11*'s final telecast from space, the day before splashdown, Aldrin said July 23 that he and his companions had "come to the conclusion that this has been far more than 3 men on a voyage to the moon, more still than the efforts of a government and industry team, more even than the efforts of one nation. We feel that this stands as a symbol of the insatiable curiosity of all mankind to explore the unknown." Aldrin said the flight recalled to him this verse from *Psalms 8*: "When I consider the heavens, the work of Thy fingers, the moon and the stars which Thou has ordained: What is man that Thou art mindful of him?"

The astronauts prepared their spaceship for the return to earth by jettisoning the service module in space. They then turned the command module (CM) so that its heat-shielded blunt end faced in the direction of flight. The 3 astronauts, strapped in their couches, flew backwards as the CM, its weight down to 10,965 pounds, smashed into the atmosphere at 12:35 p.m. EDT July 24 some 1,600 miles southwest of the mid-Pacific splash-down point.

The CM's re-entry speed of 23,900 mph. at about 400,000 feet altitude was quickly reduced by air friction, and the CM's descent was then slowed by parachutes. The capsule, with the astronauts aboard, plunged into the Pacific at 12:50 p.m. EDT at a speed of 21 mph. It came down 11 miles from the waiting recovery ship, the carrier *Hornet*, but only $1\frac{3}{4}$ miles from its aiming point, 950 miles southwest of Hawaii and 250 miles south of Johnston Island. The CM capsized immediately after splash-down but was righted by inflated balloons.

The recovery procedure that followed had been designed to fulfill the requirements of a 21-day quarantine imposed on the astronauts and all they brought with them from space. Its purpose was to cancel even the remote possibility that they might infect the earth with some disease organism from the moon. The quarantine period, which began July 21 on their departure from the moon, had 18 days left to go when they reached the earth.

Navy swimmers brought by helicopter to the floating CM attached a flotation collar and 2 rafts to the capsule. Isolation suits—green coveralls with hoods and face masks—were handed to the astronauts by one of the swimmers, and the astronauts donned these suits before emerging from the CM. The astronauts then crawled out onto the rafts, and the swimmer used decontaminants on both the astronauts and the CM hatch. Aldrin decontaminated the swimmer.

The astronauts were flown by helicopter to the *Hornet* and hustled into a quarantine trailer. (They were kept in this Mobile Quarantine Facility until the trailer was flown to Houston and they were transferred July 27 to the spacious Lunar Receiving Laboratory at the Manned Spacecraft Center.)

2 other men, who had taken up housekeeping in the trailer July 19, shared their quarantine. They were Dr. William R. Carpentier, who served as their physician, and John K. Hirasaki, a NASA engineer. Shortly after the astronauts entered the trailer Carpentier made a quick preliminary examination and then announced that all 3 astronauts had survived their lunar journey in excellent health. Hirasaki went to work on the CM as soon as it was brought aboard ship and placed near the trailer. Hirasaki removed and decontaminated various items. The lunar sample containers and photographic films were passed through a decontamination lock for early processing and study.

Pres. Nixon had been waiting aboard the *Hornet* to greet the returning astronauts. 2 hours after splash-down the astronauts were ready to chat with the President through the glass window of the quarantine trailer. "This is the greatest week in the history of the world since Creation," Nixon said, "because as a result of what happened this week, the world is bigger infinitely and also ... the world's never been closer before." (Nixon visited the astronauts at the beginning of a 12-day, 24,500-mile round-the-world tour he had started 2 days previously.)

The 3 *Apollo 11* crew members were allowed to leave the Lunar Receiving Laboratory (LRL) at the Manned Spacecraft Center in Houston, Tex. Aug. 10 after their 21-day quarantine. NASA doctors had found all 3 free of contamination by any possible lunar organism. 20 technicians and others who had shared the astronauts' quarantine also left the LRL. Several members of the Inter-Agency Committee on Back Contamination and other scientists warned that the quarantine precautions used after the *Apollo 11* flight were faulty and should be improved for future flights in which there might be more serious danger that astronauts might bring back a space disease.

At a televised press conference in Houston Aug. 12, the astronauts described their reactions to being on the moon. "It was a stark and strangely different place," Armstrong said, "but it looked friendly to me, and it proved to be friendly." He reported that "temperatures weren't high. ... We felt very comfortable in the lunar gravity. It was preferable both to weightlessness and the earth's gravity."

Aldrin said they had found it "fairly convenient" to walk "at the rate of 5 to 6 miles an hour" on the moon. "One could jump in kangaroo fashion, 2 feet at a time," he reported, "but .... I found that a standard loping technique of one foot in front of the other worked out quite well." Aldrin's feet sank further in some places than in others, he said. "Somewhere in rather flat regions, the footprint would penetrate perhaps a half an inch or sometimes only a quarter of an inch and give a very firm response. In other regions, near the edges of these craters, we could find that the foot would only sink down maybe 2, 3, possibly 4 inches."

Armstrong said that moving around on the moon was complicated by "the lunar curvature and the local roughness. It seemed ... like swimming in an ocean with 6-foot or 8-foot swells and waves. ... You never can see very far away from where you are. This was even exaggerated by the fact that the lunar curvature is so much more pronounced."

(Aldrin disclosed in an article in the Aug. 22 edition of *Life* magazine that after they had returned to the lunar module with their samples of the moon's soil and rocks, he had found "a distinct smell to the lunar material, pungent like gunpowder." He also reported that he had celebrated communion in the lunar module shortly after landing on the moon. Collins had disclosed during a CBS TV interview Aug. 17 that he planned to make no more space flights because of the difficulty he found in continuing the training needed "to do a good job" as an astronaut.)

The cost of the *Apollo 11* moon landing was officially estimated by NASA at $350 million, but Richard D. Lyons noted in the *N.Y. Times* July 22 that other sources estimated the cost at perhaps $30 billion. The latter figure included expenditures on research and development, construction of facilities and such pre-Apollo spaceflight programs as Mercury and Gemini. NASA arrived at its $350 million figure by adding up $185 million for the Saturn-5 booster rocket, $55 million for the command and service module, $40 million for the lunar module (LM) and $70 million for operations expenses. The $30 billion figure included most of the 36\frac{1}{4}$ billion NASA had spent in its 11 years of existence up to July 1969. The 11 years of expenditures provided $23 billion for manned spaceflight, $5.2 billion for scientific investigations, 3\frac{1}{2}$ billion for space technology, 2\frac{1}{2}$ billion for support operations, $1.1 billion for space applications and $900 million for aircraft technology.

(The cost of the total Apollo program so far was put at $24 billion, excluding the 2\frac{3}{4}$ billion spent on the Mercury, Gemini, Ranger, Surveyor and Orbiter programs, which provided data on manned flights to the moon and landing on the moon itself and thus made the Apollo program possible.) As an example of "staggering" research and development costs, Lyons cited $1.7 billion in NASA contracts to the Grumman Aircraft Engineering Corp. for designing and building the LM.

(Other federal agencies also had high costs for space work. The Defense Department spent more than $2 billion on space in 1968 and perhaps $20 billion on space in the past 15 years. Space expenditures totaling $143.6 million in 1968 were reported for the Agriculture, Interior and Commerce departments, National Science Foundation and Atomic Energy Commission.)

The first manned landing on the moon was made during history's 33d manned space flight. 20 of the previous space flights were made by Americans (who traveled a total of 78,955,000 miles on these flights and accumulated a total of 4,514 man hours in space). 12 of the flights were made by Russians. *Apollo 11* was the 4,039th man-made object launched into space. 6 man-made satellites were then in orbit around the moon, and about 1,743 man-made objects (about 780 vehicles and nearly 1,000 bits of debris) were in orbit around the earth.

The $4,000 Galabert International Astronautics Prize for 1969 was awarded to the *Apollo 11* astronauts July 21. The announcement was made in Paris.

Scientists Report on Moon Rocks

After nearly 2 months of "preliminary" analysis of the 48 pounds of glassy soil and rocks brought back from the moon aboard *Apollo 11*, NASA scientists asserted at a press conference in Washington Sept. 15 that the chemical composition of the moon rocks was "unlike that of any known terrestrial rock." Much of the lunar material consisted of tiny glass spheres and glass-encrusted rock.

Speaking at the Smithsonian Institution, the scientists reported the lunar material to be between 2 and $3\frac{1}{2}$ billion years old. Imbedded in the surface of the rocks were rich accumulations of the "rare gases" helium, neon, argon, krypton and xenon. They presumably had been among the particles that boiled out of the sun and shot out into space to form the "solar wind." The rocks were found to have a far higher content of such "rare" high-melting-point elements as chromium, titanium, yttrium and zirconium than rocks originating on the earth. There was no indication of such "precious" metals as gold, silver or platinum and comparatively little of such low-melting-point elements as lead, bismuth, sodium and potassium. No traces of water or indications of life were found in the lunar material.

Dr. Robin Brett, a Manned Spacecraft Center scientist, reported that the lunar rocks had a titanium oxide content of as much as 12% (compared with a maximum content of $4\frac{1}{2}\%$ for the richest terrestrial ores) and 10 times as much chromium as found in any rocks on earth.

Dr. Paul Gast of the Lamont Geological Observatory said in his summary of the scientists' preliminary findings: (a) Many of the rocks were igneous (previously molten). (b) Most of the rocks apparently had been tossed around on the surface of the moon because they were pitted on all sides, presumably by micrometeor impacts. (c) Most of the rocks were smooth and rounded, presumably because of erosion due to solar radiation and micrometeor impacts. (d) The 23 samples analyzed had virtually identical chemical composition; presumably, therefore, they were characteristic of the Sea of Tranquility area in which they had been found. (e) None of the people or laboratory animals exposed to the lunar materials had displayed reactions.

Gast said that the lunar rocks' composition, so different from that of earth rocks, gave support to the theory that the earth and moon had condensed separately from the same primordial gas cloud. The early analyses, he said, cast doubt on a 2d theory—that the moon had originally been part of the earth and had been thrown into space in some cataclysm. The analyses apparently neither supported nor contradicted a 3d theory—that the moon had been a small, wandering astronomical body formed elsewhere in space but "captured" by the earth's gravity.

William Kommerer of NASA's Marshall Space Flight Center reported that plants treated with the lunar soil had grown slightly faster and greener than untreated test plants.

After determining that no disease organisms had been brought to earth with the lunar samples, NASA officials had begun Sept. 12 to distribute 18 pounds of the lunar rock and soil to 142 scientists of the U.S. and 8 other countries for more thorough analysis.

Dr. Thomas Gold of Cornell University theorized in the journal *Science* Sept. 26 that much of the glass so prevalent in the lunar material might have been created by a solar flare 30,000 to 100,000 years ago. He held that such a flare, most likely caused by a comet or asteroid impact on the sun, would have increased the sun's heat 100-fold for 10 to 100 seconds. The heat would also have melted the material on the surface of Mercury and coated Mercury with glass, Gold said. Since most of the flare's brightness would have been in the ultraviolet range, much of it would have been filtered out by the earth's atmosphere, and the earth, therefore, would have been protected from any great blast of heat, Gold held. But he said that if a great penetrating burst of radiation had accompanied the flare, it could have had an effect on biological evolution. Gold's theory was disputed by Dr. William R. Greenwood, a geologist studying the lunar specimens. Greenwood theorized that the glass originated in splashes of molten material thrown up by the impact of meteors hitting the moon and exploding.

(NASA reported Sept. 22 that the glassy Australasian tektites, found largely in Australia, the Philippines and southeast Asia, had been identified as material that had been sprayed from the moon's surface about 700,000 years ago when an iron-nickel meteor 3 miles in diameter collided with the moon and formed the 56-mile-wide crater Tycho. Dr. Dean R. Chapman, chief of the Thermo-&-Gas-Dynamics Division at NASA's Ames Research Center in Mountain View, Calif., made the identification by relating the chemical composition of tektites to the composition of the rock on Tycho's rim, as determined by the unmanned lunar probe *Surveyor 7*, which had landed there in 1968. According to Chapman, the Australasian tektites had been ejected from the Rosse Ray, projecting for 1,000 kilometers northeast of Tycho and passing over the smaller crater Rosse. The 700,000-year-old impact was one of at least 3, NASA said, that had showered 10 to 100 million tons of lunar material on the earth. Of the others, an impact 35 million years ago had thrown material on what is now the U.S., and a 15 million-year-old impact had sprayed the area of what is now Czechoslovakia.)

After more than 2,000 unsuccessful attempts by 2 observatories to bounce a laser beam off an 18-inch-square reflector left on the moon by the *Apollo 11* crew, success had been achieved Aug. 1 by the University of California's Lick Observatory atop Mount Hamilton, Calif. Using the laser beam as a measuring tool, Dr. Carroll O. Alley of the University of Maryland reported Aug. 20 that the first precise measurements showed the reflector at the moon's Tranquility Base to have been 232,271.4064 miles away from the University of Texas' McDonald Observatory on Mount Locke, Tex. at 10:30 p.m. EDT Aug. 19. This was about 40 meters (131.2 feet) further than the Jet Propulsion Laboratory had calculated. The measurement was said to have a 4-meter (13.1-foot) chance of error. Eventually the error factor was expected to be reduced to 6 inches.

The difficulties in hitting the laser target had been due to early uncertainty as to *Apollo 11*'s exact landing point on the moon and to an inaccuracy of about $2\frac{1}{2}$ miles in the estimate of the distance between the earth and the moon. The landing site was located July 24 by Dr. Eugene M. Shoemaker of the California Institute of Technology. Shoemaker, one of the lunar surface experts at the Manned Spacecraft Center in Houston, pinpointed the site by finding identical landmarks on photos from *Apollo 11* and from unmanned Orbiter probes that the U.S. had sent into orbit around the moon.

2d U.S. Crew Walks on Moon

2 U.S. astronauts landed an Apollo lunar module (LM) on the moon Nov. 19 and made 2 walking-exploring-working excursions lasting a total of 8 hours on the lunar surface during a $31\frac{1}{2}$-hour stay on the moon. They then brought the ascent section of the LM back to the *Apollo 12* command and service module (CSM), which had been waiting in lunar orbit with the 3d member of *Apollo 12*'s crew aboard as pilot.

Once all 3 crew members were together aboard the CSM, the LM, which had been dubbed *Intrepid*, was sent crashing down to the moon. The CSM, dubbed *Yankee Clipper*, was brought out of lunar orbit the next day and headed back to earth.

The $350 million 10-day round trip to the moon was ended with a safe splash-down in the South Pacific Nov. 24.

The 2 *Apollo 12* astronauts who walked on the moon were Charles (Pete) Conrad Jr., 39, commander of the *Apollo 12* crew, and Alan La-

Vern Bean, 37, LM pilot. The 3d crew member, CSM pilot Richard Francis Gordon Jr., 40, stayed in lunar orbit on the *Yankee Clipper* while his 2 companions explored the moon. All 3 started the trip Nov. 14 with the rank of Navy commander. But Pres. Nixon, who had gone to Cape Kennedy to watch the launching, phoned them on their return to earth Nov. 24 to congratulate them and to inform them that he had promoted them all to the rank of captain.

Detailed plans for the mission had been made public long in advance, and the entire flight, from launching to splashdown, was covered "live" by TV, radio and the press. The astronauts transmitted TV reports to the public repeatedly throughout the mission and held a live televised press conference from their command module as the spaceship was returning to earth.

Apollo 12 had risen from Cape Kennedy's Launch Pad 39A at 11:22 a.m. EST Nov. 14 in the midst of a pelting rainstorm. Pres. and Mrs. Nixon and their daughter Tricia were among some 3,000 special invited guests who watched from a distance of about $3\frac{1}{2}$ miles as a 3-stage Saturn-5 booster rocket carried the spaceship up into the stormclouds.

About 36 seconds after *Apollo 12* had left the pad, power in the spaceship was cut off by overloads apparently caused by lightning. *Apollo 12*'s standby batteries automatically took over and kept most systems operating for the anxious minute or so that it took for the astronauts to reset switches and restore power from the 3 fuel cells. (Later Conrad and Bean went through connecting hatches into the 2-section LM to inspect it; they reported that, like the CSM, it had suffered no damage.)

The Saturn-5 carried *Apollo 12* into a 118-mile-high parking orbit around the earth 11 minutes 40 seconds after launching. During the 2d revolution around the earth, the booster rocket's 3d (S4B) stage was reignited for $5\frac{1}{2}$ minutes over the Pacific to increase the spaceship's speed from 17,400 mph. to 24,100 mph. and thus bring it out of orbit and into its 260,000-mile curving path toward the moon. The aim was so accurate that a potential course correction scheduled for later that day was canceled.

At 2:43 p.m. Gordon flicked a switch to separate the CSM from the S4B stage, in whose adapter the LM was encased. He then turned the CSM around and brought its docking probe into the LM's docking ring. After the CSM and LM were mated, the S4B was cast adrift and sent into orbit around the sun.

Apollo 12's course at this time was what was usually described as a "free return" trajectory because on this path, if the engines were not used, the spaceship would be pulled around the moon by lunar gravity and then flung back toward the earth. Such a path had been followed by the 3 manned Apollo spaceships that had preceded *Apollo 12* toward the moon.

At 6:15 p.m. EST Nov. 15, when the spaceship was 135,125 miles from the earth, Conrad fired the 20,500-pound-thrust service rocket for 9 seconds to take *Apollo 12* out of the "free return" trajectory and put it into a slower and riskier "hybrid" trajectory. The purpose of the change was to time the lunar landing to take place as the sun was rising over the Ocean of Storms landing site and while the Goldstone (Calif.) tracking antenna was in position to communicate with the spaceship. As a bonus, the trajectory change saved fuel. The course change was so accurate that potential corrections scheduled for Nov. 16 and 17 were canceled.

At 7:52 a.m. EST Nov. 17, when *Apollo 12* was 211,322 miles from earth and 38,944 miles from the moon, the spaceship escaped the dominance of the earth's gravity, and the moon's gravity became dominant. *Apollo 12*'s speed had been declining as it approached this point. Thereafter its speed increased as, in effect, it fell toward the moon.

Lunar gravity pulled *Apollo 12* into a curving path around the moon Nov. 17, and the spaceship passed behind the lunar disk at 10:32 p.m. EDT. While *Apollo 12* was behind the moon, Conrad fired the service rocket for 6 minutes to slow the spaceship's speed from 5,554 mph. to 3,584 mph. This deceleration put the spaceship at 10:47 p.m. into a lunar orbit with an apocynthion of 194 miles and a pericynthion of $67\frac{1}{2}$ miles. It emerged from behind the moon at 11:04 p.m., after 32 minutes in which there had been no radio contact with the earth, and the astronauts reported the successful achievement of a lunar orbit.

The orbit was made more circular at 3:06 a.m. EST Nov. 18 by means of a 17.6-second engine ignition that lowered the apocynthion to 74.6 miles and the pericynthion to 64.4 miles.

After circling the moon 10 times in orbit, Conrad and Bean crawled through the connecting hatches from *Yankee Clipper* to *Intrepid* shortly after 6 p.m. EST Nov. 18 to get *Intrepid* ready for the lunar landing. The CM and LM were then separated at 11:16 p.m.

As Conrad and Bean approached the lunar surface in *Intrepid,* the initial handling of the descent rocket was entrusted to the LM's onboard computer. At an altitude of 500 feet, however, Conrad took semimanual control of the landing vehicle and brought it down at 1:54 a.m. EST Nov. 19 exactly on target in the dry Ocean of Storms. They were then 233,450 miles from the earth. The landing site was about 600 feet from *Surveyor 3*, an unmanned U.S. probe that had soft-landed in a lunar crater in Apr. 1967. The site was 954 miles west of the spot in which *Apollo 11* had landed in the Sea of Tranquility. The coordinates of *Apollo 12*'s landing site were given as Latitude 30° 2′ 10″, Longitude 23° 25′ 5″.

Conrad reported to the control center in Houston that the landing was a difficult one, at least partly because of the dust stirred up by the landing rocket, which fired downward. "It's a good thing we leveled off high," he said later, "... because I sure couldn't see what was underneath us once I got into that dust."

Gordon, alone in *Yankee Clipper* as it passed in orbit later Nov. 19 above the landing site, reported that he could see both *Intrepid* and *Surveyor 3* through a sextant.

With Conrad and Bean clad in their bulky but protective space suits, the cabin of *Intrepid* was depressurized and the exit hatch opened. Starting the first of the mission's 2 moon walks, Conrad backed out of the hatch at 6:39 a.m. EST Nov. 19 and climbed gingerly down the 9-rung ladder.

Conrad, 6 inches shorter than Neil Armstrong (the first man to step on the moon), finally placed a booted foot on the lunar surface at 6:45 a.m. Commenting on how much harder it was for him than for Armstrong to make the final long step from the bottom rung to the moon, Conrad said: "Whoopie! Man, that may have been a small one for Neil, but that's a long one for me." Bean, following Conrad down the ladder nearly a half-hour later, stepped out on the moon at 7:14 a.m.

During their first EVA (extra-vehicular activity) on the moon, the 2 astronauts experimented with walking, collected rock samples and set up a variety of experiments.

A major disappointment of the moon walks was the failure of their color TV camera. TV transmission had started at 6:41 a.m. Nov. 19 as Conrad descended the ladder. It continued as Bean joined him but stopped at 7:25 a.m., presumably because the camera lens may have been pointed accidentally into the sun and its phosphorus-coated image tube damaged.

During their 4-hour one-minute first moon walk, the astronauts ranged as far as 1,300 feet from *Intrepid,* or about 5 times as far as the *Apollo 11* crew members had gone from their LM. They moved about 600 feet from *Intrepid* to set out a magnetometer, a lunar ionosphere detector, a lunar atmosphere detector, a seismometer and a solar wind spectrometer. These instruments formed the permanent $25 million lunar scientific station labeled ALSEP 1 (for Apollo lunar surface experiments package). They were powered by a SNAP 27 (for systems for nuclear auxiliary power) atomic generator. SNAP 27's fuel was 7 pounds of radioactive plutonium-238. As soon as the generator was set up and the instruments plugged in, the radio monitoring the instruments began transmitting data to earth.

After a busy 4 hours on the lunar surface, Conrad and Bean were back in the LM by 10:31 a.m., when they closed the hatch and repressurized the cabin.

Conrad stepped down on the lunar surface again at 11:01 p.m. Nov. 19 to start the mission's 2d moon walk. On this excursion, he and Bean clambered 155 feet down into the crater in which *Surveyor 3* had landed. After photographing *Surveyor 3* from all angles, they removed various items from the probe—including its TV camera—for study back on earth. Conrad reported that the Surveyor—originally mostly white—had turned "brown," that "some of it's even a reddish color." "It looks like something's rained on it," he said. The 2 astronauts also collected additional rocks and core samples drilled out of the lunar surface.

Returning to *Intrepid*, the astronauts retrieved a solar wind collector, loaded their samples aboard the LM and were back inside shortly before 3 a.m. Nov. 20.

After 31 hours 31 minutes on the moon, Conrad and Bean fired the ascent engine of their lunar module at 9:25 a.m. Nov. 20 and rose from the moon's surface to lunar orbit. They left on the moon *Intrepid*'s 4-legged descent section, the ALSEP 1 scientific station with its SNAP 27 atomic generator, an unfurled American flag, a plaque inscribed "Apollo 12 November 1969" and bearing the signatures of the 3 *Apollo 12* astronauts, and various items of superfluous equipment such as their backpacks, lunar boots, cameras, tools and the hammocks they had slept in aboard *Intrepid*. They brought back with them about 75 pounds of lunar surface samples (in 2 boxes and a bag), various items taken from *Surveyor 3*, films of the scenes they had photographed on the moon and miniature flags (which had been brought with them on the trip) of 136 nations, the 50 U.S. states and 4 U.S. possessions.

After a series of carefully computed orbital maneuvers, *Intrepid* docked with the orbiting command module *Yankee Clipper* at 12:58 p.m. Nov. 20 while the CM was making its 31st revolution around the moon. With the 2 vehicles locked tightly together, the hatches between *Intrepid* and *Yankee Clipper* were opened. Conrad and Bean then crawled with their lunar dust-covered moon mementos through the hatches into the CM, where they rejoined Gordon.

At 4:45 p.m. ground controllers at the Manned Spacecraft Center in Houston transmitted a radio command that sent *Intrepid* back down toward the moon. The LM crashed into the Ocean of Storms at 5:17 p.m. about 40 miles south of its initial landing site. The crash was detected by the seismometer left on the moon, and scientists monitoring the impact reported at the Houston center that the reverberations lasted for more than a half-hour. Dr. Maurice E. Ewing, director of Columbia University's Lamont-Doherty Geological Observatory, told reporters in Houston that "it is as though someone struck a bell in the belfry of a church. ... We're not sure what this means." Dr. Frank Press of MIT added that "we've never seen anything like it on earth."

After a day spent photographing the moon from orbit to get additional data on future Apollo landing sites, the astronauts ignited the *Apollo 12* service rocket at 3:49 p.m. Nov. 21 to bring the spaceship out of orbit and head it back to earth. The 2-minute 10-second engine burn, started as *Yankee Clipper* completed its 44th revolution around the moon, increased the vehicle's speed from 3,400 mph. to about 5,450 mph. It took place while the CSM was behind the moon and out of radio contact with earth.

A CSM maneuvering rocket was fired at 6:49 a.m. Nov. 22 to make a minor course correction as *Yankee Clipper* hurled earthward with ever-increasing speed. The revised trajectory was so accurate that a potential additional correction scheduled for Nov. 23 was canceled. During the return journey the astronauts saw a sight never seen by men—an eclipse of the sun by the earth.

Preparing for re-entry Nov. 24, the *Apollo 12* astronauts jettisoned the service module and then turned the command module so that its heatshielded blunt end faced the direction of flight.

Yankee Clipper smashed into the atmosphere at an altitude of 400,000 feet and a speed of more than 24,000 mph. some 1,300 miles west of its South Pacific target point. Friction with the atmosphere scorched the surface of the spaceship and reduced its speed to about 168 mph. 3 small drogue parachutes were deployed to stabilize the CM, and the main parachutes then lowered the spaceship the rest of the way.

Yankee Clipper splashed down in the South Pacific at 3:58 p.m. EST Nov. 24 some 404 miles southeast of Pago Pago and 2,651 miles south-southwest of Honolulu. It plunged into the water only 2.6 miles from its aiming point and about 3 miles from the waiting recovery carrier *Hornet*.

The module capsized almost immediately on entering the water, but the astronauts righted it by inflating flotation bags. Swimmers threw flight suits and face masks to the astronauts after the hatch was opened, and the astronauts donned them as required under quarantine procedure. They then crawled out onto a life raft, were hoisted aboard a helicopter and were flown to the *Hornet*. On the *Hornet* they immediately entered a special quarantine trailer in which they later traveled to the quarantine facilities (the Lunar Receiving Laboratory) at the Manned Spacecraft Center in Houston. Initial medical exams showed all 3 astronauts to be in excellent health. The CM was also hauled aboard the *Hornet* and placed in quarantine.

Moon Rocks Differ in Age

The lunar rock and soil brought back from the moon's Ocean of Storms by the *Apollo 12* astronauts was formed 2.2 billion to $2\frac{1}{2}$ billion years ago, it was reported in Houston, Tex. Dec. 12 at a press conference of scientists who performed preliminary analyses of the moon material. The lunar rock brought back from the Sea of Tranquility by the *Apollo 11* astronauts had been tentatively dated as 3.8 billion to 4.6 billion years old. The Dec. 12 report was made by Dr. Oliver A. Schaeffer, a physicist of the State University of New York at Stony Brook, who used the potassium-argon dating method.

Dr. S. Ross Taylor, a geochemist of the Australian National University, said that chemically the rocks from the 2 areas were similar, like "first cousins," but not identical. The Ocean of Storms material had about half the titanium of the Sea of Tranquility material, fewer refractory elements (such high-melting-point ones as zinc and yttrium) and fewer volatile (low-melting-point) elements.

Dr. Palmer Dyal, who monitored the magnetometer left on the moon by the *Apollo 12* astronauts, had revealed in Houston Nov. 25 that the moon's magnetic field, while relatively puny (30 to 40 gamma, according to various measurements made by Dyal), is many times greater than had been predicted on the basis of previous observations. Some scientists had expected it to be no more than 2 gamma. (The earth has a magnetic field of about 30,000 gamma.)

OTHER MEN IN SPACE

Soviets Dock Manned Ships

2 manned Soviet Soyuz spaceships were linked together in orbit Jan. 16, and 2 Soviet cosmonauts, dressed in space suits, crawled from one spaceship into the other. The spaceships then separated and landed safely Jan. 17 and 18.

This was the first docking of manned vehicles in space and the first transfer of personnel between orbiting spaceships. The 2 cosmonauts who made the transfer were the first men to go into space in one vehicle and to return to earth in another. The experiment was hailed in the USSR and elsewhere as an important step toward assembling permanent space stations in orbit. The name of the spaceship series—Soyuz, meaning "union"—was considered an indication of the mission's purpose.

The mission started at the Baikonur cosmodrome in Kazakhstan, Soviet Central Asia, where the spaceship *Soyuz 4* was launched Jan. 14. With Lt. Col. Vladimir Aleksandrovich Shatalov, 41, aboard as the sole occupant, *Soyuz 4* was shot into space at 10:39 a.m. (Moscow time) atop what appeared to be a 3-stage carrier rocket. *Soyuz 4* quickly achieved an orbit with an initial apogee of 225 kilometers (139.8 miles), perigee of 173 kilometers ($107\frac{1}{2}$ miles), 88.25-minute period of revolution and 51° 40' of inclination to the equatorial plane. During his 4th and 5th revolutions around the earth, Shatalov fired his vernier engine to raise the apogee to 237 kilometers ($147\frac{1}{4}$ miles), the perigee to 207 kilometers (128.6 miles) and the period to 88.75 minutes.

The mission's 2d spaceship, *Soyuz 5,* was shot into space from the Baikonur cosmodrome Jan. 15 at 10:14 a.m. (Moscow time). It carried 3 cosmonauts—Lt. Col. Boris Valentinovich Volynov, 34, the spaceship commander, Aleksei Stanislavovich Yeliseyev, 34, civilian flight engineer, and Lt. Col. Yevgeni Vasilyevich Khrunov, 35, research engineer. *Soyuz 5* went into an orbit with an initial apogee of 230 kilometers (142.9 miles), perigee of 200 kilometers ($124\frac{1}{4}$ miles), period of 88.7 minutes and inclination of 51° 40'. During the 6th revolution, Volynov fired his vernier engine to raise the apogee to 253 kilometers (157.2 miles), the perigee to 211 kilometers (131.1 miles) and the period to 88.92 minutes.

The spaceships made visual contact with each other Jan. 15 and were in radio contact as well.

Before the docking attempt started, Shatalov revised the orbit of *Soyuz 4* Jan. 16 at 10:22 a.m. (Moscow time) during the spaceship's 32d revolution around the earth. The apogee was raised to 253 kilometers (157.2 miles), the perigee lowered to 201 kilometers (124.9 miles) and the period lengthened to 88.85 minutes.

After final checking of all equipment and approval from ground controllers, Tass reported, the "automatic closing-in of the spaceships was started" at 10:37 a.m. (Moscow time) Jan. 16. The distance between the 2 craft was quickly reduced to 100 meters (328 feet).

Shatalov then took over the manual controls of *Soyuz 4* and steered it to physical contact with *Soyuz 5.* This joining of the 2 spaceships, nose to nose, took place at 11:20 a.m. Jan. 16 while *Soyuz 4* was making its 34th revolution around the earth and *Soyuz 5* was on its 18th revolution. Tass reported that "the docking was followed by the mechanical engagement of both craft, their rigid bracing and joining of electric circuits." "This was the first time that an experimental space station with 4 compartments for the crew . . . was assembled and started functioning in space," Tass said. The jubilant tape-recorded conversation between the 2 spaceship commanders on the achievement of the first manned docking included an exchange in which Shatalov said: "At last I've found you. . . . I've been hunting for you for quite some time." Volynov responded: "You raped me."

With the spaceships firmly linked together, Khrunov and Yeliseyev donned their space suits. They pulled themselves out of *Soyuz 5* and into space through the orbital compartment's hatch while their spaceship was making its 19th revolution. Using handrails to move about, they performed an hour of experimental work in space. They then entered *Soyuz 4* through the hatch of its orbital compartment, removed their space suits and took their places as new members of the *Soyuz 4* crew. The "walk-over" took place while the 4-compartment "space station" was in an orbit with an apogee of 250 kilometers (155.3 miles), perigee of 209 kilometers (129.9 miles), period of 88.85 minutes and inclination of 51° 50'.

The spaceships were uncoupled at 3:55 p.m. (Moscow time) Jan. 16 and continued in orbit independently.

The Soviet Union had given no advance notice that either of the spaceships were to be launched, and it reported their achievements only after they had taken place. It was noted, however, that these reports after the event were made with unusual speed. A videotape of *Soyuz 4*'s launching was shown on Moscow TV perhaps an hour after the spaceship had gone successfully into orbit. Similar speed was noted in news agency and TV coverage of the rest of the mission. Videotapes were shown of activities inside the 2 spaceships, of the docking, of the transfer of the 2 cosmonauts from *Soyuz 5* to *Soyuz 4* and of the separation of the spaceships.

Soyuz 4, with its augmented 3-man crew, was brought out of orbit Jan. 17 and landed safely at 9:53 a.m. (Moscow time) in Kazakhstan about 25 miles northwest of the city of Karaganda. *Soyuz 5,* with Volynov remaining as its sole occupant, came out of orbit Jan. 18 and landed safely at 11 a.m. in Kazakhstan about 125 miles southwest of the town of Kustanai. In each case, the retro-rocket was fired to bring the spaceship out of orbit, the orbital compartment was then jettisoned, and the module-descending canister, with its strapped-down crew inside, smashed into the atmosphere. The first part of the return trip was an aerodynamic flight, and the spaceship was then lowered to the ground by parachute. Waiting helicopters quickly picked up the cosmonauts.

(Sir Bernard Lovell, director of Britain's Jodrell Bank radiotelescope observatory, said Jan. 16 that the manned Soyuz docking showed the USSR to have a "4-year lead" over the U.S. in techniques for assembling equipment in space. If the Soviets continued to gain at this rate, he declared, "my guess is that the Americans will be looking rather small fry by the middle-1970s." He predicted that by 1974-5 the Russians would be using the moon to assemble observatories and spacecraft for planetary flights.)

7 Cosmonauts Fly 'Cosmic Troika'

7 Soviet cosmonauts were sent into space in 3 precision launchings from the Baikonur cosmodrome Oct. 11, 12 and 13. After circling the earth 80 times each in similar orbits, the cosmonauts and their 3 spaceships—*Soyuz 6*, *Soyuz 7* and *Soyuz 8*—were brought down safely about 100 miles from Karaganda, Kazakhstan Oct. 16, 17 and 18.

The "cosmic troika," as the Soviet press dubbed the mission, set records for the number of men and the number of manned vehicles in space simultaneously. But despite reports that a goal of the flight had been to link at least 2 of the spaceships together to form an orbiting space station, no attempt at such a feat apparently was made.

The crews of the 3 spaceships were: *Soyuz 6*—Lt. Col. Georgi Stepanovich Shonin, 34, Ukrainian, *Soyuz 6*'s command pilot; Valeri Nikolayevich Kubasov, 34, a civilian, flight engineer. *Soyuz 7*—Lt. Col. Anatoly Vasilievich Filipchenko, 41, command pilot; Vladislav Nikolayevich Volkov, 33, a civilian, flight engineer; Lt. Col. Victor Vasilievich Gorbatko, 34, research engineer. *Soyuz 8*—Col. Vladimir Aleksandrovich Shatalov, 41, commander of the 3-ship group flight and command pilot of *Soyuz 8;* Dr. Aleksei Stanislavovich Yeliseyev, 35, a civilian, flight engineer.

The only cosmonauts of the 7 to have flown in space previously were Shatalov and Yeliseyev, who had participated in January in the joint flight in which *Soyuz 4* and *Soyuz 5* had achieved the USSR's first docking of manned vehicles in space. Shatalov and Yeliseyev were the only Soviet cosmonauts to survive 2 space flights. (The only other Soviet cosmonaut to make 2 space flights, Col. Vladimir Komarov, was killed in Apr. 1967 as he completed the 2d flight.)

Soyuz 6 was launched from a rain-drenched pad at the Baikonur cosmodrome Oct. 11 at 2:10 p.m. Moscow time. Its initial orbit had an apogee given as 139 miles, perigee of $115\frac{1}{2}$ miles, period of 88.36 minutes and $51.7°$ angle of inclination. *Soyuz 7* was launched Oct. 12 at 1:45 p.m. into an orbit with the same angle of inclination, 88.6-minute period, apogee of $140\frac{1}{2}$ miles and perigee of $128\frac{1}{2}$ miles. By this time the apogee of *Soyuz 6*'s orbit was 143 miles, its perigee $120\frac{1}{2}$ miles and its period 88.6 minutes. *Soyuz 8,* launched Oct. 13 at 1:29 p.m., achieved an orbit with an identical period and angle of inclination, apogee of $138\frac{1}{2}$ miles and perigee of $127\frac{1}{2}$ miles. Repeated alterations in apogee and perigee were made in the orbits of all 3 spaceships during the flight, and Tass reported later that 31 such maneuvers were performed during the mission.

The re-entry section of *Soyuz 6*, with its 2 occupants, landed 112 miles northwest of Karaganda Oct. 16 at 12:52 p.m. Moscow time. *Soyuz 7*'s descent section and its 3 crew members came down 96 miles northwest of Karaganda Oct. 17 at 12:26 p.m., and *Soyuz 8*'s re-entry vehicle, with its 2 occupants, landed about 90 miles north of Karaganda Oct. 18 at 12:10 p.m. The orbital modules of the 3 spaceships had been jettisoned in space before the cosmonaut-carrying reentry vehicles came down. All cosmonauts were reported to have survived the flight in excellent health.

As usual in Soviet space missions, there had been no advance announcement that the launchings were planned. Each launching was disclosed after the spaceship involved was successfully in orbit, and there was no official hint that another ship would join the mission. Soviet authorities withheld all information about the specific goals of the flight. As a result, much of the Western speculation and reporting on the mission was based on what appeared to be leaks to Western newsmen through "informed sources" and on Soviet newspaper and magazine articles written ostensibly about general Soviet goals and plans but apparently based on the current mission.

Heinz Kaminski, director of West Germany's Space Research Institute at Bochum, had predicted Oct. 10 that the USSR would put a space station manned by at least 6 scientists into earth orbit before the end of 1969. He reported that recent Soviet space activity appeared directed to such a feat. UPI correspondent Henry Shapiro reported in Moscow the same day that, according to informed sources, the USSR planned to launch the first of at least 3 manned Soyuz-class spaceships the weekend of Oct. 11-12. He said they were to be used to build an orbiting space station from which Soviet spaceships could be sent to the moon and planets.

In its first announcement of the mission, made after *Soyuz 6* had gone into orbit, Tass disclosed Oct. 11 that the spaceship carried welding equipment, and Tass emphasized the "great importance" of experimenting with welding metals in space. Western observers agreed that welding might be the best method of assembling a space station in orbit. The actual welding experiment was carried out Oct. 16 during the spaceship's 77th revolution. While the cosmonauts, who flew the entire mission without donning spacesuits, remained in one section of the spaceship, the other section, containing the welding equipment, was depressurized so that the equipment was in an airless space environment. Kubasov then activated the automatic welding equipment by remote control to attempt to weld metals by 3 methods—plasma, electron-beam and arc welding. It appeared that only the electron-beam method worked.

Soviet TV science commentator Aleksei F. Konstantinov had appeared to cast doubt on the rumors of a space-station role for the *Soyuz 6-8* mission by reporting at the launching site Oct. 11 that *Soyuz 6* had no "device to secure docking," although it had "other systems." Kubasov added in a live telecast from *Soyuz 6* Oct. 12 that his spacecraft also lacked "automatic systems for maneuvering close" to other spaceships. Western observers pointed out later that the orbits were so low that the Soviet spaceships could not remain in them for much more than 2 weeks. It was also reported that the Soyuz spacecraft, weighing an estimated 14,000 pounds each on launching, had a life-support capability of only 10 days for a 3-man crew and of no more than 20 days with 2 men aboard. In a pre-launching interview, Yeliseyev said Oct. 13 that the mission's experiments would lead to the "creation of even larger orbiting stations," but he added that there "will be no need" to transfer crewmen between spaceships in the current mission.

The never-named "chief designer" (the official in overall control) of the Soyuz program said in an interview in *Pravda* Oct. 15: The object of the *Soyuz 6-8* flight was the "further improving [of] the spaceship systems and methods of their control. Controlling a group of spacecraft is far more complicated than [controlling] a single spaceship. This requires well-planned organization, which should be preceded by efficient training of the entire personnel of recording stations. This is probably the main task of the flight. The time is not far off when there will be much larger groups of vehicles ... in space."

Soviet reports emphasized the amount of manual operation and control required of the cosmonauts during the flight. They indicated that Soviet space planners foresaw a need for more dependence on cosmonauts and less on computer-backed ground controllers in future space operations. (They also reported a considerable amount of biomedical checking conducted by the cosmonauts on themselves.) Tass quoted the "chief designer" Oct. 19 as saying that the mission "made a start on the comprehensive exploration of space with the direct participation of man." *Pravda* quoted the never-named "deputy chief designer" as reporting that "close" approaches (to within 500 yards) made by *Soyuz 7* and *8*, while *Soyuz 6* observed from a distance, were made manually, in some cases with instructions from the ground control center and in some cases with calculations and observations made by the cosmonauts without ground aid.

USSR Delays Moon Landing

Prof. Mstislav V. Keldysh, chairman of the Soviet Academy of Sciences, indicated to newsmen in Stockholm Oct. 24 that the USSR had indefinitely postponed any attempt to fly men to the moon. "At the moment we are concentrating wholly on the creation of large [earth] satellite stations," he said. "We no longer have any scheduled plans for manned lunar flights."

Asked at a Moscow press conference whether the USSR intended to fly men to the moon, Keldysh replied Nov. 4: "I can only say that such operations are not planned for the coming months." He said he expected that the first permanent manned Soviet station would be put into orbit around the earth "certainly within 10 years, and I think less than 5 years, anyway, literally in the nearest future." He hinted that the main question to be answered was whether men would be harmed by spending months in space. "There is a whole lot of evidence that weightlessness does have harmful effects," he reported.

(Soviet space scientist V. V. Parin said in an inverview in *Komsomolskaya Pravda* Nov. 22 that space flight "reduces the flexibility of men's bones." He explained that "the organism of a cosmonaut dehydrates in weightlessness, and calcium leaves the bones.")

Dr. Boris N. Petrov, a member of the Soviet Academy of Sciences, indicated in an apparently authoritative article in *Pravda* Dec. 30 that the high cost of manned lunar flight was a major reason that the USSR had indefinitely postponed efforts to fly men to the moon. "Our program by no means excludes manned flight to the moon," he declared, "but at the present time we attach prime importance to lunar exploration by unmanned vehicles. The economic side is of no small importance. Unmanned vehicles are many times less expensive than manned ones."

LUNAR & INTERPLANETARY PROBES

2 Soviet Probes Reach Venus

2 unmanned Soviet spacecraft, *Venera (Venus)* 5 and *Venera 6,* reached the planet Venus May 16 and May 17, respectively. *Venera 5,* which had been launched Jan. 5, and *Venera 6,* launched Jan. 10, traveled about 350 million kilometers ($217\frac{1}{2}$ million miles) each on curving trajectories in more than 4 months. Soviet spokesmen described the flights as highly "successful" and reported that they had produced large amounts of scientific data. Each probe carried a bas-relief of Lenin and the USSR coat-of-arms to Venus.

In each case the instrument-carrying descent capsule, protected by a heat shield, was ejected from its space "station" before it plunged into Venus' hot, cloudy atmosphere. The 2 probes reached Venus about 300 kilometers (186 miles) apart. They entered on the "night" side of the planet so that their radio antennas could point directly back to earth. On entry, the speed of the capsules was first slowed by aerodynamic braking from 11.17 kilometers (7 miles) a second to 210 meters (689 feet) a second. The capsules were then eased down by parachute; *Venera 5* took 53 minutes and *Venera 6* took 51 minutes for the parachute descent.

Tass reported that the entry procedure for *Venera 5* began at 7:08 a.m. (Moscow time) May 16 when the probe, 50,000 kilometers (31,000 miles) from Venus, responded to a radio command from earth and "automatically jettisoned" its instrument capsule. At about 9:01 a.m. (Moscow time) and a distance of about 100 kilometers (62 miles) from the planet's surface, the capsule entered Venus' atmosphere. Tass said that the searing aerodynamic deceleration to which the capsule was subjected "was accompanied by a sharp increase in overloads and a considerable rise of temperature on the outer surface of the apparatus." During the parachute descent, Tass reported, the probe's instruments "measured the temperature, pressure and chemical composition of the atmosphere," and "an altimeter determined the height of the craft over the planet's surface. The on-board radio complex ensured uninterrupted transmission of these measurements." (Britain's Jodrell Bank radio-astronomy observatory reported that it had started picking up the descending capsule's signals at 9:20 a.m. [Moscow time] but had lost the signals at 9:55 a.m.)

The USSR's space communications center indicated that *Venera 6* had entered Venus' atmosphere at 9:02 a.m. (Moscow time) May 17. Tass had predicted May 16 that the capsule would enter at 9:03 a.m.

No Soviet spokesman said anything about "soft" landings or about receiving signals from either capsule after it had touched down on the surface of Venus.

Pravda May 17 quoted the "chief designer" of the Venera spacecraft as saying that "the chief goal of the experiment, the descent of 2 stations into the atmosphere of Venus, was attained." *Pravda* quoted him May 18 as saying "man will never go" to Venus because data from the Venera probes "make it clear that the planet is not fit for man's life." He said the flight proved that Venera-class spacecraft could operate much closer to the sun than previous probes. He predicted that in future Soviet space shots, "Jupiter and Mars will be approached by automatic process."

The Soviet Academy of Sciences confirmed June 3 that all of the 3 Soviet probes that had plunged into Venus' dense atmosphere had stopped radio transmissions while still high above the planet's surface. (The capsule of *Venera 4* had entered the Venusian atmosphere in Oct. 1967.) According to a release by Tass, the Soviet news agency, Soviet scientists presumed that the 405-kilogram (893-pound) capsules had been crushed by the pressure of the heavy Venusian atmosphere and that the radio transmitters had thus been destroyed.

Data from *Venera 5* and *6* indicated that Venus had mountains rearing up more than 50,000 feet above the Venusian valleys. U.S. scientists said the observations apparently were made at altitudes of 34 to 12 miles above the planet's surface. The atmospheric pressure on the Venusian surface under *Venera 5* was given as 140 times that on earth, and the temperature was reported to be nearly $1,000°F$. Under *Venera 6*, the surface atmospheric pressure was given as 60 times that on earth, the temperature as $750°F$. Soviet scientists speculated that *Venera 5* had come down over a deep basin, *Venera 6* over a plateau. *Venera 5* and *6* reported that the Venusian atmosphere was 93%–97% carbon dioxide and less than .4% oxygen.

Luna 15's Purpose Not Explained

The Soviet's unmanned *Luna 15* probe was launched from the Baikonur cosmodrome in Kazakhstan July 13, 3 days before *Apollo 11* began the flight culminating in the landing of the first men on the moon. *Luna 15*'s timing led to speculation that the USSR was trying to deflect world attention from the U.S. feat.

Luna 15 was launched at 5:55 a.m. Moscow time July 13 and put into a parking orbit around the earth. The probe was then accelerated to escape velocity and placed in a trajectory toward the moon. A course correction was made July 14. After 102 hours of flight, *Luna 15* was sent into orbit around the moon at 1 p.m. Moscow time July 17 by means of a retro-firing of its rockets.

Luna 15's initial lunar orbit had an apocynthion (high point) of $126\frac{1}{2}$ miles, a pericynthion (low point) of $34\frac{1}{2}$ miles, a 2-hour 30-second period of revolution and a $127°$ angle of inclination to the plane of the moon's equator. But its orbit was changed July 19 and again July 20. The 2d change, made at 5:16 p.m. Moscow time July 20, just before *Apollo 11*'s landing craft headed down from the command and service module toward the moon, put *Luna 15* into a course that crossed *Apollo 11*'s path.

Tass said that the orbit *Luna 15* assumed July 19 had an apocynthion of 221 kilometers ($137\frac{1}{3}$ miles), pericynthion of 95 kilometers (59 miles), 2-hour $3\frac{1}{2}$-minute period and $126°$ angle of inclination. The orbit assumed July 20, however, had an apocynthion given as 110 kilometers ($68\frac{1}{3}$ miles), pericynthion of only 16 kilometers (not quite 10 miles), one-hour 54-minute period and $127°$ angle of inclination. Western observers said that *Luna 15*'s orbit could take the probe roughly over the *Apollo 11* landing site.

Luna 15 crashed into the moon at about 6:51 p.m. Moscow time after circling the moon 52 times in orbit. Tass merely announced that "at 18 hours 47 minutes [6:47 p.m.] Moscow time ... a retro-rocket was switched on, and the station [*Luna 15*] left the orbit and reached the moon's surface in the present area." Tass said that *Luna 15*'s work had been "completed." Sir Bernard Lovell, director of Britain's Jodrell Bank radio-telescope observatory, said that *Luna 15* had probably come down in the dry Sea of Crises, about 500 miles from the *Apollo 11* landing site in the Sea of Tranquility. He said signals from *Luna 15* had ended when the craft reached the lunar surface. It was estimated that *Luna 15* might have crashed into the moon at a speed of perhaps 300 mph.

The voyage of *Luna 15* produced not only suspicion that the Russians were trying to steal the world spotlight from *Apollo 11* but also some fear that the Soviet probe might interfere with the U.S. manned landing.

Col. Frank Borman, U.S. astronaut who had visited the USSR earlier in July, phoned Moscow July 18, therefore, to inquire about *Luna 15.* Prof. Mstislav Vsevolodovich Keldysh, president of the Soviet Academy of Sciences, replied by cablegram the same day. He gave the parameters of the probe's initial orbit and said: "It is supposed that in this orbit *Luna 15* will remain for 2 days. In case of further changes in the orbit ... you will receive additional information. The orbit of probe *Luna 15* does not intersect the trajectory of *Apollo 11* spacecraft announced by you in flight program." When *Luna 15* shifted orbit July 19 and 20, Keldysh cabled the new data to Borman. (The U.S. and USSR had agreed, in the UN space treaty of 1967, to provide data of this sort, but this was the first instance in which the Russians had given the U.S. direct information about a Soviet spacecraft while it was in flight.)

Keldysh's messages provided no information about *Luna 15*'s mission. And the Soviet Union, in keeping with its usual practice, had neither announced its *Luna 15* plans in advance nor given any material facts about the probe's purpose during or after the flight. Western speculation about *Luna 15*, therefore, was based on rumors, possible information leaks and hints given by Soviet cosmonauts, scientists, officials and writers.

A Reuters dispatch from Moscow Mar. 14 had cited "usually reliable sources" as asserting that the USSR was planning a Mar. 16 shot—which might be delayed because of bad weather—to send an unmanned spaceship to the moon and back.

Lt. Col. Aleksei A. Leonov, a soviet cosmonaut, told Japanese newsmen in Moscow that the Russians planned to use an unmanned spaceship to get lunar soil in time to exhibit it at Japan's Expo 70 world's fair in Osaka in 1970.

According to Western reports, based on information from unnamed "Communist sources," unsuccessful attempts at such a mission had been made by the Soviet Union in unreported launchings from the Baikonur cosmodrome early in April and June 14. The first rocket reportedly was destroyed on the launching pad, and the 2d blew up in the air, the reports said. The Communist sources used as authority for these reports were cited in Moscow July 2 as saying also that a 3d Soviet attempt was planned for July 10.

In its vaguely worded announcement of the *Luna 15* launching July 13, Tass said that "the aim of the flight is to check the systems on board the automatic station [*Luna 15*] and to conduct further scientific exploration of the moon and space near the moon."

U.S. astronaut Borman said July 13, however, that Soviet space experts he had met in Moscow "had made references" to an early Soviet mission to bring lunar soil to the earth.

Sir Bernard Lovell said at the Jodrell Bank observatory July 15 and 17 that he considered it likely that *Luna 15*'s mission was to bring back lunar soil. "If the Russians intend to put *Luna 15* into orbit and just leave it there, the whole operation is incomprehensible," he declared July 17. Lovell speculated July 18 that *Luna 15* might have dropped a capsule secretly to the back side of the moon while the probe was out of range of radio observation from the earth.

Many U.S. and other Western space "experts" held throughout the *Luna 15* flight, however, that *Luna 15* did not have the capability of landing on the moon and returning to earth.

Much of the time that *Luna 15* was in flight the Soviet Union remained completely silent about it. Tass said nothing about its probe between the initial announcement of the launching July 13 and the announcement July 17 that *Luna 15* had become "yet another artificial satellite of the moon." In its announcement of *Luna 15*'s demise July 21, Tass said that unlike the USSR's previous Luna orbiting probes, *Luna 15* "can land in various areas of the lunar surface."

U.S. Nobel Prize-winning lunar expert Harold C. Urey, a NASA consultant, commented in Washington July 21 on the Soviet practice of withholding virtually all information about planned space missions and those in progress: "I am so proud of the United States for telling the world exactly what we tried to do," he declared. "It shows an enormous self-confidence in our effort. If it fails, we say it did. I do wish our Russian friends would come out and tell us what they are trying to do."

2 U.S. Probes Pass Mars

The unmanned U.S. space probes *Mariner 6* and *Mariner 7* flew past Mars July 31 and Aug. 5 at distances as close as 2,132 miles and 2,130 miles, respectively. They sent back close-up photos of the planet of unprecedented clarity and radioed back masses of scientific findings that led scientists to assert at a Sept. 11 press conference that Mars was probably devoid of life.

The Sept. 11 press briefing was held in Washington by the Jet Propulsion Laboratory (JPL) of Pasadena, Calif., which conducted the Mariner program for NASA. Dr. Norman H. Horowitz, California Institute of Technology (Caltech) biologist, summarizing some of the conclusions, said at the Sept. 11 session that Mars appeared to be "a cold desert by earth standards." None of the data provided by *Mariner 6* and 7 "encourages the belief that Mars is an abode of life," Horowitz declared.

Mariner 6, launched from Cape Kennedy Feb. 24, flew 241 million miles in 156 days and was $59\frac{1}{2}$ million miles from the earth July 31 when it came to within 2,132 miles of the Martian surface. *Mariner 7,* launched from Cape Kennedy Mar. 27, traveled 197 million miles in 130 days and was 61.8 million miles from the earth Aug. 5 when it flew within 2,130 miles of Mars. After passing Mars, the 2 probes continued on in orbit around the sun.

The 2 probes investigated overlapping areas on Mars although *Mariner 6* concentrated on the equatorial region while *Mariner 7*'s area was the southern hemisphere and the south polar cap. Together they provided 198 photos that included far-approach pictures of the whole plane and close-up shots of 20% of the Martian surface.

Mariner 6 began its photographic mission July 29 with the transmission of a far-approach picture of Mars taken at a distance of 770,000 miles. Each photo thereafter provided more detail as the probe came closer to the planet.

Radio contact with *Mariner 7* was lost for 7 hours July 30–31. JPL scientists speculated that the probe had been hit by a meteoroid and thrown out of orientation, its directional antenna pointed away from the earth. The presumed impact also appeared to have caused garbled transmission on about 20 of the probe's 92 telemetry channels. Signals from earth restored contact and reoriented the probe so that its solar panels were again collecting power from the sun, its navigational system was again locked on the guide star Canopus and its directional antenna again pointed towards the earth for maximum efficiency. The mishap did not damage *Mariner 7*'s photographic abilities, and the probe sent back high-quality photos of Mars beginning Aug. 2 at a distance of 1,054,000 miles from the planet.

At the Sept. 11 press conference, Dr. George C. Pimental, chemistry professor at the University of California (Berkeley) announced that he had been mistaken Aug. 7 in reporting the presence on Mars of methane and ammonia—2 gases associated with life on earth. Pimental had made his Aug. 7 announcement at a JPL press briefing in Pasadena. He said Aug. 7 that he and an associate, Dr. Kenneth C. Herr, after careful infra-red spectrometer analysis, were "confident that we have detected gaseous methane and gaseous ammonia" in the Martian atmosphere and "cannot restrain the speculation that it might be of biologic origin." He also reported data suggesting that the Martian polar cap "is composed of water ice" and that "solid carbon dioxide ... is suspended as a cloud over the polar cap," providing "protection from ultra-violet radiation" in a "hospitable region" "near the edge of the polar cap, [where] polar ice provides a reservoir of water." Confessing that he was wrong in reporting methane and ammonia in the Martian atmosphere, Pimental explained Sept. 11 that the infra-red spectrometer lines that had appeared to indicate these 2 life-associated gases actually had been the similar-appearing lines of solid carbon dioxide.

Pimental's conclusion that the polar cap was made of water ice was challenged by several scientists, most of whom said that various indicators showed the cap to be frozen carbon dioxide (dry ice). Dr. Robert B. Leighton, Caltech physicist in charge of Mariner TV operations, had said at an Aug. 5 press briefing in Pasadena that Mars' south polar cap appeared to be made of "a great thick layer" of frozen carbon dioxide. Some of Mars' many craters seemed to be partially or completely filled with solid carbon dioxide, he reported.

In his Aug. 7 report Pimental said his instruments had indicated a surface temperature of $-73°$F. at the edge of the polar cap. This temperature would be too high for solid carbon dioxide (since carbon dioxide needs a temperature of $-253°$ to freeze at the low surface pressure on Mars). But Dr. Gerry Naugebauer of Caltech, who had used an infra-red radiometer specifically designed to measure Mars' surface temperature (Pimental's instruments were not designed for this purpose), reported after Pimental's announcement Aug. 7 that 200 readings from his radiometer had indicated temperatures in the same region of about $-250°$ and had produced "strong circumstantial evidence that the polar caps are in fact predominantly made of carbon dioxide rather than of water ice."

Naugebauer had reported Aug. 2 that, in the equatorial area covered by *Mariner 6*, surface temperatures ranged from about $60°$ at noon to about $-100°$ at night. A. J. Kliore of JPL said Aug. 2 that *Mariner 6* had confirmed the findings of *Mariner 4* in 1965 that the atmospheric pressure on the surface of Mars was as light as the earth's air pressure at altitudes of 100,000 to 150,000 feet.

Dr. Bradford A. Smith, a New Mexico State University astronomer, reported Aug. 2 that the Martian phenomena popularly called "canals" appeared, from the Mariner photos, to be "diffuse, somewhat linear features" made up of "large, irregular, very-low-contrast splotches" or, in some cases, several craters roughly in alignment. Nix Olympica, a white feature whose brightness, as seen from the earth, increased in the Martian afternoon, was identified as a 300-mile-wide crater with an apparent peak in its center. Leighton noticed terracing and "avalanche chutes" in many of Mars' craters and said that "you would need something the size of an asteroid to make some of those large craters."

Dr. Charles A. Barth of the University of Colorado, who had analyzed gases in the Martian upper atmosphere by means of the Mariners' ultra-violet spectrometers, reported Aug. 2 that he had found no trace of nitrogen, an element present in all earth life forms. Naugebauer Aug. 7 confirmed the apparent absence of nitrogen. The consensus of the scientists was that Mars' atmosphere was almost entirely carbon dioxide, with some gaseous oxygen and carbon monoxide released by the breakdown of carbon dioxide and with a very small amount of water vapor.

Barth and Dr. Charles Hord of the University of Colorado reported Aug. 7 that considerable ultra-violet light from the sun reached the Martian surface. Hord said that such strong ultra-violet radiation "would destroy many of the important molecular bonds of organic compounds." "Martian life, if it exists, must be made of stronger stuff [than earth life] or have some sort of protective mechanism," he said.

Scientists at the Sept. 11 press conference reported finding on Mars what Dr. Robert Sharp, a Caltech geologist, described as a 200,000-square-mile "chaotic" region consisting of ridges, valleys and depressed areas. Leighton reported that a 1,200-mile-wide circular area named Hellas seemed to have no craters. Sharp speculated that the area and any craters in it might be covered with a light, porous, grainy material of the consistency of popcorn.

The scientists said Sept. 11 that the Mariner data had given no clue so far as to why some areas of Mars become darker in the Martian spring and summer. This phenomenon had been attributed to the growth of vegetation. But Horowitz asserted that this darkening "is the major mystery right now."

Soviet Zond Circles Moon and Returns

A 6-ton Soviet lunar probe dubbed *Zond 7*, launched from the USSR's Baikonur cosmodrome Aug. 8, circled the moon Aug. 11 and then returned to earth, where it made a soft landing Aug. 15 south of Kustanay in northern Kazakhstan.

The announcement of the launching was made by Tass Aug. 8 after the probe had gone into a successful lunar trajectory. *Zond 7* was put into parking orbit around the earth before heading toward the moon. Reuters Aug. 11 cited "unusually reliable Soviet sources" as saying *Zond 7* was on "a specific 'mapping' mission as a probable prelude" to a Soviet attempt to land an unmanned probe on the moon and to return it to earth.

Tass reported that *Zond 7* had finished its mission "successfully" after a double re-entry into the earth's atmosphere—the first entry slowing the probe and forcing it up and out of the atmosphere, the 2d entry slowing it further; parachutes brought it down the final $4\frac{1}{2}$ miles.

Soviet Moon Rocket Reported Destroyed

It was reported in Washington Nov. 17 that a 10 million-pound-thrust rocket that the USSR had been readying for manned lunar flight had exploded on its launching pad at the Baikonur cosmodrome during the summer in its first test launching.

The destruction of the rocket, to which some U.S. space experts ascribed the USSR's failure to attempt a manned lunar flight, was reported by the U.S. magazine *Aviation Week and Space Technology* and by science editor Peter Fairley of Britain's Independent Television News. A U.S. spy satellite had photographed "extensive damage" caused by the explosion, Fairley reported.

The Russians had never confirmed the existence of the rocket, which was reported to be more powerful than the Saturn-5 used by the U.S. for manned lunar flights.

U.S. PROGRAM

Space Budget Cut

As the Apollo program neared its initial landing of men on the moon, the U.S. began cutting back on appropriations for space research and technology. NASA's budget was cut from the $4.247 billion estimated for fiscal 1969 to $3.947 billion for fiscal 1970. (Space programs in the Defense Department and Atomic Energy Commission budgets, however, were scheduled to bring the fiscal 1970 space outlays total up to $6.1 billion.) Since the "most expensive developmental phase" of the Apollo moon-landing program was completed, the Apollo budget was reduced from $2\frac{1}{2}$ billion in fiscal 1969 to $1.9 billion in fiscal 1970. The Apollo and Apollo applications programs (manned flights in earth orbit and to the moon) were to be continued with 5 flights a year scheduled in calendar 1970-1. Outlays for it were to rise from $242 million in fiscal 1969 to $465 million in fiscal 1970.

Space science projects were budgeted at $479 million—for developmental work to begin in calendar 1970 on the Planetary Explorer program (unmanned flights to the planets). Space technology outlays were to decline by $23 million from fiscal 1969 figures to a total of $358 million, which included funds for development of Nerva-1, a nuclear-powered rocket engine for the Saturn-5 booster.

(A White House report informed Congress Jan. 17 that U.S. spending on space activities in the 11 space-age years through 1968 totaled $56,727,500,000, of which $17 billion was spent by the Defense Department and $38 billion by NASA. The report had said that "the USSR [space] effort is vigorous and appears to be on a trend of increasing investment and increasing accomplishment. At existing rates, the Russians could overtake and possibly surpass the U.S." In the 11 years through 1968, however, the U.S. had launched 606 space vehicles and the USSR only 358, the report declared.)

11 Moon Landings & 'Space Shuttle' Requested

Dr. George E. Mueller, associate NASA administrator for manned space flight, revealed at the Manned Spacecraft Center in Houston Mar. 13 that NASA was asking Congress to finance 11 manned U.S. landings on the moon. The proposed landings would be spaced at 2 or 3 a year and would end in 1974.

Mueller asserted that the cost of the additional 10 landings, after the first, would be relatively cheap—starting at about $100 million in fiscal 1970—because the additional Saturn-5 rockets and Apollo spaceships were already bought or on order. They had been contracted for because NASA had not expected the Apollo program to be so successful that the first moon landing might be attempted on the 5th Apollo flight. It had been assumed that at least several more Saturn-5s and Apollo spaceships would be required for pre-lunar landing tests. The additional money would be needed for the scientific instrumentation and for the development of "a new constant-volume space suit in which an astronaut can walk normally and move freely about the surface of the moon."

The program would consist of 2 phases: (1) landing, in which "we expect to develop the capability to pinpoint-land at points of scientific interest," and (2) exploratory, in which astronauts would land at points of "great interest scientifically" (6 such spots had already been picked) to make seismic, temperature and photographic records and gather samples.

Mueller had warned at a press conference in New York Jan. 28 that the U.S. would be "out of the manned spaceflight business" if Congress failed to provide money for projects beyond the initial moon landing. He said that Soviet space expenditures were "about 50% greater than ours." "We will find the Russians doing many spectacular things in space," he predicted, and he asserted that they would "probably [be] moving ahead of us" in the 1970s.

NASA disclosed May 7 that it had established 2 task groups to oversee efforts to develop a manned space station and a "space shuttle." The space station would be a permanently orbiting base in which scientists and other space personnel would work. They would commute to and from the station in the space shuttle, a reusable spacecraft that would not duplicate the wasteful—but currently necessary—practice of using a new booster rocket for every new space flight.

Charles W. Mathews, deputy associate NASA administrator for

manned space flight, headed the space station effort; Dr. Mueller, associate NASA administrator, was in charge of the space shuttle program. LeRoy E. Day, ex-director of Apollo Test, headed the Space Shuttle Task Group under Mueller's supervision.

2 "lifting bodies," wingless possible precursors of the space shuttle, had already been test flown at Edwards Air Force Base, Calif. in a joint Air Force-NASA test program: NASA's HL-10 was tested in powered flight; the Air Force's X-24A in non-powered flight. Carried aloft by B-52 bombers, both demonstrated after being dropped that their odd, flattened shapes gave them enough aerodynamic lift for flight.

NASA Administrator Thomas Paine said at a press conference in London June 3 that the U.S.' first permanent orbiting laboratory might be manned by an international team of scientists. If the program were approved, he said, the project should start in 1975 with a small laboratory in orbit, but additional sections would be added each year until there would be working room for about 50 scientists.

John Discher, deputy director of the Apollo applications program, had said Mar. 11 that the U.S. planned to orbit an experimental space station in the burned-out body of a Saturn rocket during 1971. The plan called for 3 missions in the rudimentary space station: a first mission of 28 days during which the station would be put together; a mission of 28 days during which a doctor would live with and observe 2 other men in the station, and a 3d mission of 56 days during which medical observation would continue and a telescope would be set up in the station.

Later in 1969, the Air Force indicated that it, too, was interested in a space shuttle. Lt. Gen. Samuel C. Phillips, new commander of the Air Force Space & Missile Systems Organization in Los Angeles, said in a Washington Post interview published Oct. 15 that his top priority program was the development of a reusable space shuttle. Phillips estimated that launchings by means of reusable shuttles could cut the cost of space shots from $1,000 a pound to perhaps less than $100.

The Air Force plan called for developing the space shuttle in cooperation with NASA. NASA July 22 had announced plans to launch its first orbital workshop in 1972 and to use the first 2 stages of a Saturn-5 rocket as the disposable launching vehicle. The space station, which would include an observatory to study the sun and stars, would be assembled on earth before the launching. Its 3-man crew would go up to it later in an Apollo spacecraft launched atop a Saturn-1B rocket. The crew would return to earth in the Apollo capsule after nearly a month, and new crews would go up by the use of new Saturn-1Bs and new Apollo spaceships.

Slowdown Urged

Sen. Edward M. Kennedy (D., Mass.) urged May 19 that the U.S. space program be slowed after the "lunar landing and exploration are complete" and that "a substantial portion of the space budget . . . be diverted" to "pressing problems" at home.

Speaking in Worcester, Mass. at the dedication of Clark University's $5.4 million Robert Hutchings Goddard Memorial Library, Kennedy said that "we need a dedication not only to the national security . . . [but] to social justice that will eliminate the causes of the divisions and the violence that beset our land." At the library, built in honor of the U.S. rocket and space pioneer, Kennedy said: "We need not try to get to Mars or Venus merely because the Russians might get there first. . . . I am for the space program. But I want to see it in its right priority, one which will let it continue into the future and not have to be cut back or abandoned because the nation that supports it is hobbled by internal disorder. We should develop a plan for an orderly programmed exploration of outer space, but we no longer need an accelerated program."

NASA Administrator Thomas Paine said at the Manned Spacecraft Center in Houston May 20 that he was "disappointed" that Kennedy had presented "such a dispiriting vision of the nation's vigor and destiny in space." Kennedy "is wrong," Paine declared. The U.S. "should not weakly yield technological supremacy in space to the Soviets. We should not ground our astronauts after Apollo."

Vice Pres. Spiro T. Agnew, replying in Washington May 21 "to those who would denigrate our space effort," asserted that "this Administration has already demonstrated its belief in the strength and potential of America's space program." He quoted the late Pres. John F. Kennedy's assertion that space is there "and we're going to climb it." But he reported that the Administration was taking steps to "evaluate the costs and alternatives available to us in extending the program once man has been placed on the moon and returns."

Apollo 11's spectacular success in landing men on the moon added fuel to the controversy over demands that the U.S. reorder its national priorities. Many critics of the Johnson and Nixon Administrations asserted that the nation's treasure was being wasted on space and military uses, particularly on the war in Vietnam and on reaching the moon. By comparison, they held, too little money was being spent to combat poverty, hunger, discrimination, deprivation, disease and on other problems on earth.

A Louis Harris poll indicated July 14, however, that there had been a shift in public opinion in favor of landing a man on the moon. In February, Americans had opposed the moon landing by 49% to 39%, but when "a carefully drawn cross-section of 1,607 adults" was questioned June 12-23, only a month before *Apollo 11* was launched, the lunar project was favored by 51% to 41%. Harris held that "basically the change in public attitude can be attributed to the feeling 'if we have gone this far, we ought to finish the job and actually land on the moon.'" He added, however, that, according to the same survey, "by 56% to 37%, the same American people simply do not think the space program 'is worth the $4 billion a year which has been spent on it.'"

Senate Democratic leader Mike Mansfield (Mont.) and Senate Democratic whip Edward M. Kennedy (Mass.) July 16 rejected Vice Pres. Agnew's call for a national commitment to land a man on Mars by the end of the century. "The needs of the people on earth, and especially in this country, should have priority," Mansfield declared. "When we solve these problems, we can consider space efforts," Mansfield said. Kennedy, agreeing with Mansfield's assertion, said: "The Apollo program is for landing a man on the moon and exploration and should take another one to 2 years. I think after that the space program ought to fit into our other national priorities."

House Democratic whip Hale Boggs (La.), disagreeing with the 2 Senate Democratic leaders, said July 16 that following Mansfield's advice "would be like telling the Wright brothers to stop after they flew a couple of hundred feet." He said the U.S. should investigate the feasibility of such projects as a colony on the moon and manned flights to other planets.

Health, Education & Welfare Secy. Robert H. Finch and Housing & Urban Development Secy. George W. Romney July 28 joined the chorus urging a shift of priorities toward domestic problems. Speaking at a Western Governors Conference in Seattle, Romney said that "having shown we can put people on the moon, we're at the point where we ought to give greater priorities to domestic needs on earth." Romney cited the need for housing: "60% of the people in the U.S. can't afford to buy a $20,000 home," he said, "and houses are deteriorating faster than we can replace them."

Agnew disagreed with the Finch and Romney argument, which he described July 28 as "specious" reasoning to try to prove that "we cannot afford to venture into space." "We do not need a transfer of dollars from the space program to other programs," he said. "We need a transfer of its spirit—an infusion of American dedication to purpose and hard work."

Nixon Accepts Mars-Landing Goal

Pres. Nixon Sept. 15 indicated his acceptance of a "balanced space program" that included the "goal" of landing men on Mars by the year 2000 but opposed an expensive effort to get men to Mars as soon as possible.

The balanced program, detailed in a report entitled "The Post-Apollo Space Program: Directions for the Future," was presented to the President Sept. 15 by a cabinet-level Space Task Group headed by Vice Pres. Agnew. The report was made public Sept. 17 at a White House press conference attended by Agnew and the other members of his task group—Dr. Robert C. Seamans, Air Force secretary; Dr. Thomas O. Paine, national aeronautics and space administrator, and Dr. Lee A. DuBridge, the President's science adviser.

White House Press Secy. Ronald L. Ziegler, announcing Nixon's acceptance of the report, was asked Sept. 15 whether he would call the President's action a commitment to land a man on Mars. Ziegler replied, "Yes, I think so."

The report recommended that the U.S. "accept the long-range option or goal of manned planetary exploration with a manned Mars mission before the end of the century as the first target." The task group rejected these 2 "extreme options": (1) a crash program to put men on Mars as soon as possible regardless of cost and (2) the virtual abandonment of manned space flight after the current lunar explorations.

3 optional programs for landing men on Mars were proposed in the report: (1) a big-budget program to make the first Mars landing in 1983; (2) an effort in which annual expenditures would be lower and men would land on Mars in 1986, and (3) a program in which the decision on a landing date might be deferred until after 1990.

Officially, the task group called all 3 options acceptable and did not favor any one of them. Agnew, however, said he preferred the option for the 1986 landing. Costs under this option were estimated at about $24 billion; annual spending would remain below $5 billion until fiscal 1975, when it would rise to $5\frac{1}{2}$ billion, and spending would peak at about $8 billion a year in the early 1980s. Paine was known to prefer the plan for a 1983 landing, a program under which spending would amount to $7.7 billion in fiscal 1975 and reach $9.4 billion in 1980. A decision on trying for a 1986 landing would not have to be made until 1976, when such preliminaries as equipment development and unmanned landings on Mars would have provided a sound-enough basis for such a determination. A decision on attempting a manned landing in 1983 would have to be made not later than 1974.

The task group reported that these 4 items would have to be developed in preparation for manned flight to Mars: (1) a manned space station in permanent orbit around the earth; (2) a reusable space shuttle to ferry men and equipment from earth to space station and back, (3) a nuclear rocket for space flights of several years' duration and (4) a nuclear-powered tug to move space stations and big space vehicles around in space.

The "balanced space program" proposed by the task group called for continued manned exploration of the moon and the eventual establishment of semi-permanent scientific stations there. The program would include a variety of unmanned interplanetary probes. Several of these vehicles would make the essential preliminary unmanned landings on Mars. Others would go on "grand tour" trips in which a single probe would fly past several of the outer planets—Jupiter, Saturn, Uranus, Neptune and Pluto—on the rare occasions when the planets are in favorable alignment in their orbits. The program provided for the development of additional communications, navigation, weather, scientific and other special-purpose satellites.

NASA plans for manned landings on Mars had been discussed in Washington Aug. 6 by Dr. Paine and Dr. George Mueller, associate NASA administrator for manned flight. The plans called for 2 spaceships, each carrying up to 12 astronauts, to be sent to Mars together. The 2-ship expedition would make a rescue possible if necessary. 6 of the astronauts would spend 4 to 6 weeks on the Martian surface while the mother ships remained in orbit around Mars. The round trip was expected to take about 2 years.

A commitment to an early manned landing on Mars was opposed, however, by many scientists and members of Congress. Dr. William H. Pickering, director of the Jet Propulsion Laboratory in Pasadena, Calif., had said July 26: "Now that Apollo [the manned landing on the moon] has been accomplished, rather than set another ambitious goal we should have a period of consolidation" during which "the balance should be increased toward unmanned effort." Chairman George P. Miller (D., Calif.) of the House Science & Astronautics Committee asserted Aug. 11 that a commitment to a manned Mars landing would be "premature" for "5, perhaps 10 years." Miller said he was "not against going to Mars or elsewhere when the time to begin such a great undertaking has come." Before then, however, he called for a "balanced program" fully exploiting "the great potential of unmanned" as well as manned spacecraft. In a Gallup poll of 1,517 adults in more than 300 areas across the U.S. Aug. 6, 53% opposed and 39% favored "setting aside money" for a manned Mars landing.

Unmanned "grand tours" of the outer planets had been proposed in a report issued Aug. 3 by the 23-member Space Science Board of the National Academy of Sciences. Orbital conditions available for such flights in the mid-1970s will not recur for 180 years and therefore should not be wasted, the board said. It suggested that: (a) A probe be sent to pass Jupiter in 1974 and drop an instrumented probe into the planet's atmosphere; (b) a probe be sent into orbit around Jupiter (which has 13 natural satellites) in 1976; (c) a probe be sent to pass Jupiter, Saturn and Pluto in 1977 and possibly drop a radio beacon to the surface of Saturn; (d) another probe be sent on a grand tour of Jupiter, Uranus and Neptune in 1979; (e) a probe be sent in the early 1980s to pass Jupiter and Uranus and to drop a probe into Uranus' atmosphere. The board also proposed a study of the feasibility of (1) sending a spacecraft to an asteroid to bring back samples of the asteroid's surface material and (2) sending a probe to Halley's Comet. The reporting board was headed by 2 co-chairmen: Dr. James A. Van Allen of the University of Iowa and Dr. Gordon J. F. MacDonald of the University of California Santa Barbara.

Fund Authorization Drops, Mol Canceled

A bill authorizing $3,715,527,000 for the space program in fiscal 1970 was passed by the House Nov. 6 and Senate Nov. 7 and was signed by Pres. Nixon Nov. 18. It was $297,846,000 less than the amount authorized in 1968. Among the bill's provisions were ones to bar funds from college individuals participating in campus disruptions, to require registration and data from top-level employes transferring between NASA and space contractors, to express the "sense of Congress" that consideration be given to geographic distribution of research contracts and to require that only the U.S. flag be placed on the surface of the moon or other planets on U.S. missions.

Among the casualties of the belt-tightening in space projects was the Air Force's work on a manned orbiting laboratory (Mol). Deputy Defense Secy. David R. Packard announced June 10 that the Mol project was being ended because of the cut in defense spending. $1.3 billion had been spent in the effort to place a 15-ton, 2-man spaceship in orbit for reconnaisance and other military missions. Mol's end came after a bipartisan drive against the project in Congress.

Solar Probe Fails

A 148-pound solar probe designated *Pioneer 10* was sent aloft from Cape Kennedy Aug. 27 but deliberately destroyed by ground controllers moments after lift-off because its 3-stage Delta booster rocket was veering dangerously off-course. The last of the Pioneer series, *Pioneer 10* had been scheduled to orbit the sun in a course that would have brought it alternately inside and outside the earth's orbit. Its mission had been to provide data on plasma, energetic particles and magnetic and electric fields radiating outwards from the sun.

Also destroyed was a 45-pound "hitchhiker" satellite designated as *TETR 3* (test and training satellite). *TETR 3*'s purpose was to aid in the training of personnel and the testing of the equipment of NASA's Manned Space Flight Network.

U.S. Satellites

The U.S. launched a number of scientific and military satellites during the year. The launchings, in chronological order:

Jan. 22—The 641-pound scientific satellite *Oso 5* (for orbiting solar observatory) was sent into a 350-mile-high circular orbit by means of a 3-stage Delta rocket launched from Cape Kennedy, Fla. at 11:48 a.m. EST. The $12 million satellite was designed to study solar radiation and the sun's influence on near-earth space. NASA said that data gathered by *Oso 5* was expected to help in predicting solar flares and in providing advance warning of intense solar activity that could affect the scheduling of manned space flights.

Feb. 5, Apr. 12, Apr. 15, June 3, July 31, Sept. 22—A secret satellite was launched Feb. 5 into polar orbit by means of a Thor-Agena booster rocket fired from Vandenberg Air Force Base, Calif. by an Air Force-industry team. A secret 26-foot-long Air Force reconnaissance satellite was put into a 300-mile-high polar orbit Apr. 12 by means of an Atlas-Agena booster rocket launched from Cape Kennedy; the satellite's mission was to get information to be used in the later-canceled manned orbiting laboratory. Secret satellites were sent into polar orbit by means of 2 Titan-3B/Agena-D rockets launched by the Air Force from Vandenberg Base Apr. 15 and June 3 and by means of 2 Thor-Agena rockets sent up from Vandenberg Base by an Air Force-industry team July 31 and by the Air Force Sept. 22.

Feb. 9—The non-Communist world's biggest communications satellite, a 1,600-pound military spacecraft dubbed *Tacomsat* (tactical communications satellite), was carried into a synchronous orbit over the Pacific by means of a 3-stage Titan-3C booster rocket launched from Cape Kennedy at 4:09 p.m. After the Air Force booster's first 2 stages exhausted their fuel and dropped away, the Titan-3C's 3d stage put *Tacomsat* initially into a 100-mile-high parking orbit. A 2nd firing of the 3d stage's engine then brought the satellite into a higher transfer orbit, and a 3d ignition carried it to its post above the equator, where it seemed to hang motionless 22,300 miles above the Galapagos Islands. In an equatorial orbit at that altitude, a satellite's speed matches the speed at which the earth turns beneath it, and it therefore remains constantly above the same geographical point. The $30 million experimental *Tacomsat* was orbited for the Defense Department to test the possibility of using a network of such satellites to keep in constant communication with U.S. troops, naval vessels and military aircraft in all parts of the world. *Tacomsat*, built by the Hughes Aircraft Co., was capable of relaying 10,000 2-way phone conversations simultaneously.

Feb. 26—The 320-pound U.S. weather satellite *Essa 9* was sent into an 887-mile-high polar orbit by means of a Delta rocket launched by NASA at Cape Kennedy. The drumlike satellite, 42 inches in diameter, carried 2 cameras on a mission that called for it to photograph the cloud cover of the entire earth once every 24 hours. It was built by RCA under the direction of NASA's Goddard Space Flight Center in Greenbelt, Md.

Apr. 14—The 1,269-pound *Nimbus 2* weather satellite was sent into a 690-mile-high polar orbit by means of a Thorad/Agena-D rocket launched by NASA and the Air Force's 6595th Aerospace Test Wing from Complex 2 of the Western Test Range at Lompoc, Calif. The 10-foot-high satellite carried a record 7 meteorological experiments and 2 nuclear isotope SNAP-19 generators. One of the experiments was to follow the wanderings of an elk named Moe (for meteorology, observation and ecology) in Yellowstone National Park. His movements were radioed to the satellite by means of a 7-pound radio that had been attached to the animal. Should the experiment succeed, the method was to be used to study the migration of whales, polar bears, elephants, sea turtles and other animals. The launching was also used to orbit a "hitchhiker" geodetic satellite, the 13th of the Army's Secor (for sequential collation of range) satellites.

May 23—2 755-pound Vela radiation-detection satellites and 3 small military research satellites were orbited by means of a Titan-3C launched by the Air Force at Cape Kennedy at 3:57 a.m. EDT. The two Velas were stationed 180° apart so that they could keep both hemispheres under observation simultaneously. Their mission was to detect any violation of the treaty banning nuclear explosions and to check on Communist China's atomic tests. The 3 research satellites were to provide data on radiation.

June 5—The 1,393-pound *Ogo 6* (for orbiting geophysical observatory) was sent into near polar orbit by means of a Thorad-Agena rocket launched from Vandenberg Air Force Base, Calif. at 7:43 a.m. The last in the series of large Ogo observatory-type spacecraft, *Ogo 6* achieved an orbit with an apogee of 683 miles, perigee of 248 miles, period of 100 minutes and 82° angle of inclination to the equatorial plane. It carried 25 experiments to get data on the upper atmosphere and ionosphere, the auroral regions around the poles and the edges of the regions of trapped radiation surrounding the earth. It was expected to measure the intensity of solar flares during a period of maximum sunspot activity.

June 21—The 174-pound *Explorer 41*, the 7th satellite in the interplanetary monitoring platform (IMP) program, was sent into near polar orbit June 21 by means of a 3-stage Delta rocket launched from Vandenberg Base at 1:40 a.m. while an electrical power failure blacked out most of the rocket base. *Explorer 41*, its 12 experiments designed to provide data on cislunar space, solar plasma, magnetic fields and cosmic rays, achieved an orbit with an apogee of about 135,000 miles, perigee of perhaps 215 miles, period of about 4 days and inclination of 85°. It was to support the *Apollo 11* mission by providing solar radiation data.

June 28—A 3-year-old 14-pound pigtail monkey (*macaca nemestrina*) named Bonny was sent into space in *Biosatellite 3* in what had been scheduled as possibly a 30-day orbital flight. But the monkey, an adolescent male, became so "sluggish" in space that he was brought down July 7, after only 9 days in orbit. Bonny died July 8, about 12 hours after returning to earth. According to a preliminary autopsy report July 9, he probably died of heart failure caused by weightlessness, immobility and cold. The purpose of the $92 million NASA experiment was to provide data about the mental, emotional and physiological processes in a man-like mammal during an extended space flight.

The 31-inch Thailand-born Bonny, his brain-wave pattern, blood pressure and other reactions monitored by sensors implanted in his brain, arteries and other organs, had been strapped to a form-fitting couch late June 28 in the 1,536-pound *Biosatellite 3*, a spaceship built by the General Electric Co. at an estimated cost of $55 million. The space vehicle was sent aloft by means of a 2-stage Delta rocket launched from Cape Kennedy at 11:16 p.m. It achieved an orbit with an apogee of 245 miles, perigee of 225 miles and period of 92 minutes.

According to telemetry data, Bonny appeared to be in good health and performed well during his first 2 days in orbit. He drank water by sucking from a stainless-steel nipple; he ate at designated times after pulling a lever to get food pellets, and he won additional food by performing 2 tasks—matching symbols on a display panel and aligning 2 rotating disks. When he slept, which he did more fitfully than normal, he dreamed, the monitored data revealed during the early days of the flight. But Dr. W. Ross Adey, the experiment's chief scientist, reported July 1 that Bonny had stopped performing the tasks that yielded the extra food. By July 4, it was reported, Bonny's efficiency had decreased considerably, and he showed little interest in his work.

Faced with a threatened "serious deterioration" in the monkey's condition, space authorities decided July 7, during his 130th revolution around the earth, to bring him out of orbit. A radio signal detached his 315-pound capsule from the orbiting *Biosatellite 3*, and the monkey-carrying capsule splashed down safely in the Pacific about 1 p.m. Hawaii time. It came down about 25 miles north of the island of Kauai, Hawaii, after an Air Force plane was foiled by heavy clouds in an attempt to seize the parachuting capsule in midair. A helicopter then fished the capsule out of the water, and it was flown to Hickam Air Force Base, Hawaii.

Bonny died at Hickam Base at 1 p.m. Hawaii time July 8. Dr. Adey disclosed in Washington Oct. 22 that Bonny had succumbed to the effects of weightlessness. Adey, anatomy and physiology professor at UCLA, said astronauts had suffered mildly from the same ailment. He warned that it was "not realistic to go ahead with plans for a major space system [in which men would be subjected to extended periods of weightlessness] without information on man's ability to perform at a high level on a continuing basis in space." Adey said that during the flight, Bonny had lost 20% of his launching weight, mainly through fluid loss due to excessive perspiration and urination.

Aug. 9–The 640-pound scientific satellite *Oso 6* (for orbiting solar observatory) was sent into near circular orbit by means of a 2-stage Delta-N rocket launched from Cape Kennedy at 3:52 a.m. EDT. The $12 million satellite achieved a 350-mile-high orbit, inclined at an angle of 33° to the equatorial plane and with a period of 95 minutes. The satellite, sent up to continue the work of earlier Oso satellites, was designed to study evolutionary changes in various features of the sun at a time of near-peak activity in the current 11-year solar cycle.

The Delta's 2d stage carried into orbit a package attitude control (Pac) system as a "hitchhiker." The Pac experiment was a test of a long-life low-power, 3-axis earth stabilized control system designed to make it possible for the Delta 2d stage to be used as an experiment platform.

Aug. 12–The 1,987-pound U.S. satellite *ATS 5* (for applications technology satellite) was sent aloft by means of a 2-stage Atlas-Centaur rocket launched from Cape Kennedy at 7:01 a.m. EDT. On separating from its booster rocket, however, the satellite began to wobble. Ground controllers, therefore, fired the satellite's apogee engine as soon as

ATS 5 reached apogee the first time. The object of this ignition was to put *ATS 5* into a synchronous orbit 22,300 miles above the equator about 1,100 miles west of Quito, Ecuadro. The apogee engine put *ATS 5* into an orbit with an apogee of 22,927 miles, perigee of 22,221 miles, period of 24 hours 24 minutes and 2.6° angle of inclination. *ATS 5* then began to drift toward its planned station from a position originally over the Bay of Bengal. The satellite, separated from its apogee engine Sept. 5, continued to spin backwards at a rate of 65 revolutions a minute.

ATS 5's primary mission was to test a gravity-gradient stabilization system consisting of 4 124-foot booms forming an X with the satellite in the center. The system's function was to keep the satellite pointed toward the earth so that its experiments always pointed in the proper direction. A major *ATS 5* experiment was to test an L-Band communications and air-traffic control system.

Borman Visits USSR

Col. Frank Borman, 41, who had commanded *Apollo 8* in the first manned orbiting of the moon, visited the USSR July 2-10 on a trip during which he stressed the desirability of U.S.-Soviet cooperation in space projects. Borman, accompanied by his wife, Susan, and his sons Frederick, 17, and Edwin, 15, was a guest of the Soviet-American Relations Institute. Although his visit was unofficial, Borman conferred with Soviet Pres. Nikolai V. Podgorny in the Kremlin July 9 and later said that the meeting "was very encouraging and beneficial when we think of cooperation in space."

Borman, the first U.S. astronaut to visit the USSR, was greeted on his arrival in Moscow July 2 by 3 Soviet cosmonauts—Lt. Col. Gherman S. Titov, Maj. Gen. Georgi T. Beregovoi and scientist Konstantin Feoktistov.

Visiting Leningrad July 3, Borman said at a press conference: "I am now working on a program under which, in the mid-1970s, we hope to launch a big orbiting space station or manned laboratory... I hope and can foresee a time in that program when U.S. and Soviet spacemen will fly together."

Borman July 5 visited Zvezdny Gorodok (Stellar Town), the cosmonaut community 25 miles northeast of Moscow, an area normally closed to foreigners. He and his family, guided by Titov, flew July 6 to Simferopol in the Crimea, where he became the first American (and the 3d foreigner of any nationality) to tour the Yevpatoria space tracking station. They flew July 8 to Novosibirsk, in Siberia, where Borman visited Academgorodok (Academic Town) and talked to the elite Soviet scientists stationed there. Dr. Feoktistov, who accompanied them, helped describe scientific and technological achievements of the center. They then flew back to Moscow late July 8.

After meetings with Podgorny and leading Soviet scientists in Moscow July 9, Borman left the USSR with his family July 10 to return to the U.S. At a press conference in Cape Kennedy July 12, Borman said Podgorny had told him that the Soviets were planning to send large manned space stations into orbit around the moon.

(Borman, who had voluntarily removed himself from space-flight status Jan. 9, had been appointed in May as field director of NASA's Space Station Task Group, whose mission was to develop plans for manned orbiting space stations.)

NASA Personnel Changes

Walter M. Schirra Jr., 45, the only astronaut to fly in Mercury, Gemini and Apollo spaceships, disclosed in Houston Mar. 11 that he had agreed to become president of the Regency Corp. of Denver, an industrial equipment-leasing organization. In resigning from the Navy and the space program, Schirra said, he had rejected offers of a NASA desk job or of promotion to an admiralcy.

NASA announced May 7 that Navy Capt. Alan B. Shepard Jr., 45, the first American to fly in space, had been restored to eligibility for space flight. An operation in 1968 had corrected an inner ear disorder that had grounded him in 1963 after his single space flight.

Col. James A. McDivitt, 40, who had commanded the *Apollo 9* earth orbital flight, retired from active space flying June 25 and immediately stepped into a new job as manager for lunar landing operations at the Manned Spacecraft Center in Houston, Tex. A NASA spokesman explained that McDivitt would be "responsible for planning lunar landing missions subsequent to the first landing." McDivitt Sept. 25 was appointed manager of the Apollo spacecraft program to succeed George M. Low, who had been appointed deputy NASA administrator.

Dr. George E. Mueller, 51, resigned Nov. 10, effective Dec. 10, as associate NASA administrator and director of the manned space-flight program.

Rocco A. Petrone, 43, director of launch operations at the John F. Kennedy Space Center, had been appointed Aug. 22 to be director of the Apollo program. He succeeded Lt. Gen. Samuel Cochran Phillips, who returned to the Air Force Sept. 1 as commander of the Air Force Space & Missile Systems Organization. Walter J. Kapryan, 49, deputy director of launch operations at the space center, succeeded Petrone as director.

Dr. Abe Silverstein retired Oct. 31 as director of NASA's Lewis Research Center in Cleveland. Bruce T. Lundin of NASA's Advanced Research & Technology Office succeeded Silverstein at the Lewis Center, effective Nov. 1.

Dr. Gene Simmons, a geophysics professor at MIT, was named Oct. 22 as chief scientist at the Manned Spacecraft Center in Houston.

NASA announced Aug. 14 that 7 aerospace research pilots from the Air Force's canceled manned orbiting laboratory (Mol) were being reassigned as NASA astronauts. The newly named astronauts were: Maj. Karol J. Bobko, 32, of Seaford, N.Y., the first U.S. Air Force Academy graduate to become a NASA astronaut; Navy Lt. Cmndr. Robert L. Crippen, 32, or Porter, Tex; Air Force Maj. Charles G. Fullerton, 31, of Portland, Ore.; Air Force Maj. Henry W. Hartsfield Jr., 35, of Birmingham, Ala.; Marine Maj. Robert F. Overmyer, 33, of Westlake, Ohio; Air Force Maj. Donald H. Peterson, 35, of Winona, Miss., and Navy Lt. Cmndr. Richard H. Truly, 32, of Meridian, Miss. An 8th Mol research pilot, Air Force Lt. Col. Albert H. Crews, 40, of Alexandria, La., was assigned to the Flight Crew Operations Directorate at the Manned Spacecraft Center in Houston.

Dearth of Scientist-Astronauts Charged

NASA Aug. 21 denied a report by Victor Cohn in the Aug. 21 *Washington Post* that every scientist-astronaut except geologist Harrison (Jack) Schmitt had been dropped from training for landing on the moon. Cohn had reported that even Schmitt would have to wait more than 2 years to get a chance at a lunar flight and that the remaining 12 scientist-astronauts had been assigned to train for the earth-orbiting Apollo applications program, which was to start launchings in 1972. Cohn reported in the Aug. 22 *Washington Post* that despite the NASA denial, "unofficial NASA sources" had confirmed that Schmitt was considered the only officially designated scientist-astronaut with enough flying ability to be acceptable for a moon landing.

Even Dr. Don L. Lind, a nuclear physicist who was not officially listed as a scientist-astronaut because he had qualified for astronaut service as a pilot, had been designated not for lunar landing training but for training for an Apollo applications flight. The irony of the Lind assignment was that he was trained in geology, had spent the past 2 years working on scientific problems involved in lunar exploration and was considered the "expert" among the astronauts on lunar surface experiments. Dr. George Mueller, associate NASA administrator for manned flight, said Aug. 21, however, that "Lind is high on our list" for assignment to a lunar landing mission and that "we hope to include a scientist on or before *Apollo 17*," which he indicated might be sent to the moon by mid-1971. NASA mission planners had pointed out that lunar landings were "tricky" flight problems and would require good pilots for "some time to come."

Dr. F. Curtis Michel, 35, a physicist, had announced Aug. 5 that he was resigning as a scientist-astronaut in order to devote more time to science. His departure was reported to be another indication of the dissatisfaction of many scientists with the space program's failure to emphasize science in manned flight. This failure was said to be a major reason for the resignation Aug. 1 of Dr. Wilmot N. Hess, director of the Science Operations Office at the Manned Spacecraft Center in Houston, and the resignation of Dr. Elbert King, curator of the Lunar Receiving Laboratory at the same center. Hess and King had complained that too much stress was placed on engineering to the detriment of concentration on science.

Court OKs Prayers in Space

U.S. District Court Judge Jack Roberts in Austin, Tex., dismissed a lawsuit by Mrs. Madalyn Murray O'Hair, an atheist, to ban the broadcast of prayers and Bible readings by U.S. astronauts on future space flights, it was reported Dec. 4. In her suit against the NASA, Mrs. O'Hair said the religious statements by astronauts on the *Apollo 8* and *Apollo 11* missions had violated constitutional provisions forbidding establishment of a state religion. Roberts said the astronauts had spoken as individuals, not as U.S. government representatives, and that to have prohibited their statements would have violated their religious rights. (The U.S. Supreme Court Mar. 8, 1971 upheld Judge Robert's ruling.)

Air Force Ends UFO Study

U.S. Air Force Secy. Robert C. Seamans Jr. announced Dec. 17 that the Air Force had ended its 21-year investigation of UFOs (unidentified flying objects, or "flying saucers"). Investigators for the inquiry, codenamed "Project Blue Book," had checked 12,618 UFO sighting reports, Seamans said, but had found no evidence that any reported UFO represented an extra-terrestrial spaceship or constituted a threat to national security. Seamans held that a continuation of the inquiry "cannot be justified on the ground of national security or . . . the interest of science."

The end of the Air Force inquiry was applauded by a number of people enthusiastic over research into UFOs. They indicated that better progress could be made by investigators less skeptical than the Air Force. Stuart Nixon, secretary-treasurer of the National Investigations Committee on Aerial Phenomena (NICAP), a non-governmental organization, said at NICAP headquarters in Washington Dec. 17 that "UFOs can now be given the serious scientific attention they require."

A symposium on UFOs was held in Boston Dec. 26-27 in conjunction with the annual meeting of the American Association for the Advancement of Science. Astronomers, physicists and social scientists who participated in the symposium appealed to the Air Force to preserve the secret records compiled in the Blue Book inquiry. They said the records could be helpful in studying the physical or psychological phenomena that might have contributed to UFO sightings. The principal organizer of the symposium was Dr. Thornton Page, director of the Van Vleck Observatory at Wesleyan University and a visiting scientist at the Manned Spacecraft Center in Houston. Another organizer was Dr. Walter O. Roberts, president of the American Association for the Advancement of Science.

Boeing Gets Moon-Car Contract

NASA announced Oct. 28 that the Boeing Co.'s aerospace group in Huntsville, Ala. had been selected to design and build "4 flight-qualified lunar roving vehicles" to be used by astronauts when they explore the moon. As then planned, the 4-wheeled electric cars were to weigh no more than 390 pounds each, be capable of carrying 2 astronauts and 500 pounds of equipment at a speed of 10 mph. and have a range of 8 miles although they probably would be used for distances of only 3 miles—their communication range. Total cost of the 4 cars was estimated at $19 million. Boeing won the contract in competition with Bendix Corp., Chrysler Corp. and Grumman Corp.

INTERNATIONAL DEVELOPMENTS

Soviet Launchings

Soviet space authorities orbited at least 55 Cosmos satellites—*Cosmos 263* through *Cosmos 317*—during 1969.

The West German Institute for Space Research in Bochum reported Apr. 24 that *Cosmos 279*, sent up Apr. 15, was a Soyuz space capsule and had made a soft landing on Soviet territory that day. *Cosmos 279*'s initial orbit had an apogee of 173.9 miles, perigee of $120\frac{1}{2}$ miles, period of 89.1 minutes and $51.8°$ angle of inclination to the equatorial plane. The West German institute reported June 6 that *Cosmos 284*, launched May 29, had landed in the USSR earlier June 6. The institute reported Aug. 27 that *Cosmos 294*, launched Aug. 19, had landed in the USSR early that morning. Richard D. Lyons had reported in the *N.Y. Times* June 29 that *Cosmos 286*, which was landed in the USSR June 14, was a spy satellite (as was a secret U.S. Air Force satellite brought down over the Pacific June 23). George C. Wilson reported in the *Washington Post* Sept. 17 that *Cosmos 298* was an unarmed FOBS (fractional orbital bombardment system) missile, which, like all FOBS missiles, was brought to earth before completing a full orbit. Previously, Wilson said, the USSR had conducted 13 FOBS launchings between Sept. 17, 1966, when the first took place, and Oct. 2, 1968. *Cosmos 305*, launched Oct. 22, was also brought down after less than a full orbit.

Informed sources in Washington had reported Jan. 30 that an unmanned Soviet space vehicle sent up the previous week had aborted due to failure of its 2d stage. The spacecraft burned up on re-entry into the atmosphere, and the USSR did not reveal the launching. More than one unannounced Soviet launching was made during the period, an informed source in Washington asserted Jan. 30.

Peter Grose reported in the *N.Y. Times* Mar. 28 that an unmanned Soviet space probe bound for Mars was "believed to have failed shortly after blast-off" early Mar. 27 at the Baikonur cosmodrome.

2 meteorological satellites dubbed Meteor were launched by the USSR Mar. 26 and Oct. 6. The first achieved an orbit with an apogee of 427 miles and perigee of 386 miles. The 2d went into an orbit with an initial apogee of 690 kilometers (429 miles), perigee of 630 kilometers ($391\frac{1}{2}$ miles), period of 97.7 minutes and an $81.2°$ angle of inclination to the equatorial plane. The 2-Meteor system relayed weather data at 6-hour intervals.

The USSR's 11th and 12th Molniya-1 communications satellites were orbited Apr. 11 and July 23.

Soviets Orbit International Satellites

Intercosmos 1, a satellite carrying instruments made in the USSR, Czechoslovakia and East Germany, was sent up from an undisclosed Soviet launching site Oct. 14. The satellite, whose announced mission was to study solar radiation, achieved an orbit with an apogee of about

400 miles, perigee of 160 miles, period of 93.3 minutes and 48.4° angle of inclination. Tass reported that an "operational group" of Soviet scientists was directing the flight. The launching was observed by representatives from Bulgaria, East Germany, Hungary, Poland, Rumania and Czechoslovakia, and observatories in the latter 6 countries and the Soviet Union were cooperating in the use of the satellite for a joint study.

Intercosmos 2, the 2d Soviet-made scientific satellite carrying instruments produced by other Soviet-bloc countries, was launched into orbit Dec. 25 by means of a Soviet rocket. Its instruments were designed by East German, Soviet, Bulgarian and Czechoslovak specialists and built in East Germany and the USSR. Its mission was to study the ionosphere.

Space Debris Hits Japanese Ship

Japanese delegates to the UN Committee on Outer Space revealed in Geneva July 4 that falling wreckage from a Soviet space vehicle had badly damaged the 3,000-ton Japanese freighter *Dai Chi Chinci* June 5 and had seriously injured 5 crewmen. The incident took place as the ship cruised in international waters between the Soviet island of Sakhalin and the Siberian province of Khabarovsk. The Japanese informed Western delegates that 2 Soviet ships had collected most of the debris shortly after the ship was hit but that the Japanese ship's captain had picked up enough of the debris for his government to identify it as wreckage of a Soviet space vehicle. The incident was disclosed during an unsuccessful attempt by the committee to agree on provisions of a proposed treaty on liability for damage caused by space objects.

(The U.S. State Department had disclosed May 7 that it had given to the Soviet embassy in Washington May 6 a 14-inch metal sphere dropped from a Soviet space vehicle and washed up on the Alaskan coast late in 1968. This was the first such action under a treaty, which took effect in Dec. 1968, that provided for the rescue of astronauts and the return of fallen space objects.)

Quebec-France Link Dispute

Canada's federal government continued its attack on agreements under which France and Quebec planned a joint communications satellite system. The signing of the agreements without prior notification of the Canadian federal government was condemned by Canadian External Affairs Min. Mitchell Sharp Jan. 24 and 31. He told the House of Commons Jan. 31 "that the participation of the federal government will be required in any satellite scheme, and therefore consultations will of necessity be involved." Sharp said that Quebec should have been "willing to take into its confidence the government of the country and to disclose its intentions before making them known and including them in letters of intent to the government of another country."

Quebec Education Min. Jean-Guy Cardinal, who had signed the agreements in Paris, asserted through an aide Jan. 31 that Sharp's "statement in no way changes the position of Quebec. Education is an exclusive provincial jurisdiction.... Quebec maintains that telecommunications is merely a means of education, just as chalk and blackboards are. It would have been neither preferable nor useful to consult with Ottawa." Cardinal had returned to Quebec from his European trip Jan. 28.

Quebec Premier Jean-Jacques Bertrand defended his regime's actions Feb. 6 and said his government was "being reproached for taking part in studies whose effect could be the improvement of the level of education and culture in Quebec." Bertrand accused Ottawa of "unjust fiscal sharing" that forced Quebec to try "to the best of its ability to meet its constitutional responsibilities and to prepare for its future." Bertrand asked how the federal government, which could not be trusted in the field of tax distribution, could be trusted in matters of technical means to influence culture and education. He rejected any suggestions that Quebec's attempts to raise overseas capital were unconstitutional.

U.S. Launches Canadian Satellite

A Canadian-built scientific satellite dubbed *Isis 1* (international satellite for ionospheric studies) was launched by NASA from the Western test range at Lompoc, Calif. at 1:47 a.m. Jan. 30. The 532-pound *Isis 1*, sent into space by means of a 3-stage thrust-improved Delta booster rocket, achieved a near-polar orbit with an apogee of 3,522 kilometers (2,184 miles), perigee of 574 kilometers (356 miles), period of 238.3 minutes and 88.5° angle of inclination. Essentially a spheroid, 50 inches in diameter by 42 inches high, *Isis 1* was the 3d Canadian ionospheric satellite to be launched by the U.S. in a cooperative program. It was to continue the topside-sounding program, started with the Canadian satellite *Alouette 1*, of studying the ionosphere from above.

No Intelsat Agreement

Representatives and observers from about 100 nations held a 4-week conference at the U.S. Department of State in Washington

Feb. 24–Mar. 21 in an unsuccessful effort to reach agreement on a permanent program for managing the 68-nation International Telecommunications Satellite Consortium (Intelsat).

The Soviet observers were headed by Vladimir Minashin, chief of the Soviet Communications Ministry. Minashin declared at the conference Feb. 27 that 3 principles should be basic to an international space communications system: (1) under a system encompassing domestic, regional and global networks, any country should have the right to use whatever networks met its needs; (2) every participating nation should have equal representation and rights in the governing bodies; (3) expenses of and income from the systems should be shared among participants on the basis of their use of the facilities.

The major U.S. objection to the Soviet position was the U.S. stand that only major users of the system should be represented on the governing board. The U.S., which had paid nearly $60 million of Intelsat's $100 million in outlays so far, managed the program through its privately owned Communications Satellite Corp. (Comsat) and had about 53% of the vote.

Several Intelsat members had indicated a desire to create an international body to take over some of Comsat's functions and thus diversify control. French delegates said that since Intelsat satellites had fixed positions over different geographic locations, they were regional in character and should be controlled regionally by their regional users. The U.S., which had borne the major expense in developing the system, held that it had to control equipment in which it had such a heavy public investment.

Intelsat's 2d of a new series of 322-pound communications satellites had gone into orbit Feb. 5. It was lofted by a Delta rocket launched from Cape Kennedy at 7:39 p.m. NASA charged Comsat $5 million for launching the $6 million satellite. The 3d of 6 satellites in Intelsat's Intelsat-3 series, but only the 2d to achieve a successful orbit (rocket failure destroyed the first Intelsat-3), the new satellite went initially into an elliptical orbit. But its small on-board motor was ignited Feb. 7, and it pushed it into a synchronous orbit in which it appeared to hang motionless above the equator 22,300 miles over the Gilbert Islands. On reaching its station above the Pacific, the satellite was named *Pacific 3*.

Pacific 3, capable of handling 1,200 2-way phone conversations or 4 TV programs simultaneously, went into commercial operation Mar. 3 on the conclusion of its testing period. (The Intelsat-2 satellites currently operating had a capacity of 240 phone conversations or one telecast each.) *Pacific 3* was put to use relaying communications between

North America and Hawaii, the Philippines, Australia, Japan and the Asian mainland.

Because *Pacific 3* was plagued by interruptions in service, a new $6.7 million Intelsat-3 was sent into orbit May 21 to replace it. The new satellite was launched by NASA at Cape Kennedy. By radio signal from the ground May 23, the engine aboard the new Intelsat-3 was ignited to start moving it to the old satellite's station above the Pacific. The old satellite was shifted 7,000 miles west to a new station over the Indian Ocean, where there was less communications business. (The synchronous satellites over the Indian, Pacific and Atlantic oceans gave Intelsat the first global communications satellite network.)

Another Intelsat-3 was launched from Cape Kennedy at 10:06 p.m. EDT July 26 by a 3-stage Delta rocket. The satellite failed to achieve its proper transfer orbit, however, and was located by radar July 28 in an orbit with an apogee of only 2,963 miles instead of the scheduled 23,000 miles. The orbit was too low for the satellite to be shifted into the intended synchronous orbit. The launching was conducted by NASA on behalf of Comsat. Its purpose was to orbit a new satellite to replace the Intelsat-3 that had been put above the Atlantic in Dec. 1968 but which had blacked out June 29 because of antenna difficulties. Comsat announced Aug. 8 that it had corrected the latter satellite's trouble and had restored the satellite to service.

U.S.-Soviet Communications Merger Urged

A panel of communications experts from 10 nations recommended Nov. 16 that the U.S. and the Soviet Union merge their communications satellite systems into an internationally controlled network open to all nations. The panel, whose recommendations were made as individuals rather than as government representatives, said the system should be governed by a board whose voting rules would give special recognition to the U.S. and the Soviet Union.

The panel conceded that "obstacles undoubtedly exist" to setting up such a vast system, but it held that the U.S.-run Intelsat network and Russia's proposed Intersputnik system could be linked without technical difficulties. The report said that "all nations should have access to a global, integrated communications satellite system" regardless of political affiliations.

The study, issued by the Carnegie Endowment for International Peace and the Twentieth Century Fund, summarized a conference held at Talloires, France Sept. 21-25.

German & British Satellites Launched by U.S.

West Germany's first satellite, the 157-pound *Azur*, was sent into near-polar orbit Nov. 7 by a 4-stage U.S. Scout rocket launched by the U.S.' NASA from Vandenberg Air Force Base, Calif. The scientific satellite, built to study the earth's radiation belt, the aurorae and solar particles, achieved an orbit with an apogee of 1,920 miles and perigee of 228 miles. Under a cooperative agreement between NASA and West Germany's Ministry for Scientific Research, the U.S. defrayed the entire cost of the 1\frac{1}{2}$ million rocket and launch facilities, the Germans paid for and provided the satellite and its instrumentation.

The 535-pound *Skynet*, Britain's first military communications satellite, was sent toward a synchronous orbit by means of a U.S. Delta rocket launched by U.S. Air Force personnel at Cape Kennedy at 7:37 p.m. A refiring of the Delta final stage Nov. 24 put *Skynet* into station 22,300 miles above the Indian Ocean at the equator. The satellite, designed to provide communications between London and points as far away as Singapore, was compatible with the U.S. military communications satellite system.

(Under an accord signed in Washington Sept. 18, the U.S. agreed to orbit a satellite that would beam farming, birth-control and other instructional TV programs to community receivers in about 5,000 Indian villages. The satellite, to go up in 1972, would be the first to send TV signals directly to users without the use of expensive relay stations on the ground. The satellite would be *ATS 6* [applications technology satellite]. It would be placed in synchronous orbit within range of the Indian TV sets. The programs would be produced and transmitted to the satellite by Indian officials.)

4th Japanese Failure

University of Tokyo scientists Sept. 22 failed for the 4th successive time in an attempt to put a Japanese satellite into orbit. The 18-inch prospective satellite was sent up in the nose of a 55-foot 4-stage Lambda-4S4 rocket launched at Uchinoura. The first 3 stages fired and separated successfully, but the rocket then veered off course, and the satellite burned up by friction with the atmosphere. Observers blamed the 4 failures on lack of control and guidance equipment. The pacifistic University of Tokyo scientists in charge of the project had refused to develop guidance and control technology for fear that it might be used for military rocketry.

The Japanese government Space Development Committee Oct. 1

unified all Japanese space efforts under a new part-government Space Development Corp. The corporation's capital of 539 million yen (1\frac{1}{2}$ million) was made up of 500 million yen from the government and 39 million yen from industry. The space committee said the corporation would be responsible for a program that would include the orbiting of an ionosphere observation satellite in 1972 and of a synchronous communications satellite in 1974. These 2 satellites had previously been scheduled for launching in 1971 and 1973, respectively. The Japanese government had decided that the satellites would be launched with the use of U.S. guidance and control technology.

Europa Shot Fails

The European Launcher Development Organization (Eldo) launched a 3-stage *Europa-1* rocket from the Woomera test range in South Australia July 3, but the German-built 3d stage failed to ignite and fell into the Pacific carrying a test payload that Eldo scientists had hoped to put into orbit. The first-stage British-built Blue Streak and the 2d-stage French-built Coralie had operated perfectly, Eldo officials reported. (Only the last of 10 scheduled Eldo launchings at Woomera remained to be conducted before Eldo moved to its new launching site in French Guiana.)

1970

The U.S.' only manned space flight of 1970 ended in near-disaster. *Apollo 13*, whose mission was to make the 3d landing of astronauts on the moon, developed a serious malfunction in flight and was forced to return to earth without a moon landing. But the astronauts were able to land on earth safely. 2 Soviet cosmonauts set an endurance record for space flight, circling the earth for almost 18 days. The USSR also sent 2 successful unmanned probes to the moon; the most spectacular feat was the landing of a robot car that toured a small area of the moon while controlled by radio from the earth. The U.S. continued to cut back on space expenditures, and the remaining Apollo series of planned moon landings was cut from 6 to 4.

MAN IN SPACE

Crippled Apollo 13 Fails to Land on Moon

3 U.S. astronauts, the oxygen and power of their *Apollo 13* command module suddenly reduced to a dangerously low level Apr. 13 by the rupture of an oxygen tank, used oxygen and power from their lunar module (LM) landing craft to keep themselves alive. The propulsion system of this improvised LM "lifeboat" then took them around the moon Apr. 14 and back to earth, where they splashed down safely in the Pacific Apr. 17.

The accident forced the cancellation of a planned landing on the moon; and for $3\frac{1}{2}$ days of the 6-day flight, which had begun Apr. 11, there appeared to be good reason for fear that the mishap might also cost the lives of the astronauts. This was the first in-flight failure in the 22 manned flights—5 of them to the vicinity of the moon—made in the U.S. space program so far.

All information on the *Apollo 13* mission—including the details of the accident and of the all-out effort that brought the astronauts back home safely—was made public as soon as U.S. space officials themselves knew the facts. In some instances, the public and the U.S. space agency learned of events simultaneously as data was radioed to earth by the 3 astronauts struggling for their lives in space. It had been reported that the public had appeared bored by the early stages of the flight—after 2 successful U.S. lunar landings. The accident and the resulting danger to the astronauts transformed the boredom into deep concern for the safety of the 3 Americans. Messages and offers of aid were sent to the U.S. from leaders of the USSR and most other countries, and prayers were offered by Pope Paul VI and religious leaders on every continent.

Apollo 13's crew was headed by Navy Cmndr. James Arthur Lovell Jr., 42, the first man to reach the vicinity of the moon twice (although he had never landed on the moon). Lovell was the current record-holder for number of space flights and hours in space. He had been the navigator of *Apollo 8* in Dec. 1968 on the first manned flight around the moon. The *Apollo 13* flight, his 4th space flight, added about 142 hours to the record 572 hours he had already spent in space. Lovell, who had said Feb. 4, 9 weeks before the *Apollo 13* launching, that the flight would be his last trip in space, told reporters in Houston, Tex. Apr. 21, 4 days after his safe return to earth, that "if the agency [NASA] wants this crew to go back . . . , we'll be glad to go."

As *Apollo 13*'s lunar module (LM) pilot, Fred Wallace Haise Jr., 36, a civilian, had been scheduled to land on the moon with Lovell before the oxygen-tank breakdown caused the lunar landing to be canceled. The *Apollo 13* mission was his first space flight.

John Leonard Swigert Jr., 38, a civilian, was a last-minute substitute as pilot of *Apollo 13*'s command module (CM). Originally assigned to the back-up crew, he was moved up to the first crew after it was discovered that the *Apollo 13* astronauts had all been exposed to German measles and that the original CM pilot, Navy Lt. Cmndr. Thomas K. Mattingly 2d, was the only one of them who had not developed immunity to the disease. The *Apollo 13* flight was Swigert's first in space. Swigert was the first bachelor to fly in a U.S. spaceship.

The fully assembled *Apollo 13* spaceship consisted primarily of 4 detachable segments—2 forming the command and service module (CSM) and the other 2 forming the lunar module (LM).

The CSM consisted of: (a) the heat-shielded command module (CM), a 12,365-pound conical capsule, $11\frac{1}{2}$ feet high and 12 feet 10 inches at the base, containing the crew compartment and most of the flight controls and instruments; (b) the service module (SM), a 51,105-pound cylindrical container, 12 feet 10 inches in diameter, carrying the spacecraft's main 20,500-pound-thrust engine and its propellents, oxygen tanks and fuel-cell power plants.

The LM segments were: (a) the 4-legged 33,325-pound descent stage, with a rocket engine designed to ease the LM down to the surface of the moon; (b) the ascent stage, carrying the ascent engine and containing the pressurized cabin designed to carry 2 astronauts on the trip down to the moon and on the flight back up from the moon to rejoin the CSM, which was to have remained in orbit around the moon.

Lovell Mar. 14 had named the CSM *Odyssey*, and he had given the name *Aquarius* to the LM.

The flight program had called for Lovell and Haise to make a risky descent in *Aquarius* to a relatively narrow, preselected valley in the moon's Fra Mauro Hills about 120 miles east of the area in which *Apollo 12* had made its manned landing. The *Aquarius* site, on the eastern shore of the moon's dry Ocean of Storms, had been picked because the Fra Mauro Hills were believed to contain some of the oldest material on the moon. While Lovell and Haise were to have been making their 33-hour expedition on the moon, Swigert had been scheduled to remain in orbit around the moon in *Odyssey*, which was to have picked up the moon-walkers after they had completed their activities on the lunar surface and had blasted off in the LM's ascent section.

Apollo 13, carrying its 3 astronauts strapped to form-fitting couches in the CM, was sent up from Cape Kennedy's Launch Pad 39A at 2:13 p.m. Apr. 11 atop a 3-stage Saturn-5 launching rocket. Spectators at the launching included Vice Pres. Agnew and West German Chancellor Willy Brandt.

The Saturn-5's first stage, operating perfectly, boosted the spacecraft-rocket assembly to an altitude of 45 miles, then cut its engines and fell away from the rising *Apollo 13*. The 2d stage's center engine malfunctioned: it discontinued ignition 2 minutes 7 seconds early. But the on-board Saturn-5 computer compensated for this by burning the 2d stage's 4 outboard engines an additional 34 seconds and by adding 10 seconds to the ignition of the 3d (or S4B) stage.

Apollo 13, still attached to its 61-foot-long S4B stage, went into a 117-mile-high orbit at 2:24 p.m. The spacecraft circled the earth twice in orbit while the astronauts and ground controllers checked their instruments and control displays to make sure that everything was operating properly. At 4:48 p.m. Lovell reignited the S4B engine to increase *Apollo 13*'s speed to the escape velocity of 25,000 mph. and to head the spaceship toward the moon, then some 246,500 miles away.

Shortly after achieving a satisfactory translunar trajectory, Lovell started the procedure of attaching *Odyssey*, the CSM, to *Aquarius*, the LM, which had gone into space encased in the cylindrical S4B. Lovell first fired explosive bolts to detach *Odyssey* from the S4B. After this he moved the CSM about 60 feet forward of the S4B and turned the CSM around so that its pointed nose was aimed at the open end of the S4B. He then slowly moved *Odyssey* back toward the S4B and the enclosed *Aquarius* until the docking probe in the CSM's nose was thrust into the LM's docking collar. After *Odyssey* and *Aquarius* were tightly locked together, Lovell carefully backed the assembled CSM-LM away from the S4B and thus pulled *Aquarius* out of the S4B.

Apollo 13 continued to draw away from the S4B until the 2 vehicles were speeding toward the moon at a distance of about 40 miles from each other. At this point a radio signal directed the S4B to vent its remaining oxygen. The jet of oxygen increased the S4B's speed, and the S4B hurtled toward the moon at a velocity higher than that of *Apollo 13*. The S4B smashed into the lunar surface at 8:09 p.m. EST Apr. 14 in the Ocean of Storms about 87 miles west-northwest of a seismometer and ion detector that had been left on the moon by the astronauts of *Apollo 12*. The force of the impact was comparable to the force created by the explosion of 11 tons of TNT. The seismometer recorded a series of moonquakes beginning about 30 seconds after impact, and the reverberations of the crash continued for 3 hours 20 minutes. The ion detector, whose main function was to record the flow of particles from the sun and to detect any evidence of lunar atmosphere, recorded a patter of particles 22 seconds after impact; this indicated that the crash had thrown up debris to a distance of at least 87 miles.

A mid-course correction, made by means of a brief burst of the service rocket at 8:45 p.m. Apr. 12, moved *Apollo 13* slightly from a trajectory that would have taken it into a 291-mile-high orbit around the moon into a revised trajectory calculated to put the spaceship into a 69-mile-high orbit.

During the joshing that had become traditional in radio conversations between astronauts and their ground controllers, Swigert suddenly recalled Apr. 12 that he had failed to file his federal income tax return (due Apr. 15), but he was assured that Manned Spacecraft Center officials would get an extension for him.

The unexplained accident that threatened the lives of the astronauts apparently happened at 10:07:42 p.m. EST Apr. 13 as *Apollo 13*, already 205,000 miles from the earth, was approaching the vicinity of the moon. It was determined later that Oxygen Tank 2, one of the 2 oxygen tanks in the SM, had exploded. Oxygen from the tanks was used both for breathing by the astronauts in the CM and for generating power for the CM through 3 fuel cells in the SM.

The destruction of the oxygen tank took place about an hour after Lovell and Haise had crawled through the tunnel connecting *Odyssey* with *Aquarius* to inspect the instruments and equipment of the landing craft. They returned to the CM shortly after 10 p.m. after reporting everything in the LM to be in proper working order. Less than a minute after the accident to the oxygen tank, Swigert radioed to the ground controllers at the Manned Spacecraft Center in Houston that "we've got a problem." Haise reported that "we had a pretty large bang," and Swigert said there had been a complete loss of power on an electrical line hooked into a fuel cell that used oxygen from the tank later determined to have ruptured. A 2d line was losing power quickly. Lovell said after looking out of a window that "we're venting something into space; it's a gas of some sort."

As the power readings declined to zero, Mission Control in Houston informed the astronauts that "we're starting to think about the LM lifeboat." But it was Lovell who radioed to the ground controllers the decision made by the astronauts. "We've been talking it over," he reported. "It looks like we're going to have to go to a LEM lifeboat." Under this procedure, one of numerous emergency steps worked out in detail for the Apollo flights, Lovell and Haise moved back into the LM to switch on its equipment and use its oxygen and power supply. Swigert remained briefly in the CM to shut off its equipment and use the last waning fuel cell to charge the batteries needed for re-entry;

while in the CM he also used oxygen from the LM. The lunar landing was tacitly canceled.

One of the ground controllers explained in radio conversation with the *Apollo 13* astronauts that the LM "lifeboat" procedure in the current situation called for the use of the LM descent engine, "the big engine of the LM, to propel the entire spacecraft stack to higher velocity" as it rounded the moon. This would give *Apollo 13* enough speed "to come back to earth a day earlier than a normal free-return trajectory," he said. During the rest of the trip, the astronauts were to use *Aquarius*' limited but adequate oxygen and power, the controller disclosed. The LM's relatively small water supply also had to be used—and conserved—to cool the spaceship's vital electronics systems. Although 2 astronauts at a time might sleep in the CM, he said, at least one astronaut would always have to stay awake in the LM to monitor its controls and equipment. As it turned out, however, most of the time 2 or all 3 of the astronauts rode in the cramped and crowded crew compartment of the LM. The heat turned down to save power, *Odyssey* and *Aquarius* quickly became so uncomfortably cold—about 45° to as low as 38°F.—that the astronauts found it almost impossible to sleep or eat.

There was no question about whether or not the crippled *Apollo 13* should continue on around the moon before heading back to earth. Rounding the moon—and using the moon's gravity to whip the spaceship back towards the earth—would take less power and bring the astronauts back sooner than would an attempt to turn back before reaching the vicinity of the moon.

At 3:43 a.m. EST Apr. 14 *Aquarius*' descent engine was fired for 30 seconds to increase *Apollo 13*'s speed by about 25 mph. and send the spaceship around the moon at a distance of 137 miles instead of the 62-mile approach previously planned. *Apollo 13* passed behind the moon at about 7 a.m. and emerged less than an hour later to begin the journey back to earth. Course-correcting blasts of the descent engine were fired at 9:41 p.m. EST Apr. 14 and at 11:31 p.m. Apr. 15. The final course-correction maneuver was made with an ignition at 7:53 a.m. Apr. 17 as *Apollo 13*, only 48,000 miles from the earth, was nearing its home planet at a speed of 6,600 mph.

Approaching the earth Apr. 17, the astronauts jettisoned the SM at 8:15 a.m. EST, and they were then able to see the damage done by the oxygen-tank explosion. "There's one whole side of the spacecraft [the SM] missing," Lovell reported. "Right by the high-gain antenna the whole panel is blown out, almost from the base to the engine. It looks like the explosion stained it a dark brown streak. It's really a mess."

The astronauts took still and motion pictures of the damage to the service module.

Lovell and Haise then joined Swigert in *Odyssey*, the hatches between *Odyssey* and *Aquarius* were closed, and the LM was jettisoned at 11:43 a.m. at an altitude of 12,000 miles. "Farewell, *Aquarius*, and we thank you," said Cmndr. Joseph P. Kerwin, a fellow astronaut, at Mission Control in Houston.

With the 3 astronauts strapped to their couches in the CM, the sole remaining segment of the spacecraft, *Apollo 13* smashed into the earth's atmosphere at a speed of 24,680 mph. at 12:53 p.m., and the buildup of ionic particles around the capsule blacked out radio communication with the astronauts until 12:59 p.m. The spaceship's speed was quickly reduced and its heat-shield and sides seared by friction with the atmosphere. The 2 16-foot-diameter drogue parachutes were deployed at 1:02 p.m., and the 3 main parachutes came out a minute later to ease *Odyssey* and its crew down to the surface of the Pacific.

The spaceship splashed down at 1:07:44 p.m. Apr. 17 only $3\frac{1}{2}$ miles from the waiting recovery carrier *Iwo Jima*, about 800 yards from the aiming point and some 610 miles southeast of American Samoa.

Helicopters from the recovery ship picked up the 3 astronauts about 35 minutes later and brought them to the carrier. Physicians aboard the carrier gave them a quick initial physical exam and reported them to be in relatively good health. Dr. Keith Baird of NASA said that Haise had been found to have a mild urinary tract infection—probably due to shortage of water—and a mild fever of 100.6°. All 3 astronauts "were more tired than the other crews I have been associated with," Baird said, but "except for being tired, I think they are all in good health." After the astronauts had returned to Houston, Dr. Charles A. Berry, chief physician and medical researcher at the Manned Spacecraft Center, disclosed Apr. 21 that Lovell had lost 14 pounds, the most weight ever lost by an American in space flight. Swigert lost 11 pounds, Haise $6\frac{1}{2}$ pounds.

(A SNAP-27 nuclear generator loaded with 8.6 pounds of radioactive plutonium fell from the jettisoned LM and splashed into deep water in the Pacific Ocean Apr. 17 northeast of New Zealand.)

Pres. Nixon flew Apr. 18 to Honolulu, where he met the 3 astronauts and presented a Medal of Freedom to each. En route to Honolulu, the President had stopped in Houston, where he had awarded a Medal of Freedom to the *Apollo 13* mission operations team. The medal was accepted by Sigurd Arnold Sjoberg, 50, director of flight operations at the Manned Spacecraft Center.

NASA & 2 Contractors Blamed

In a detailed report released by NASA June 15, NASA itself and 2 contractors—Beech Aircraft Corp. and North American Rockwell Corp. —were blamed for the near-fatal oxygen tank explosion that aborted the *Apollo 13* mission.

The *Apollo 13* review board's 2-month investigation, in which 300 scientists, engineers and technicians performed 100 tests, was headed by Edgar M. Cortright, director of NASA's Langley Research Center at Hampton, Va. The report concluded that "the accident was not the result of a chance malfunction in a statistical sense but, rather, resulted from an unusual combination of mistakes, coupled with a somewhat deficient design." These mistakes started with the installation by Beech Aircraft, the oxygen tank's manufacturer, of the wrong safety switches in the tank, and neither of the 2 companies nor NASA detected the mistake. This made the tank's protective thermostats inoperative, and the chain of circumstances that followed presumably resulted in the explosion.

NASA Administrator Thomas O. Paine told the Senate Aeronautics & Space Science Committee June 30 that changes to be made in *Apollo 14* as a result of the *Apollo 13* accident might cost as much as $15 million and would cause a delay in the *Apollo 14* launching.

About 645 pounds of safety features were added to *Apollo 14*, retired Navy Capt. Chester J. Lee Jr., ground director of the mission, announced in Washington Dec. 21. Lee said the mission's duration had been cut from 10 days to 9 as a further safety precaution. *Apollo 14*'s new safety features included redesigned oxygen tanks, the addition of a 3d oxygen tank, an extra battery and 40 more pounds of water.

Cosmonauts Set Endurance Record

2 Soviet cosmonauts—Col. Andrian Grigoryevich Nikolayev, 40, and civilian flight engineer Vitali Ivanovich Sevastyanov, 34—set a space-flight endurance record of 17 days 16 hours 59 minutes June 1-19 during an orbital mission in the spaceship *Soyuz 9*.

The principal objective of the flight was to test the ability of men to work and remain healthy during the protracted periods of weightlessness that would be experienced by crews of orbiting space laboratories. Preliminary examination of the cosmonauts after their landing showed that their training and the daily exercises they took in space had apparently enabled them to emerge in good condition from their recordbreaking flight—although it took nearly a week for their strength to return to pre-flight level.

(Nikolayev, pilot-commander of the spaceship and chief of Soviet cosmonauts, was a veteran space flier who had gone into orbit in the one-man *Vostok 3* in 1962. After his 1962 space flight, Nikolayev was married to Soviet cosmonaut Valentina Nikolayeva Tereshkova, the only woman to make a space flight. Apparently in honor of Nikolayev, *Soyuz 9* was code-named *Sokol* [*Falcon*], the code name that had been used for Nikolayev's first spaceship. Sevastyanov had never flown in space before.)

Soyuz 9 was launched at 10 p.m. Moscow time June 1 from the Baikonur cosmodrome in Kazakhstan by means of a multistage booster rocket. 9 minutes after launching, the spaceship was in orbit.

As usual with Soviet space operations, no advance announcement had been made about the mission, and the public disclosure of the launching did not take place until about a half-hour after liftoff, when Soviet space officials had determined that *Soyuz 9* was safely circling the earth.

Tass reported that by the time the spaceship had achieved its 3d revolution, it was in orbit with an apogee of 220 kilometers (136.7 miles), perigee of 207 kilometers (128.6 miles), period of 88.59 minutes and $51.7°$ angle of inclination. An orbital maneuver executed manually by Nikolayev June 2, during the 5th revolution, raised the apogee to 267 kilometers (165.9 miles), the perigee to 213 kilometers (132.53 miles) and the period to 89.05 minutes; the angle of inclination remained $51.7°$. A 2d orbital maneuver June 2, during the 17th revolution, dropped the apogee to 266 kilometers (165.3 miles), raised the perigee to 247 kilometers (153.5 miles) and increased the period to 89.5 minutes.

During the flight the cosmonauts had busy 16-hour workdays. They ate "natural" foods (ham, bread, coffee, milk, borshch and other soups, veal, currant juice, cottage cheese, etc.) rather than specially prepared space foods, and they used a stove to heat food. They exercised regularly, wearing "special load suits." They took repeated biomedical observations of each other, made careful photos of different parts of the earth and of various space phenomena, performed experiments, practiced navigation by different methods and piloted the spaceship through various maneuvers. Biological items they took along for research included specimens of drosophyllum, potato tubers, seeds of wheat, barley, onion and arabidopsis, chlorella and algae.

The cosmonauts June 13 participated in a joint meteorological observation program by photographing cloud formations over the Indian Ocean as the Soviet research ship *Akademik Shirshov* took observations from the surface and by means of balloons and while a Soviet Meteor weather satellite took TV photos from higher altitude.

Nikolayev and Sevastyanov were shown on Soviet TV June 14 as they cast ballots for candidates in the Soviet parliamentary election. Speaking from their *Soyuz 9* spacecraft, which served as their polling booth, they urged Soviet citizens on the ground to vote.

At 4:45 p.m. June 15 Moscow time the 3 cosmonauts broke the 13-day 18-hour 35-minute space endurance record set by U.S. astronauts Frank Borman and James A. Lovell Jr. in the spaceship *Gemini 7* in Dec. 1965.

An orbital correction made June 15, during the 208th revolution, shifted the orbit's apogee to 231.4 kilometers (143.9 miles), the perigee to 215.1 kilometers (133.6 miles) and the period to 88.8 minutes.

Nikolayev and Sevastyanov played a 6-hour chess match against their ground controllers during their 141st to 144th revolutions around the earth June 16. The game ended in a draw on the 36th move.

The cosmonauts brought *Soyuz 9* out of orbit just after 2 p.m. Moscow time June 19 during its 288th revolution around the earth. Detached from the spaceship's orbital compartment, which remained in orbit, the command/descent section smashed into the atmosphere and was slowed aerodynamically.

About 43 minutes after the descent had started, with the spaceship down to an altitude of about 30,000 feet, the parachutes were deployed. *Soyuz 9* made a soft landing a quarter-hour later about 47 miles west of Karaganda in Kazakhstan at 2:59 p.m.

The descending spaceship had been spotted by aerial recovery teams, and the cosmonauts were taken aboard a recovery helicopter shortly after they landed. The disclosure of the landing, which had not been announced in advance, was broadcast about 45 minutes after the spaceship touched down and after Soviet space authorities had assured themselves that the mission had ended successfully.

Tass reported June 20 that the cosmonauts were experiencing some minor difficulty in readjusting to gravity after nearly 18 days of weightlessness. Nikolayev and Sevastyanov were reported to be complaining about their "heavy bodies," but Tass added that "what was heavy an hour ago is becoming light." Tass reported June 23 that they still had not recovered completely.

At a Star City press conference restricted to Soviet newsmen, Nikolayev noted June 24 that he found recovering from the *Soyuz 9* mission "more difficult" than his readjustment after the 4-day space flight he had made in 1962. But the never-named "chief designer" of the Soyuz program said at the press conference that the *Soyuz 9* flight had showed that "man can work in space for a long time, for at least a month."

At a press conference at Moscow State University July 9, Nikolayev described the cosmonauts' readjustment to earth: "After 18 days of weightlessness, the entire body (arms, legs and head) suddenly [on landing] felt very heavy. The sensation was as if we were in the centrifugal stand, under the influence of small stress. The first day [after returning to earth] it seemed that the stress was about 2G [twice that of gravity] or perhaps slightly more. And in the days that followed it gradually decreased, disappearing completely on the 5th or 6th day. Now we feel quite well and are absolutely healthy."

Prof. Mstislav V. Keldysh, chairman of the Soviet Academy of Sciences, reiterated at the July 9 press conference that "the main task which we [the Soviet space authorities] have assigned to ourselves within the next few years is that of building orbital stations and studying near-earth space and the solar system using automatic apparatuses. At the same time we shall use space flights for the national economy— communications, meteorology, navigation, etc." "It is my opinion," he said in answer to a question, that a permanent Soviet space station could be expected "within the next few years." He said it was still "difficult to fix the maximum time" that a cosmonaut could remain in an orbital space station without physical harm. But he speculated that "this apparently may be about a month, and this is quite sufficient for a long-term station" in which crews would be relieved periodically.

Asked whether "the Soviet Union [had] at present given up plans for a manned landing on the moon," Keldysh replied: "We never announced such a program, thus, actually, we are not giving up anything. In the near future . . . we are not planning manned flights to the moon." In answer to another query, Keldysh said that the problem of "manned interplanetary flights lasting for several months . . . most probably . . . will not be solved within this decade."

(At a Kremlin reception July 3, Nikolayev was promoted from colonel to major general.)

SOVIET LUNAR & VENUS PROBES

Unmanned Mission Gets Moon Rocks

The U.S. sent no unmanned probes to the moon or planets during 1970, but the USSR continued its instrumented explorations. 2 unmanned Soviet probes visited the moon, one returning to earth with rock samples and the other leaving a robot car on the lunar surface for exploration and experimentation.

The unmanned rock-gathering probe, *Luna 16*, landed on the

moon Sept. 20, scooped up some lunar rock samples and brought them back to the USSR Sept. 24. This voyage of *Luna 16* was the first in which an unmanned space vehicle lifted itself off the moon (or any other astronomical body) and returned to earth. (2 U.S. spaceships had returned from the moon's surface, but both were manned.)

Luna 16's trip began with a launching from the Baikonur cosmodrome at 4:26 p.m. Moscow time Sept. 12. After being carried into parking orbit around the earth by its multi-stage booster rocket, the probe was injected into a lunar trajectory, and it continued on toward the moon for $4\frac{1}{2}$ more days. *Luna 16* then went into orbit around the moon Sept. 17.

After 2 orbital corrections Sept. 18 and 19 and after circling the moon 41 times, *Luna 16* started toward the lunar surface Sept. 20. The main propulsion unit was ignited at 8:12 a.m. to bring the probe out of orbit; then ignition was suspended until the probe was down to an altitude of about 600 meters (1,970 feet). At this point the engine was again ignited, and the thrust was varied under a computer program that took into account preset data and incoming data on altitude and speed of descent. The main engine was cut out at an altitude of about 20 meters (65 feet), and 2 low-thrust engines then took over. These engines were cut out when the altitude decreased to two meters ($6\frac{1}{2}$ feet).

Luna 16 landed softly on the moon's dry Sea of Fertility at 8:18 a.m. Moscow time Sept. 20. Tass reported that the location of the landing site was: Latitude 0° 41′ S.: Longitude 56° 18′ E.

Luna 16 stayed on the moon for 26 hours 25 minutes. Its most spectacular achievement before returning was its use of an electric drill to dig some 350 millimeters (about 14 inches) into the moon's surface and to collect rock samples. Tass confirmed that the drill was "controlled from the earth." The samples were sealed in an airtight container. The probe also took temperature, radiation and navigational measurements.

Using its landing stage as a launching platform, *Luna 16* blasted off from the moon at 10:43 a.m. Moscow time Sept. 21 for its return voyage to earth. As the probe approached the earth Sept. 24, the reentry capsule separated from the ascent stage. The capsule then reentered the earth's atmosphere and made a parachute-slowed descent.

Luna 16's capsule landed at 8:26 a.m. Sept. 24 in Kazakhstan some 50 miles southeast of Dzhezkazgan and about 250 miles from its Baikonur launching point. A helicopter recovered the payload, and the lunar rocks were sent to the Soviet Academy of Sciences in Moscow.

As was customary in the Soviet space program, *Luna 16* was launched without prior announcement. The disclosure of the launching was made after Soviet space officials ascertained that the probe had successfully entered a lunar trajectory. Similarly, the effort to bring back lunar rock was disclosed only after the rock was aboard the probe and the probe had been successfully launched from the lunar surface for the return trip. The only departure from this policy was a Tass announcement Sept. 23 that *Luna 16*'s "recovery capsule will land in ... Kazakhstan at 8:20 a.m. Moscow time Sept. 24." The actual landing took place 6 minutes later than the announced time.

In one of the first reports on the lunar material, *Izvestia* said Sept. 26 that the Soviet sample of lunar dust "seems to resemble somewhat dry black earth—a gray dust with a brownish tint." It added that "when a ray of light falls on it, it looks greenish and even a little reddish." *Pravda* reported Oct. 4 that "the sample ... in its main mass consists of fine-grain mineral particles." The article noted "its crumby structure and the presence of noticeable cohesive forces between the particles."

The flight of the unmanned *Luna 16*, following 2 U.S. manned landings on the moon, led to renewed speculation about the comparative value of the 2 different methods of exploring space.

The never publicly identified "chief designer" of *Luna 16*, interviewed in *Pravda* Sept. 26, paid tribute to "the tremendous value" of the U.S. manned flights. But, he said: "In our opinion, it is more expedient in the present stage of space exploration to work with automatic devices."

Georgi A. Petrov, director of the Soviet Space Research Institute, had said in an *Izvestia* interview Sept. 22: "The 'share' of man and automatic devices in space research will constantly vary with the development of technology. I consider that the time has arrived for man to board automatic space stations, their orbits covering even near-moon space. But these stations must ... be just visited by man. For some time he can work there, replace instruments, adjust them, pick up the materials of some investigations and return to the earth, while the station operates in the automatic regime till the next visit of man. ... As for moon research, short visits by man ... would be probably quite useful. ... In the exploration of planets, automatic devices are and must be running ahead of man. ... "

Boris Petrov, chairman of the Soviet Academy of Sciences, agreed in *Pravda* Sept. 24 that there was a role for manned space flight missions. But "automatic devices are now designed to conduct the main part of the exploration of outer space," he declared. He estimated that "the flight of an unmanned craft compares in cost to a manned one by a factor of one to 20 or 50."

Soviets Put Robot Car on Moon

An 8-wheeled robot car dubbed *Lunokhod 1* was landed on the moon Nov. 17. *Lunokhod* (which means, roughly, moon rover) *1* was carried to the moon by the unmanned Soviet spacecraft *Luna 17*, which had been launched from the Baikonur cosmodrome Nov. 10.

The solar-powered *Lunokhod 1*, the first wheeled vehicle to travel on the surface of the moon, was guided by radio signals from earth controllers for 5 days in a program of exploration and experiments. Its TV cameras and instruments were then shut off Nov. 21 as the vehicle was prepared for what Tass described as "hibernation" during the cold, 14-day lunar night from Nov. 24 to Dec. 8.

The *Luna 17* spacecraft, with *Lunokhod 1* aboard, was launched at 5:44 p.m. Nov. 10 (Moscow time) and put into parking orbit around the earth for less than a full revolution. The engine of the carrier rocket's final stage was then reignited and the velocity increased to send *Luna 17* into its lunar trajectory. The launching was not disclosed until the following morning, and the Soviet announcement did not reveal at that time that a lunar landing was planned or that the probe carried a wheeled vehicle.

After 2 course corrections, *Luna 17* reached the vicinity of the moon Nov. 15. It went into a circular lunar orbit at an altitude of 52 or 53 miles; the orbit's period was 116 minutes.

On radio instructions from earth, *Luna 17* dropped out of lunar orbit the morning of Nov. 17 and made a soft landing at 6:47 a.m. Moscow time on the western edge of the moon's arid Sea of Rains (or Sea of Showers).

After telemetry had indicated that all systems aboard the probe were in working order, a radio signal from earth activated a mechanism that lowered a ramp from the probe. The explosive bolts that had held *Lunokhod 1* tightly in place aboard the spacecraft were then fired by command from earth, and the freed lunar car rolled down the ramp to the surface of the moon at 9:28 a.m.

Soviet authorities waited until *Lunokhod 1* had tested its TV cameras and other instruments and had moved a distance away from *Luna 17* before announcing toward evening Nov. 17 that *Luna 17* had landed and had disgorged a wheeled vehicle. Soviet officials revealed later that they had no plans to try to bring any part of *Luna 17* or *Lunokhod 1* back to earth.

The early Soviet announcements revealed that *Lunokhod 1* was equipped with a French laser reflector placed aboard under the Franco-Soviet space-cooperation agreement. The French instrument was usable as a moving beacon for location purposes and for making very precise measurements of distances, sizes and shapes on the moon and between the moon and the earth.

Soviet authorities disclosed that *Lunokhod 1*'s control team on earth—"the full crew"—consisted of "a commander, navigator, engineer, radio operator and a driver." In manipulating the vehicle from a distance of some 240,000 miles, the earth-based crew was guided by various instruments, including the lunar car's TV cameras, which were faced in 4 directions. The scientific apparatus included a digging instrument and X-ray spectrometer to analyze the moon's soil. An X-ray (or "roentgen") telescope was used to investigate cosmic rays and radiation sources.

Tass reported Nov. 22 that in its 5 days of activity before being shut down for its first lunar night, *Lunokhod 1* had traveled 646 feet. Tass said that it had "carried out experiments to study the mechanical properties of lunar soil and to determine the chemical composition of the surface layer of lunar rocks."

Lunokhod 1 resumed its activities for a 2d lunar day Dec. 10. The hibernation period, however, was not entirely without incident. Although it remained motionless because it relied on energy from the sun for propulsion, *Lunokhod 1* retained enough power in its batteries for 2 communications sessions. During these sessions the vehicle radioed telemetric data to its Soviet controllers on earth. The temperature on the moon's surface dropped to below $-130°C.$ ($-200°F.$) during the lunar night, but *Lunokhod 1* reported that the temperature inside its instrument compartment was maintained at about $15°C.$ ($59°F.$) by means of an isotope "furnace" that heated circulating gas.

Lunokhod 1's earth controllers Dec. 9 transmitted radio signals that started the recharging of the solar batteries in preparation for the resumption of the vehicle's activities. *Lunokhod 1* was then set in motion Dec. 10 by signal from earth, and it traveled a distance of 4,854 feet before being "parked" again Dec. 22 for its 2d lunar night (Dec. 23–Jan. 7, 1971).

During *Lunokhod 1*'s first lunar night, the Crimean astrophysical observatory of the Soviet Academy of Sciences had used the French-made laser reflector aboard the inactive lunar car Dec. 5–8 as a target in a successful experiment in lunar site location. But when French experimenters aimed a laser beam from the Pic du Midi d'Ossau observatory at the spot designated by the USSR, they failed to receive a reflection.

Lunokhod 1's "parking" site for its first lunar night had been announced by the Soviets as Latitude 38° 17' N., Longitude 35° W., in the Sea of Rains south of the Bay of Rainbow. Thomas O'Toole reported in the *Washington Post* Dec. 23, however, that "a topic of gossip in scientific circles" was the "possibility that the Soviets misled their French colleagues about the [vehicle's] precise location." A U.S. scientist was quoted as saying that "the French were given the same lunar coordinates . . . as we were—and they weren't the right ones."

It was reported that the French finally found the car and their laser reflector themselves after a 2-week search.

Zond 8 Circles Moon and Returns

In a lunar mission immediately preceding the *Luna 17* flight, the unmanned Soviet probe *Zond 8* was launched Oct. 20 into a parking orbit around the earth and then sent into lunar trajectory. The launching was not disclosed until Oct. 21, but Tass then announced that the program called for *Zond 8* to round the moon Oct. 24 and to return to earth Oct. 27. The Tass announcement differed from usual Soviet practice: normally the Soviet authorities never disclosed specific goals for one of their spacecraft until the space vehicle had actually achieved them; if the mission failed, its goal was not usually revealed.

Keeping to the announced schedule, *Zond 8* whipped around the moon Oct. 24 at a minimum distance of 690 miles and then headed back to earth. *Zond 8* splashed down Oct. 27 in the Indian Ocean 450 miles southeast of the Chagos Archipelago. It was recovered by the Soviet oceanographic ship *Taman*.

Soviet Probe Reaches Venus

The USSR's unmanned spacecraft *Venera (Venus) 7*, launched at 8:38 a.m. Moscow time Aug. 7, reached the planet Venus Dec. 15 after a 320 million-kilometer (200 million-mile) flight. It then apparently was destroyed 55 minutes after plunging into Venus' flaming atmosphere. The 1,180-kilogram (2,601-pound) probe had been described by Tass after launching as "an improved version" of the USSR's 6 previous Venus probes.

According to Tass, *Venera 7* entered the planet's atmosphere at

8:02 a.m. Moscow time, and its instrument capsule was then separated from the spacecraft. The capsule was slowed to a descent rate of 250 meters (820 feet) a second by aerodynamic braking, and a parachute system was then deployed. "Signals from the descent craft were received for 35 minutes," Tass said, and the data received "are being processed and studied."

Venera 7 was the 4th Soviet probe to enter Venus' atmosphere. The first 3 sent back signals for 96 minutes, 53 minutes and 51 minutes, respectively. U.S. observers speculated, however, that *Venera 7*'s shorter period of transmission might have been caused by a faster rate of descent and that the probe, therefore, could conceivably have descended as deeply as or even more deeply than its predecessors before being destroyed.

(Tass reported Jan. 26, 1971 that radio signals from *Venera 7* "were received for . . . 23 minutes after landing." The announcement noted that this was "the first time that scientific information was relayed directly from the surface of another planet. . . . " Tass and the Soviet government newspaper *Izvestia* reported that after the landing, "the volume of the [radio] signal was about 100 times less than during the descent." *Izvestia* said this "is most probably explained by the axis of the antenna being diverted from the direction of the earth after the landing." The weak signals were analyzed and deciphered by a "special method" that required the use of computers, *Izvestia* reported. The signals took 3 minutes 22 seconds to travel the 36.3 million miles that then separated Venus from the earth.

(Data received from *Venera 7* disclosed that the surface temperature on Venus in the location of the probe's landing site was $475°C$. [allowing a margin for error of $20°$ in each direction; the Fahrenheit equivalent—$847°$ to $923°$]. The surface atmospheric pressure was given as 90 times that on earth [with an allowance of 15 atmospheres either way for error]. The atmosphere's density was recorded as about 60 times that on earth.

(*Venera 7* weighed about 220 pounds more than previous Venera probes that had penetrated Venus' atmosphere because of modifications made to allow the spacecraft to withstand pressures of up to 180 atmospheres and temperatures of up to $986°$ F. Aerodynamic braking when the probe smashed into Venus' atmosphere reduced its speed in relation to the planet from 7.1 miles a second to 656 feet a second. At an altitide of about 37 miles, when the external pressure was about .7 atmospheres, the parachute opened. The parachute, which had been redesigned after the previous Venera flights, allowed *Venera 7* to drop through the atmosphere much faster than previous Venera probes. The landing then took place at 22 seconds after 8:34 a.m. Dec. 15, 1970 Moscow time.)

U.S. PROGRAM

Moon Landings Reduced, Budget Cut

NASA Sept. 2 "reluctantly" canceled 2 of the remaining 6 manned moon landings that had been scheduled for the Apollo program. Dr. Thomas O. Paine, NASA administrator, said at a Washington press conference Sept. 2 that the cancellation, which would distinctly diminish the scientific rewards originally expected from the lunar program, had been forced on the agency by budget cuts. Because of the curtailments, "there has been a tremendous reduction in the national space capacity," he declared. "Any further [spending] cuts would have devastating effect" on U.S. space plans, Paine warned.

Only 3 of the 10 originally planned manned Apollo landings on the moon had been attempted so far (2 successfully, one unsuccessfully), and a 4th had been canceled previously. The 4 remaining planned Apollo lunar landings were rescheduled for January and July 1971 and January and June 1972. None of the rescheduled flights was to include a scientist-astronaut.

NASA's revised manned space-flight program called for Apollo to be followed by the Skylab project, in which 3-man teams would operate in an orbiting workshop for periods of up to 56 days. Skylab was scheduled to start with a Nov. 1972 launching, and it was to be ended by mid-1973. No further manned U.S. space flights were scheduled for the next 3 or 4 years until the first launchings of the projected space shuttle.

NASA's budget for fiscal 1971 had been cut to $3.27 billion from the 1970 budget of $3.75 billion. Originally NASA has requested $3.6 billion for fiscal 1971. But Pres. Nixon, when he submitted his budget to Congress Feb. 2, had requested only $3.3 billion, the lowest figure since the early 1960s. The scaling down, or stretching out, was to be applied to moon-landing and unmanned planetary flights, basic research and the training of space scientists. Current moon-landing and earth-orbital programs would be stretched out until 1974 and new endeavors postponed until the late 1970s. Employment within NASA and on space contracts was expected to decline by a total of about 45,000 to about 144,000.

But the President spoke of actions to make it possible "to begin plans for a manned expedition to Mars." $110 million was budgeted for initial work on a reusable space "shuttle" and a permanent space station. The shuttle and station, Nixon said, "could evolve into a new space capability that can fully exploit and use space for the benefit of all mankind and at the same time substantially reduce the cost of space operations."

Funds were budgeted also for continued development of a nuclear engine capable of producing 75,000 pounds of thrust in its initial version and, eventually, 200,000 pounds of thrust.

By the time NASA's appropriation had moved through Congress and had been signed by the President, the total was $3,268,675,000—$64.3 million less than the President had requested and $300 million less than NASA's original request.

The cancellation of the 2 lunar landings reduced NASA spending by $40 million, and an additional $20 million was cut from other manned space-flight operations.

The reductions in expenditures and operations brought additional dismissals of employes of NASA and of private space contractors, in both cases after drastic earlier personnel cutbacks. Morale in the U.S. space establishment was reported to have declined seriously.

Nixon Sets Goals for '70s

2 "grand tour" probes of the outer planets by unmanned spacecraft were proposed by Pres. Nixon Mar. 7 as part of a series of new U.S. space goals for the 1970s. These plans, largely based on recommendations made by the President's cabinet-level Space Task Group Sept. 15, 1969, were explained in greater detail Mar. 7 by Dr. Thomas O. Paine, NASA administrator.

The President listed "3 general purposes [that] should guide our space program" and 6 "specific objectives" to "work toward."

The first general purpose was "exploration." "A great nation must always be an exploring nation if it wishes to remain great," Nixon said. "A 2d purpose . . . is scientific knowledge—a greater systematic understanding about ourselves and our universe," and "a 3d purpose . . . is that of practical application—turning the lessons we learn in space to the early benefit of life on earth."

"We must realize that space activities will be a part of our lives for the rest of time" and should plan them "on a continuing flexible basis," the President said. "We must also realize that space expenditures must take their proper place within a rigorous system of national priorities. . . . The space budget which I have sent to Congress for fiscal year 1971 is lower than the budget for fiscal year 1970. . . ."

Nixon stated these specific objectives: "(1) We should continue to explore the moon. ... (2) We should move ahead with bold exploration of the planets and universe. ... As a part of this program, we will eventually [but not in the 1970s] send men to explore the planet Mars. (3) We should work to reduce substantially the cost of space operations. ... (4) We should seek to extend man's capability to live and work in space. The experimental space station (XSS)—a large orbiting workshop—will be an important part of this effort. ... (5) We should hasten and expand the practical applications of space technology [in such fields as meteorology, communications, navigation, air traffic control, education and national defense and to aid in such tasks as surveying crops, locating mineral deposits and measuring water resources]. ... (6) We should encourage greater international cooperation in space. ..."

Paine said that the first grand-tour launching, scheduled for 1977, would take a probe (possibly 2 unmanned spacecraft) past Jupiter in 1981, past Uranus in 1985 and past Neptune in 1988. The 2d grand-tour launching, in 1979, was to put its unmanned spacecraft into a more direct route passing Saturn in 1980 and reaching Pluto, 3.7 billion miles from the earth, in 1985. Nixon intended to request $10 million for fiscal 1972 for grand-tour planning. 2 unmanned orbital flights should map 70% of Mars' surface in 1971, Paine said, and plans called for the landing of unmanned spacecraft on Mars in 1976. An unmanned probe toward Jupiter was to be launched in 1972. In 1973, Paine said, an unmanned Mariner probe would be sent to pass Venus and then "give us our first close-up view of the planet Mercury." Paine noted that 4 manned landings on the moon were scheduled for 1971-2 and that such Apollo landings were to continue through 1974. A reusable space shuttle should make its first flight in 1976, Paine said, and Project Skylab—the construction of an orbiting space station—should then begin. "We hope to have the first launch of our nuclear propulsion rocket" in about 1978, Paine revelaed.

(NASA Feb. 18 had invited 6 aerospace companies to submit designs for the projected space shuttle.)

Paine, who had recently visited Canada, Japan and Australia to discuss space cooperation, said that people from non-Communist countries would probably be invited to fly in U.S. spacecraft before Soviet and U.S. spacemen go up together.

Germs Survived on Moon

The *Washington Post* reported May 23 that NASA's Lunar Receiving Laboratory in Houston, Tex. had found in Nov. 1969 in a piece of scientific equipment colonies of microbes that had survived on the moon for $2\frac{1}{2}$ years.

The germs, which were uncovered in a rim of plastic that protected the lens of a TV camera carried by *Surveyor 3*, were identified as *streptococcus mitis*, a benign bacteria that exists in the nose, throat and mouth of humans. The TV camera and the microbes were brought back to earth by the *Apollo 12* astronauts.

The germs had been carried to the moon by the Surveyor craft launched in Apr. 1967. The colonies had survived for 950 days on the lunar surface despite the moon's extreme heat and cold, almost vacuum-like atmosphere and constant radiation.

Scientists at the NASA facility hypothesized that the microbes had originally been deposited in the camera's protective rim inside droplets of mucus in a workman's breath while the camera was being repaired in California.

U.S. Orbital Launchings

NASA and the U.S. military launched fewer satellites in 1970 than in previous years. Details of the launchings:

Apr. 8—The 1,366-pound weather satellite *Nimbus 4* was sent into polar orbit by means of a booster rocket launched by NASA at Vandenberg Air Force Base, Calif. at 12:18 a.m. The same rocket also orbited a small Army mapping satellite—the 14th of the Secor (for sequential collation of range) satellites. *Nimbus 4*'s orbit had an apogee of 683 miles, a perigee of 673 miles and a period of 107 minutes. A highly advanced meteorological monitor, it carried instruments to measure the atmosphere's temperature and moisture content at different altitudes. (*Nimbus 3* had been launched Apr. 14, 1969.)

The Air Force used a single Titan-3C rocket to launch 2 Vela nuclear detection satellites from Cape Kennedy's Launch Pad 40. The Velas, each weighing 770 pounds, went up at 5:50 a.m. in a carefully designed program to place the monitors in circular 69,000-mile-high orbits on opposite sides of the earth from each other. The satellites, each equipped with 28 sensors to detect any violation of the treaty banning nuclear tests in the atmosphere or space, were to circle the earth once every $4\frac{1}{2}$ days. 10 Vela satellites had been launched in pairs previously, but only 6 of the earlier ones were still operating.

Apr. 15—An Air Force-industry team used a Titan-3-B-Agena rocket combination to launch a secret satellite from Vandenberg into polar orbit.

June 19—The first in a series of secret operational U.S. spy satellites was sent aloft by means of an Atlas-Agena booster rocket launched by the Air Force from Cape Kennedy at 7:39 a.m. It was said that the satellite was to hover in a near synchronous orbit about 22,300 miles above Southeast Asia in a position in which it could keep the USSR, China and North Vietnam under constant surveillance. The satellite was equipped with a TV camera and with infra-red and X-ray sensors that could detect the exhaust of rockets being launched. It was reported that the satellite would monitor Soviet FOBS (fractional orbital bombardment system) and submarine missile tests. 2 earlier-model U.S. spy satellites, launched Aug. 6, 1968 and Apr. 19, 1970, were already on station in the area.

Nov. 6—An 1,800-pound Air Force payload was launched from Cape Kennedy in an unsuccessful attempt to station over the Indian Ocean a 2d satellite capable of giving early warning of Soviet or Chinese missile attack. The satellite, referred to as 647 (or Project 647), was sent up in the nose of a Titan-3C booster rocket fired by Air Force personnel from Launch Complex 40 at 5:35 a.m. The press was barred from the launching. Pentagon sources disclosed Nov. 28 that the missile-warning satellite had gone into orbit but that, because of rocket malfunction, the orbit was too low for the satellite to monitor its assigned area more than about $\frac{1}{3}$ of the time. The satellite had been designed to operate in a near-synchronous orbit at an altitude of 22,300 miles above the equator. In such an orbit, it would appear to move in a figure-8 path over the Indian Ocean. It would have the USSR and China under constant surveillance, would be able to detect any rocket launchings and could give warning to the U.S. 30 minutes before a Soviet or Chinese rocket could reach U.S. territory. Current radar detection provided only 15 minutes' warning.

Nov. 30—An $83 million Stargazer astronomical satellite was lost after its Atlas-Centaur booster rocket, launched from Cape Kennedy at 5:40 p.m., failed to put it into orbit. The 4,680-pound payload, the biggest and most expensive scientific satellite the U.S. had ever built, was the 2d in the orbiting astronomical observatory program to fail.

(The U.S.' first satellite, the 12-year-old *Explorer 1*, dropped out of orbit Mar. 31 and burned up in the atmosphere over the South Pacific at 5:57 a.m. EST.)

INTERNATIONAL DEVELOPMENTS

China Orbits First Satellite

Communist China sent its first satellite into space Apr. 24. The 173-kilogram ($381\frac{1}{3}$-pound) payload achieved an orbit with an apogee of 2,384 kilometers (1,480 miles), perigee of 439 kilometers ($272\frac{3}{4}$ miles), period of 114 minutes and a 68.5° angle of inclination. The satellite's transmitter broadcast alternately the music of *Tang Fang Hung* (*The East Is Red*), a popular Communist Chinese song, and telemetric signals.

In its initial announcement of the successful launching, Hsinhua (the New China News Agency) Apr. 25 called the Chinese feat "a great victory for the Mao Tse-tung thought, a great victory for Chairman Mao's proletarian revolutionary line and another fruitful result of the great proletarian cultural revolution."

China was the 5th nation to orbit a satellite with its own rocket. (The first 4 were the USSR, the U.S., France and Japan.)

The various Chinese announcements gave no hint as to the satellite's launching site or details about the booster rocket. John W. Finney noted in the Apr. 26 *N.Y. Times* that China's missile-launching center had been unofficially reported to be at Shuang-cheng-tze, in Inner Mongolia's western desert 400 miles northwest of Lanchow. Finney reported speculation that the launching rocket might have been either the first stage of the rocket China had developed for its medium (1,000-mile) range ballistic missile (MRBM) or the first stage of the rocket it was still developing for its intercontinental ballistic missile (ICBM). Stanley Karnow reported in the Apr. 26 *Washington Post* that it was thought the satellite might have been launched from a test site in the Lop Nor region of northwestern Sinkiang province, where ICBM launching facilities had been under construction since 1965. Stuart Auerbach reported in the Apr. 26 *Washington Post* that NASA experts speculated that a 2-stage rocket might have been used to launch the satellite, whereas a U.S. Air Force scientist cited early data as indicating that the Chinese used an MRBM.

Chairman George P. Miller (D., Calif.) of the House Space Committee said Apr. 25 that "apparently it [the Chinese satellite] is a rather simple satellite, and they have a long way to go."

Despite the few details supplied by the Chinese announcements, the orbiting of the first Chinese satellite was covered extensively in the Western press. In the USSR, however, it was reported curtly in a 10-word item at the bottom of page 5 of *Izvestia* Apr. 26 under the heading "Announcement of the Hsinhua news agency." There were no early reports of the Chinese achievement by the Czechslovak, Bulgarian or Hungarian news agencies, but Rumania, the first East European country to inform its people of the satellite, congratulated the Chinese Apr. 26 on their "fresh successes."

Soviet Orbital Launchings

The USSR launched 72 unmanned Cosmos satellites—*Cosmos 318* through *Cosmos 389*—during 1970. 8 of the satellites—*Cosmos 336* through *343*—were put into orbit by a single Soviet rocket Apr. 25. Tass, which disclosed neither the mission nor weight of the satellites, said Apr. 27 that they were in orbits with apogees of 931 miles, perigees of 870 miles, periods of 115 minutes and 75° angles of inclination.

George C. Wilson reported in the *Washington Post* Jan. 7 that the number of reconnaissance (spy) satellites launched by the USSR had increased from 5 in 1962 to 22 launched in 1967, 29 in 1968 and 32 in 1969. He reported that the Soviets seemed to be shifting from spy satellites that stay up 8 days to those that remain up about 12 days. 5 of those launched during 1969 were of the longer lasting series, and the last one, *Cosmos 313*, was sent up from the military space center in Plesetsk instead of from the Baikonur cosmodrome in Tyuratam, the launching site for the previous ones. Wilson reported that Defense Department research chief John S. Foster Jr. had told the Senate Appropriations Committee that the USSR used its SS-9 (Scarp) ICBM (intercontinental ballistic missile) to launch its FOBS (fractional orbital bombardment system) missiles, which were brought down before completing a complete orbit around the earth. The Soviets had test-flown 11 FOBS vehicles between Sept. 17, 1966 and the end of 1967, Wilson reported, but only 2 in 1968 and one in 1969.

Richard D. Lyons reported in the *N.Y. Times* Feb. 6 that, according to U.S. and British "space experts," the USSR had successfully tested a satellite capable of hunting down and destroying other satellites. The feat was reported accomplished with *Cosmos 248*, which intercepted and destroyed *Cosmos 249* and *252*, the experts said. *Cosmos 248* was still in orbit.

West Germany's Space Research Center in Bochum reported Feb. 18 that *Cosmos 323*, launched Feb. 10, had made a soft landing earlier Feb. 18 on Soviet territory.

Western observers expressed the belief that *Cosmos 354*, launched July 29, was also sent up in a FOBS test. It traveled in an orbit inclined at an angle of 50° to the equatorial plane, and it came down before completing a full revolution. *Cosmos 365*, which went up Oct. 25 and also did not complete a full revolution, was likewise presumed to be a FOBS test craft. And Tass reported Oct. 30, the day *Cosmos 375* was launched, that the satellite had already completed its mission.

Kenneth Gatland, vice president of the British Interplanetary Society, said in London Nov. 2 that *Cosmos 373, 374* (launched Oct. 20 and Oct. 23, respectively) and *375* were apparently launched to test a satellite system for destroying U.S. military satellites. Supporting his deductions with data from the Royal Aircraft Establishment at Farnborough, Gatland speculated that *Cosmos 373* was the target of the other 2 satellites. He noted that *Cosmos 374* had been blown up, presumably on signal from earth-based controllers. George Wilson reported in the *Washington Post* Nov. 6 that all 3 satellites in the presumed test had been launched from the Soviet military site at Tyuratam by means of SS-9 Scarp rockets. Both hunter satellites (*Cosmos 374* and *375*) "presumably were blown up after they made their inspection pass" at the target satellite, Wilson said. (Wilson noted that the U.S. Air Force had been developing a "Saint" satellite inspection system and a "Bambi" satellite killer program in the 1950s but that both programs had been canceled.)

Tass reported Oct. 8 that *Cosmos 368*, launched Oct. 8, carried "scientific instrumentation to test experimental systems which ensure vital activity of test animals, further to study the influence of space flight factors on living organisms."

The USSR orbited 5 unmanned Molniya-1 communications satellites—the 13th through 17th of the series—during 1970. Their mission, Tass reported, was "to ensure the functioning of longrange telephone and telegraph communications and relaying ... [Soviet central TV programs] to the points on the Orbita television network situated in ... the Far North, Siberia, the Far East and Central Asia." A Molniya-1 launched Feb. 19 went into an orbit with an apogee of 39,175 kilometers (24,342 miles) in the Northern hemisphere, perigee of 487 kilometers ($302\frac{1}{2}$ miles) in the Southern hemisphere, period of 11 hours 43 minutes and 65.3° angle of inclination. The 14th Molniya-1 was launched June 26 into an orbit with an apogee of 39,280 kilometers (30,407 miles) in the Northern hemisphere and perigee of 470 kilometers (292 miles) in the Southern hemisphere. The 16th and 17th Molniya-1s went into orbit Nov. 27 and Dec. 25.

3 Meteor weather satellites were launched Apr. 28, June 23 and Oct. 15. The launchings raised to 5 the number of such satellites in orbit. The Meteor sent up Apr. 28 achieved an orbit with an apogee of 736 kilometers (457.3 miles), perigee of 637 kilometers (395.8 miles),

period of 98.1 minutes and an 81.2° angle of inclination. The satellite launched June 23 went into an orbit with an apogee of 906 kilometers (563 miles), perigee of 863 kilometers (536.2 miles), period of 102 minutes and 81.2° angle of inclination. The Meteor launched Oct. 15 achieved an orbit with an apogee of 674 kilometers (419 miles), perigee of 633 kilometers (393 miles), period of 97.5 minutes and 81.2° angle of inclination. Tass reported that the satellites carried equipment to photograph cloud formations and the snow layer on the shadowed as well as illuminated side of the earth, instruments to collect data on thermal energy "reflected and irradiated by the earth and atmosphere" and equipment to assure the satellite's "constant orientation to the earth."

Intercosmos 3 and *Intercosmos 4*, the 3d and 4th in a series of unmanned satellites launched from a Soviet site by the USSR in cooperation with other East European countries, were sent into orbit Aug. 7 and Oct. 14. During the preparation for *Intercosmos 3*'s launching, Czechoslovak specialists joined their Soviet colleagues, Tass reported, "in assembly operations and tests of the research equipment." The satellite, designed to study radiation, achieved an orbit with an apogee of 1,320 kilometers (820 miles), perigee of 207 kilometers (129 miles), period of 99.8 minutes and 49° angle of inclination. The data from the satellite was coordinated with ground data assembled by Polish, Bulgarian, East German, Czechoslovak and Soviet scientists. *Intercosmos 4* was launched on the first anniversary of the launching of *Intercosmos 1*. *Intercosmos 4* went into an orbit with an apogee of 668 kilometers (415 miles), perigee of 263 kilometers (163 miles), period of 93.6 minutes and 48.5° angle of inclination. The satellite carried instruments provided by the USSR, East Germany and Czechoslovakia.

Japan's First Success

Japan's first satellite, a 50.82-pound radio-equipped payload, was sent into orbit Feb. 11 by means of a 4-stage solid-fueled Lambda-4S5 rocket launched at 1:25 p.m. from Uchinoura on the Ohsumi Peninsula. The successful launching, after 4 successive failures by Japan in attempts to orbit a satellite, made Japan the 4th nation to send a satellite into space. (The first 3 were the USSR, the U.S. and France.) It was reported that it had cost a total of $52,700,000 for the 16 years of research and development that finally put the first Japanese satellite up.

The satellite was launched by scientists of Tokyo University's Institute of Space & Aeronautic Science, who dubbed the payload *Ohsumi*. (The satellite was also called *Rising Sun Satellite 1*, and it was listed scientifically as "Artificial Satellite 1970-11-A.")

Ohsumi achieved an orbit with an apogee of 5,100 kilometers (3,169 miles), perigee of 350 kilometers (217 miles), period of 110 minutes and inclination of 31° to 32°.

The satellite was sent aloft without the usual guidance equipment. The university scientists who developed and launched it had refused to develop guidance devices for fear such equipment might later be used for weapons. Instead, a unique device sent *Ohsumi* into orbit by a system described as "unguided gravity turn." The satellite did not use the usual solar batteries for power, and its initial power source failed within hours after launching. By Feb. 12 the satellite's radio was no longer functioning.

A Japanese effort to put a scientific satellite into orbit failed Sept. 25 because the 4th stage of its carrier rocket did not ignite. The 137-pound payload, launched at Uchinoura, went up in the nose of the first of a new series of MU4SI rockets.

Franco-German Project

The 264-pound West German satellite *Dial* was sent into a near-equatorial orbit Mar. 10 by means of a French-built 3-stage Diamant-B rocket launched by French personnel at 9:21 a.m. from the new French space center at Kourou, French Guiana. *Dial* achieved an orbit with an apogee of 1,760 kilometers (1,090 miles) and perigee of 339 kilometers ($210\frac{1}{2}$ miles). This was France's 5th successful satellite launching, the first from the Kourou site. All 4 previous French shots had been made at the Hammaguir center in Algeria, which had been given up in 1967.

A French satellite was sent into near-circular orbit Dec. 12 by a carrier rocket launched from France's space center at Kourou. The satellite's mission was to provide measurements of the earth from altitudes of between 465 and 530 miles.

British Skynet & Orba Fail

A 2d British Skynet communications satellite was sent into a preliminary orbit aboard a U.S. rocket Aug. 19 in an unsuccessful plan to put the military payload into synchronous orbit 22,300 miles over the Indian Ocean. The $5\frac{1}{2}$ million 535-pound satellite was lost Aug. 22, however, after its onboard rocket failed as the U.S. Air Force control center at Sunnyvale, Calif. was signaling it to shift the satellite into its designated position. The satellite, built by the Philco-Ford Corp. of Newport Beach, Calif., had been launched from Cape Kennedy by Air Force personnel, and the U.S. charged Britain $4.6 million for the launching.

A British 3-stage Black Arrow booster rocket was launched from the Woomera range in South Australia Sept. 2 in an unsuccessful effort to put a 180-pound Orba satellite into orbit. The 30-inch gold-plated sphere, designed to measure air density from near-polar orbit, failed to achieve orbital velocity because ignition of the rocket's 2d stage ended 13 seconds too soon.

Italy Orbits U.S. Satellite

In the first launching of a U.S. satellite by another country, Italian engineers Dec. 12 used a 4-stage U.S.-made Scout rocket to send a U.S.-built X-ray astronomy satellite into an equatorial orbit. The 315-pound unmanned spacecraft was launched from Italy's San Marco platform in Formosa Bay 3 miles off the coast of Kenya near Malindi. The launching was conducted by the University of Rome's aerospace research center, which was reimbursed by the U.S. NASA.

The satellite, known as X-ray Explorer, achieved an orbit with a 340-mile altitude and 90-minute period of revolution. NASA and the Italian team dubbed the satellite *Uhuru* (Swahili for "freedom") because it was launched on Kenya's independence day. The mission of the satellite, the 42d spacecraft in the Explorer series of unmanned scientific probes, was to provide information on high-energy X-ray sources both inside and outside the galaxy. The satellite was part of the U.S.' small astronomy satellite program, which was managed by NASA's Goddard Space Flight Center at Greenbelt, Md.

U.S. Firm Gets Canadian Contract

Canada's federal cabinet July 31 awarded the contract for Anik, Canada's first communications satellite, to Hughes Aircraft Co. of California. The decision followed months of industrial and public lobbying over the amount of Canadian content that would be included in the production of the satellite.

An original bid of $55 million by RCA Ltd. of Montreal, which was based on 65% Canadian content, had been withdrawn July 6 after RCA learned that Hughes had bid $30 million. RCA asked for a chance to refigure the contract on a modified form of cost-plus arrangement. In a statement July 8, RCA management charged that no Canadian company could compete in price with the U.S. aerospace industry and said it had bid on the program with the understanding that the development of a Canadian national space program was a government priority. RCA Pres. John Houlding said in a July 10 press conference that the year-old Canadian satellite-building program would be ended if the Anik contract went to a U.S. firm. RCA discussed another offer for the contract—reportedly $35 million plus an incentive fee—with communications ministry officials in Ottawa July 13. This proposal was rejected, according to a July 31 report.

The cabinet decision was based on an undisclosed firm price bid by RCA and 3 Hughes proposals, ranging from $28 million to more than $30 million, depending on the amount of Canadian content.

Hughes' side in the controversy was supported by Eric Kierans and his communications department, and Telesat Corp., the public-private corporation established to operate the satellite system. Favoring RCA were Labor Min. Bryce Mackasey, the Department of Industry, Trade & Commerce, and Science Council of Canada Chairman O. M. Solandt. Solandt had charged July 6 that it would be "madness" for Canada to award the contract to the U.S. and "throw away" the abilities Canadian scientists and engineers had built up in satellite technology.

The cabinet decision, however, was at least partly based on the argument that Canada would benefit more if it developed expertise in specific subsystems in an ongoing relationship with a major U.S. company. The decision called for additional Canadian content to be negotiated in subcontracts.

The total cost of the satellite program was set at $90 million, including launching costs of $7 million per satellite, of which 2 were to be launched by U.S. space teams at Cape Kennedy. Some 34 earth receiving stations were to be built under terms of the program.

Intelsat Orbits 2 Satellites

The International Telecommunications Satellite Consortium (Intelsat) launched 3 communications satellites during 1970. One of the launchings was unsuccessful.

The first launching took place Jan. 14 when a 644-pound Intelsat-3 was sent into preliminary orbit by means of a 4-stage Delta booster sent up from Cape Kennedy by NASA at 7:16 p.m. The Delta's first 3 stages put the satellite into an orbit with a 22,700-mile apogee and 165-mile perigee. A radio signal from a Comsat ground station at Andover, Mass. Jan. 16 ignited the Delta's 4th stage and sent the satellite into its permanent synchronous orbit 22,300 miles above the South Atlantic. In this orbit it appeared to hang motionless over the equator. Its owner, Intelsat, put it into operation some 10 days later to relay phone and TV signals back and forth between the U.S., Europe, Africa and North and South America.

The satellite was the 6th in the Intelsat-3 series but only the 4th to be operating. The other 3 operating Intelsat-3s were in synchronous orbit above the Atlantic, Pacific and Indian Oceans. 2 Intelsat-3s had failed to achieve orbit.

Like the previous Intelsat-3s, the new satellite was launched by NASA on contract to Comsat (Communications Satellite Corp.), the U.S. member and operating manager of Intelsat. The new satellite cost Intelsat about 11\frac{1}{2}$ million, including the price of the satellite and rocket and the expenses of the launching. Comsat, charged for 53% of the cost, paid about $6.1 million of the total, but it insured its share for an $872,000 premium with Lloyd's of London and other insurers, and it would have collected $4.57 million if the satellite had failed to go into its stationary orbit. The other nations did not insure their 43% in this first insured satellite.

The next Intelsat-3 was sent into space Apr. 22 by means of a Delta launched by NASA at Cape Kennedy. It achieved a preliminary orbit with an apogee of 22,700 miles and perigee of 165 miles. A radio signal from the Comsat station at Andover, Me. Apr. 24 ignited the 4th-stage motor to start the satellite toward its permanent synchronous orbit 22,300 miles above the South Atlantic. Its mission was to relay phone and TV signals back and forth from various points in North and South America, Europe and Africa.

The 8th Intelsat-3 was blasted into space from Cape Kennedy July 23 in an unsuccessful effort to put it into a synchronous orbit over the Pacific. The 332-pound satellite was launched by NASA on contract with Comsat, which acknowledged July 25 that the satellite was lost in space. The satellite, not insured, was the 3d lost by Comsat. The satellite had achieved a preliminary orbit July 23, but contact with the satellite was lost July 24 after Comsat controllers, by means of a radio signal, fired the on-board engine to start shifting the satellite to its planned Pacific station.

U.S. & Soviet Experts Confer

A 5-member delegation of U.S. space officials met in Moscow Oct. 26-27 with a delegation of Soviet space experts for the first of a planned series of talks on U.S.-Soviet cooperation in space rescue and possibly other space matters. The discussions were confined almost completely to such technical matters as linking vehicles (docking) in space and transferring crew members between space ships. The 2 delegations agreed that all spacecraft should have compatible docking systems so that any space ship could rescue occupants of any other space ship, regardless of nationality.

The conference produced an agreement providing for future meetings by 3 committees of space technicians from the 2 countries. The agreement was signed in Moscow Oct. 28 by Robert S. Gilruth, head of the U.S. delegation and director of the U.S. Manned Spacecraft Center, and by Boris N. Petrov, head of the Soviet delegation and chairman of the Soviet Council for International Cooperation in the Exploration & Use of Space. Other delegation members also signed.

Arnold W. Frutkin, assistant NASA administrator for international affairs and a member of the U.S. delegation, said at a press conference in Washington Oct. 29 that the Soviet delegates had been frank and open in describing their space-docking methods. "They told us a great deal that we didn't know," he declared. While U.S. astronauts had been transferred repeatedly through a "tunnel" between docked and closed spaceships, all Soviet transfers of personnel between spaceships, Frutkin noted, had been made by EVA (extra-vehicular activity), in which a space-suited cosmonaut went out into space and then entered the other vehicle.

Members of the U.S. delegation were taken on a tour of Zvezdny Gorodok (Stellar Town, or Star City), the carefully guarded cosmonaut and space scientist community near Moscow, where they were shown spacecraft and other Soviet space hardware. They were reported to be the first foreigners invited to see the lunar soil brought back to earth by the unmanned Soviet lunar probe *Luna 16*. (Dr. Mstislav V. Keldysh, president of the Soviet Academy of Sciences, disclosed Oct. 28 that the Soviet probe had brought back a little more than 100 grams [about $3\frac{1}{2}$ ounces] of lunar material.)

The Moscow meeting took place shortly after 2 Soviet cosmonauts, Maj. Gen. Andrian G. Nikolayev and Vitali I. Sevastyanov, had arrived in New York Oct. 18 for a 10-day U.S. visit. The cosmonauts toured the Manned Spacecraft Center in Houston, where they spoke Oct. 22 at the concluding session of the 4-day 7th annual meeting of the American Institute of Aeronautics & Astronautics. (Although previous meetings of the institute had drawn from 2,500 to 6,000 engineers and scientists, the current meeting, taking place after severe personnel and spending reductions in the U.S. aerospace industry, had only 1,200 registrants.)

The cosmonauts, whose 18-day flight in *Soyuz 9* in June had set an endurance record, said at a press conference in Houston Oct. 22 that the USSR was planning even longer flights to check into what Nikolayev described as the "very serious" physical disabilities the cosmonauts had suffered from their long period of weightlessness. He reported that their general post-flight weakness and disorientation had continued for several days after their return to earth. The cosmonauts indicated that the principal immediate Soviet goal in manned space flight was to develop an orbiting space station.

U.S. astronaut James Lovell said Oct. 6 at an International Astronautics Congress in Constance, West Germany that he believed no "American astronaut would object to flying with a Soviet crew" provided "the language problems could be solved and there were sufficient training beforehand." Soviet cosmonaut Nikolayev the previous day had indorsed the possibility of mixed Soviet-American flights, saying: "The history of human development has many examples of mixed teams . . . in travel and exploration." At the congress, the USSR agreed Oct. 8 to participation by 4 Soviet specialists in a scheduled meeting of the space rescue committee of the International Academy of Astronautics. 2 American aerospace officials Oct. 7 proposed the creation of a fleet of international space laboratories under UN control. The laboratories, to be built largely with U.S. and Soviet assistance, would house 40 to 50 people. The proposal came from George W. Morgenthaler and Robert B. Demoret of the Martin Marietta Corp.

1971

4 more U.S. astronauts landed on the moon during 1971 in 2 missions and conducted the most ambitious lunar exploration program so far. In the 2d mission, the astronauts drove a small electric car $17\frac{1}{2}$ miles while exploring the lunar surface. The USSR conducted 2 manned earth orbital flights in both of which cosmonauts docked their spacecraft with unmanned satellites. The 2d mission ended in tragedy June 30 when the 3 cosmonauts were found dead in their capsule after its return to earth. Both the U.S. and USSR continued unmanned probes of the moon and nearby planets. The U.S. achieved the first orbiting of another planet when it put a probe into orbit around Mars.

EXPLORING THE MOON

3d U.S. Landing on Moon

2 astronauts from the spaceship *Apollo 14* landed their lunar module (LM) *Antares* on the moon Feb. 5. The 3d 2-man team to set foot on the earth's natural satellite, they left the LM twice for $4\frac{1}{2}$-hour working and exploring sessions on the lunar surface during the record $33\frac{1}{2}$ hours they remained on the moon. They then returned in the LM's ascent stage to *Apollo 14*'s command-and-service module (CSM), the *Kitty Hawk*, which had remained in lunar orbit with *Apollo 14*'s 3d crew member aboard.

The *Apollo 14* astronauts had started their half-million-mile journey Jan. 31 with a launching at Cape Kennedy, Fla. They ended it Feb. 9 with a safe splash-down in the Pacific.

The 2 *Apollo 14* astronauts who reached the moon's surface were: (a) the spaceship commander, Navy Capt. Alan Bartlett Shepard Jr., 47, one of the U.S.' original 7 astronauts and the first American to make a space flight (a sub-orbital mission in the one-man Mercury capsule *Freedom* 7 May 5, 1961); and (b) the lunar module pilot, Navy Cmndr. Edgar Dean Mitchell, 40, who had never flown in space before. The 3d *Apollo 14* crew member was the command module pilot, Air Force Maj. Stuart Allen Roosa, 37, making his first space flight. Roosa remained aboard the CSM *Kitty Hawk* in orbit around the moon while Shepard and Mitchell were on the lunar surface.

The *Apollo 14* mission cost the U.S. an estimated $400 million—or at least $25 million more than any previous Apollo flight. The 3-stage Saturn-5 booster rocket alone cost $185 million. The spaceship cost $95 million—$55 million of the total for the command module and $40 million for the landing module. The $95 million in operations expense included the wages of some 10,000 launching team members. The high (and growing) cost of manned spaceflight had made the Apollo program a subject of controversy and was a major reason for widespread demands that manned spaceflight be replaced with less costly unmanned space projects.

Because of several bomb threats in the months before the launching, extra guards had been posted at Cape Kennedy and additional security restrictions imposed at the space center.

Apollo 14 with its 3 astronauts aboard was sent up from Cape Kennedy's Launch Pad 39A at 4:03 p.m. EST Jan. 31 after a rainstorm had caused a 40-minute delay in the launching.

The first stage of the Saturn-5 booster, its 5 rocket engines developing a total of 7,591,215 pounds of thrust, carried the 4,949,100-pound rocket-and-spaceship assembly up and over the Atlantic, firing for some $2\frac{1}{2}$ minutes before exhausting its fuel, separating from the moon-bound space vehicle and dropping into the ocean. The 2d stage then ignited and fired for about 7 minutes, pushing *Apollo 14* further upwards, before it too exhausted its fuel, separated from the spaceship and dropped back. The 3d (S4B) stage then ignited and boosted the spaceship into a 117-mile-high orbit around the earth.

For the next $2\frac{1}{2}$ hours the astronauts and ground controllers checked instruments and the performance of the spacecraft, and they found all operating properly. At 6:33 p.m., therefore, as *Apollo 14* sped over the Pacific after circling the earth $1\frac{1}{2}$ times, the astronauts ignited the still-attached S4B stage again. The S4B, firing for 5 minutes, increased the spaceship's speed to 25,000 mph and injected *Apollo 14* into a translunar trajectory.

Trouble developed at 7:11 p.m. EST Jan. 31 after the astronauts had begun the next major maneuver. This required them to detach the command-and-service module (CSM) from the S4B stage, move the CSM ahead of the S4B, turn it around, bring it back to the S4B, dock (link) it nose-to-nose with the landing module (LM), which remained in the open front of the S4B, and then extract the LM from the S4B.

With Roosa at the controls, all went well until the CSM's docking probe had been thrust into the LM's docking collar. The probe's 3 capture latches then failed to attach themselves to the LM's funnel-like drogue. This attachment, or "soft" docking, is normally followed by a "hard" docking, the attachment of the 12 docking latches in the CSM's docking ring. After the first failure, Roosa moved the CSM backward, aimed again at the LM's docking collar and failed in a 2d docking attempt.

During a period of nearly 2 hours the astronauts made 5 unsuccessful attempts to dock by the prescribed method. In each case the capture latches failed.

On their 6th attempt, using an unorthodox method, they finally docked. Success was achieved at about 9 p.m. by retracting the probe just before contact, by-passing the use of the capture latches and executing the docking procedure by employing only the docking latches.

At 9:52 p.m. the LM, attached firmly to the CSM, was withdrawn from the S4B, and the 2 assemblies were separated.

(The S4B was later speeded up by radio signal and sent crashing into the moon at 2:41 a.m. EST Feb. 4 in the arid Ocean of Storms where it dug a 33-foot crater. The crash was monitored by a seismometer that had been left on the moon by *Apollo 12*'s astronauts.)

The docking problem for awhile caused NASA officials to consider canceling the manned landing on the moon. The reason for debating cancellation was the fact that the LM's ascent stage, on returning from the surface of the moon with Shepard and Mitchell aboard, would normally dock with the CSM to facilitate the transfer of the astronauts and their lunar spoils from the LM to the CSM prior to their eventual return to earth. The astronauts, however, spent about $4\frac{1}{2}$ hours early Feb. 1 in dismantling and testing the docking system under directions from controllers on the ground. They found nothing wrong, and the flight controllers decided to continue with the landing mission. It was pointed out also that even if the docking system failed to work on the return from the moon, the astronauts could still transfer from LM to CSM by a "space walk," an emergency method they had practiced. This would require that they exit in space suits from the LM and use a clothesline-like tether to bring their lunar luggage aboard the CSM.

The astronauts fired the CSM's 20,500-pound-thrust main rocket for 10 seconds at 10:38 p.m. EST Feb. 1 to correct their spaceship's course. Without the correction *Apollo 14* would have missed the moon by 2,419 miles. The correction put the spaceship into a trajectory that would take it to a point about 70 miles from the moon.

While the cabin of the spaceship was in darkness Feb. 2, the astronauts reported seeing brief flashes of light as previous Apollo astronauts had seen. Scientists had speculated that these flashes were caused by cosmic rays. (The astronauts noticed similar flashes when they darkened the cabin on the return trip Feb. 8.)

By a .7-second ignition of the CSM engine at 9:01 p.m. Feb. 3 the astronauts again refined their course with the objective of putting *Apollo 14* into lunar orbit at an altitude of about 66 miles above the lunar surface.

Pulled by the moon's gravity, *Apollo 14* whipped around the moon at 1:48 a.m. EST Feb. 4. An ignition of the CSM engine at 1:59 a.m. then put the spaceship into a lunar orbit with an apocynthion (high point) of $195\frac{1}{2}$ miles above the moon's surface and pericynthion (low point) of 69 miles. On completing the 2d 2-hour revolution around the moon, the astronauts fired their engine at 6:14 a.m. and put *Apollo 14* into a lower (descent) orbit with an apocynthion of 66.7 miles and pericynthion of $11\frac{1}{2}$ miles.

Preparing for their excursion on the moon, Shepard and Mitchell struggled into their cumbersome space suits at about 9 p.m. EST Feb. 4. They then opened the hatch between the command ship (*Kitty Hawk*) and the landing module (*Antares*) and crawled through the connecting tunnel into the LM. Roosa remained alone in the CSM. Assuming their posts in *Antares*, Shepard and Mitchell checked the LM's equipment and reported everything in working order.

At 11:50 p.m., as *Apollo 14* was making its 12th trip around the moon, the astronauts uncoupled *Antares* from *Kitty Hawk*, and the 2 spaceships continued on in separate lunar orbits.

Antares' descent was started during the 14th revolution at about 4:05 a.m. EST Feb. 5 as the LM was in the low point of its orbit. Dropping a distance of 45,000 feet in 12 minutes 46 seconds, they carefully selected a relatively level spot in their target area in the moon's Fra Mauro Hills.

The 4-legged *Antares* touched down at 4:18 a.m. EST Feb. 5 only 87 feet from the chosen landing spot on an 8° slope. The coordinates of their landing site were given as 3.66°S. and 17.478°W.

The astronauts examined the surface of the moon from the windows of their LM. Mitchell, looking through the right window, said: "We seem to be sitting in a bowl. It slopes toward us from the west. It's rather choppy.... The skyline is quite undulating. There is a large old depression ... to the north of us which forms another bowl very similar to the one that we appear to be sitting in."

The astronauts again checked their spaceship's equipment and found everything apparently in working condition. They started their first meal on the moon at 6 a.m. EST Feb. 5, then adjusted their equipment and their backpack PLSS (portable life support systems) for operation outside the landing craft on the surface of the moon. They depressurized the LM's cabin at 9:52 a.m. by venting its oxygen, and Shepard opened the hatch at 9:49 a.m.

Shepard, clad in his space suit and burdened with his clumsy backpack, stepped out of the hatch at 9:50 a.m. EST Feb. 5 to the "porch" atop the LM's ladder. As he backed down the ladder a few minutes later, he pulled a lanyard that opened an equipment bay near *Antares'* base and started a color TV camera that recorded his descent and his first movements on the surface of the moon.

At 9:57 a.m. Shepard stepped off the last rung of the ladder and became the 5th man to set foot on the moon. A ground controller radioed to Shepard enthusiastically as he watched him by TV from a distance of 230,000 miles: "Al, beautiful.... It looks like you're about on the bottom step—and on the surface. How's that for an old man?"

"OK, you're right," Shepard replied. "Al is on the surface. It's been a long way, but we're here." "Now I can see the reason we [*Antares*] have a tilt is we landed on a slope," Shepard added. "The surface in which the forward footpad landed is extremely soft. As a matter of fact, it's a small depression. The soil is so soft that it comes up all the way to the top of the footpads."

Shepard tested his ability to walk in the light gravity (1/6 the gravity of the earth) and over the dusty, rubble-strewn surface.

Mitchell then backed down the ladder and became the 6th man to walk on the lunar surface.

During the $4\frac{1}{2}$ hours they spent in this first of their 2 working-exploring sessions on the moon's surface, one of the most important tasks Shepard and Mitchell performed was to set up a nuclear-powered ALSEP (for Apollo lunar scientific experiment package) research station. Because the ground near the landing site was so bumpy, they had to move some 300 feet west of the LM to find a spot smooth enough for their research station. In carrying their instruments and rock samples, they used a 2-wheeled cart.

Mitchell deployed 3 geophones (vibration detectors) along a cable at distances of about 10, 160 and 310 feet from the ALSEP station. He then tried to fire 21 explosive charges at 15-foot intervals along the cable to create seismic waves for the geophones, but only 13 of the charges went off. The equipment the astronauts deployed included another seismic experiment linked to the geophones. This experiment used a mortar package designed to fire 4 explosive projectiles on radioed command from the earth after the astronauts' departure.

Earlier, after Shepard had set up an American flag, the ground controllers had relayed to the astronauts a telephoned message of congratulations from Pres. Nixon. It was revealed that Mitchell had placed on the moon a package containing the first verse of *Genesis* in 16 languages and a microfilm copy of the complete bible.

Completing their first working session on the moon, the astronauts returned to the LM with 43 pounds of lunar surface samples. They then closed the hatch at 2:29 p.m. Feb. 5.

After sleeping restlessly, Shepard and Mitchell opened the LM's hatch again at 3:20 a.m. EST Feb. 6 and climbed outside for a 2d $4\frac{1}{2}$-hour working-exploring period on the lunar surface.

Pulling their 2-wheeled, equipment-laden cart, the astronauts started walking in the direction of a major objective—Cone Crater, in which scientists had hoped that the astronauts would find specimens of the oldest rocks on the moon. The climb over difficult terrain proved to be longer and more exhausting than anticipated. Because their time was growing short, and on the radioed advice of ground controllers and space physicians, the astronauts gave up after covering about a mile.

Abandoning the effort to reach Cone Crater, they collected, instead, surface samples from closer sites. Some of these samples, they believed, had been thrown up when Cone Crater was formed.

During this working session, Mitchell used a portable magnetometer to take the first lunar magnetic-field readings ever taken by an astronaut.

Before returning to the LM, Shepard used a secretly prepared golf club (a 6-iron) and 3 special heat-resistant golf balls to try golf shots in the airless, light-gravity conditions of the moon. He missed the first shot, connected on the 2d and 3d and estimated that the balls went "miles and miles and miles" (possibly a half-mile).

As their mission on the moon ended, the astronauts stowed their 2d batch of lunar samples aboard the LM and closed the hatch at 7:43 a.m.

Using the LM's 4-legged descent section as a launching pad, Shepard and Mitchell blasted off in *Antares'* ascent section at 1:47 p.m. EST Feb. 6 after spending $33\frac{1}{2}$ hours on the moon.

Antares rose to orbital altitude and was still on its first revolution around the moon when it was docked at 3:36 p.m. with the CSM, or *Kitty Hawk.* The docking was handled by Roosa, who had been riding in orbit in *Kitty Hawk,* while his colleagues were busy on the moon. The docking mechanism worked without trouble on the first attempt.

Shepard and Mitchell then crawled through the connecting tunnel from the LM into the command module and brought their lunar mementos with them.

Antares was separated from *Kitty Hawk* at 5:58 p.m., and the LM was sent crashing down on the moon by radioed command about 2 hours later. The seismometer left by the astronauts on the moon recorded the impact.

Apollo 14 was brought out of lunar orbit and headed back toward the earth by a $2\frac{1}{2}$-minute firing of the CSM's main rocket at 8:38 p.m. Feb. 6 while the spaceship was behind the moon on its 35th lunar revolution.

A minor course correction was made by means of a brief burst of the maneuvering rockets at 1:37 p.m. Feb. 7.

As *Kitty Hawk* approached the earth with ever-increasing speed, the astronauts jettisoned the 24-foot-long service module by firing explosive bolts at 3:37 p.m. EST Feb. 9. This took place at an altitude of 150 miles over the western part of the Indian Ocean.

At 3:51 p.m. the command module, its blunt heat shield facing in the direction of flight, plunged into the atmosphere at a speed of 24,500 mph. Friction with the air heated the outside of the spaceship to a temperature of more than 5,000° F., and the buildup of an envelope of ionized air around *Kitty Hawk* caused a radio blackout for more than 3 minutes.

Its descent slowed first by friction and then by parachutes, *Apollo 14* dropped out of the clouds and splashed safely into the Pacific at 4:05 p.m. EDT Feb. 9. It came down 5 miles from the waiting recovery carrier *New Orleans*, only .7 mile from its aiming point and 880 miles south of Samoa.

Navy frogmen dropped by helicopter quickly placed a flotation collar around the capsule and handed the astronauts clean coveralls and masks to wear for their brief helicopter trip to the carrier and short walk to the mobile quarantine trailer aboard the carrier. The quarantine period was 3 weeks from their departure from the moon. (The astronauts had spent 21 days in isolation at Cape Kennedy before the the start of the flight to minimize the possibility of infections that might interfere with their mission. The quarantine after their return was a precaution against the remote possibility that they had brought back some harmful microbe from the moon.)

The selection of 47-year-old Alan Shepard as *Apollo 14* commander had been criticized in London Feb. 4 by ex-U.S. astronaut L. Gordon Cooper Jr., who had once been considered a candidate—and probable choice—for the assignment. Cooper, 43, told reporters that he was "still in good physical condition" and "considerably younger than Shepard," who had been grounded from 1963 to 1969 by Meniere's disease, an ear affliction that caused vertigo, a loss of hearing and buzzing sounds in the ear. (Shepard's ear condition was corrected by an operation in 1968, and he was restored to flight status.) Cooper, who admitted that "I have my own feelings about him [Shepard]," said: "I don't feel that a man who is not fully qualified to fly an airplane without having a copilot with him and who has various physical problems really should bump more qualified people."

Astronauts Drive Car on Moon

In the 2d manned lunar landing of 1971, 2 astronauts drove a 460-pound electric car on the moon July 31–Aug. 2 during 3 record-breaking days of scientific exploration and experimentation on the lunar surface.

The astronauts and their 4-wheeled *Lunar Rover 1* made the 240,000-mile trip to the moon aboard the spaceship *Apollo 15* on a mission described by NASA sources as the most successful from a scientific standpoint of all U.S. manned space flights so far. The 12-day flight started with a perfect liftoff from Cape Kennedy's Launch Pad 39A July 26 and ended with a safe splashdown in the Pacific Aug. 7.

The astronauts who reached the moon, both of them Air Force officers, were Col. David Randolph Scott, 39, the flight commander, and Lt. Col. James Benson Irwin, 40, the pilot of *Apollo 15*'s lunar module (LM), which they had dubbed *Falcon*. While they explored the moon from its surface, *Apollo 15*'s 3d crew member, Air Force Maj. Alfred Merrill Worden, 39, pilot of the spaceship's command module, *Endeavour*, remained in lunar orbit making scientific observations aboard *Endeavour*. Scott and Irwin became the 7th and 8th men to set foot on the moon. The mission was Scott's 3d in space but his first to the moon. It was the first space flight for both Irwin and Worden.

All advance data on the $445 million mission had been made public long before the flight took place, and the flight itself was covered "live" by TV and radio from the pre-launch countdown to the splashdown.

Apollo 15, with the 3 astronauts aboard, was launched from Cape Kennedy at 9:34 a.m. EDT July 26 atop a 363-foot Saturn-5 rocket assembly. The booster's first and 2d stages were fired for periods of $2\frac{1}{2}$ and nearly 7 minutes, respectively, and were discarded in turn before the 3d (or S4B) stage, firing a bit less than 3 minutes, put the spaceship into a 105-mile-high orbit about 12 minutes after liftoff. Consisting of the S4B stage and the command, service and lunar modules of the *Apollo 15* spaceship proper, the assembly weighed a total of 107,500 pounds and was the heaviest man-made object sent into orbit so far. It traveled in orbit at a speed of 17,300 mph.

For nearly 3 hours, as *Apollo 15* circled the earth in orbit, the astronauts and ground controllers checked instruments and equipment. At 12:24 p.m., as the spaceship was flashing across the Pacific on its 2d circuit around the earth, Scott restarted the S4B engine for a 6-minute "burn" that brought the S4B-*Apollo 15* assembly out of orbit and into a translunar trajectory at an initial speed of 24,218 mph.

Shortly after the spaceship had achieved its lunar trajectory, Worden separated the command and service modules (CSM, or *Endeavour*) from the Saturn-5's S4B stage, which held the lunar module (LM, or *Falcon*) and lunar rover. He moved the CSM ahead of the S4B, turned it around and then brought it back to the S4B. By maneuvering the CSM, Worden nudged the command ship's docking probe into the LM's docking collar. The 2 modules were locked to each other in this position just after 1 p.m. EDT, and Worden drew the LM out of the S4B by backing the CSM-LM assembly away from the S4B. The discarded S4B was then speeded on toward the moon. (The 30,836-pound S4B crashed into the moon July 29 at a distance of 220 miles from a seismometer left on the surface during the *Apollo 12* mission. The impact had a force equivalent to that created by the explosion of 10 tons of TNT, and it sent shock waves 62 miles below the lunar surface.)

After pressurizing the cabin of the LM, Scott and Irwin crawled through the tunnel between the CSM and LM at 7:31 p.m. EDT July 26 to check out the landing vehicle's equipment. Except for a non-critical glass cover found "shattered" on an instrument dial, everything was reported to be in good shape.

The astronauts and ground controllers confirmed July 27 that there was a short circuit in a Klixon toggle switch needed in one of 2 banks of electrical circuits when the 20,500-pound-thrust main engine was fired by computer direction. A 2d bank, however, was in perfect order. The engine thus could be fired either manually on the affected bank or by computer direction on the unaffected bank. The short had been indicated by a warning light that had begun to flash erratically on the cockpit instrument panel 2 hours after liftoff. Worden successfully fired the main engine by use of a manual switch July 27 to prove that the short circuit would not jeopardize the mission and to make a minor mid-course correction.

The crew was warned by ground controllers July 27 to be sure to exercise regularly to overcome the weakening that otherwise would result from long hours of weightlessness in space. Since Worden was not scheduled to land and work on the moon, he was considered especially vulnerable to such weakness, so special stress was placed on instructions he received to exercise.

In a cosmic-ray experiment July 28, the astronauts darkened their cabin, covered their eyes with light-proof masks and counted a total of 55 flashes or streaks in a one-hour period from about 1 p.m. to 2 p.m. The flashes or streaks were believed to be caused by cosmic rays passing through their eyeballs.

Apollo 15 reached the vicinity of the moon at about 4 p.m. EDT July 29 and was whipped around the moon by lunar gravity. At 4:05 p.m. the astronauts fired their main rocket to slow the vehicle and put it into orbit around the moon. The initial orbit had an apocynthion (high point) of 195 miles above the lunar surface and pericynthion (low point) of 70 miles. At 8:14 p.m. the engine was fired again to lower the orbit to a high point of 67 miles and a low point of $10\frac{1}{2}$ miles. The purpose of the lowering of the pericynthion, in preparation for the lunar landing, was to conserve LM fuel.

Scott and Irwin prepared for their descent to the moon July 30 by donning space suits and crawling through the tunnel from *Endeavour* into *Falcon*.

The LM, with Scott and Irwin aboard, was separated from the CSM at 2:15 p.m. EDT July 30. Worden remained aboard *Endeavour*.

Falcon's descent engine was fired at 6:04 p.m., and the LM landed on the moon at 6:16 p.m. on the dusty Marsh of Decay (Palus Putredinis) at the edge of the dry Sea of Rains. The spider-like 4-legged landing vehicle came down not far from its target site, perhaps a mile east of the 1,200-foot-deep Hadley Rille and $1\frac{1}{2}$ miles from the Apennine Front, one of the highest mountain ranges on the moon.

The 2 astronauts checked over *Falcon*'s instruments after landing and then depressurized the cabin. About 2 hours after touchdown, Scott opened the top hatch, leaned out, took motion pictures of the area and described the landing site and surroundings in detail to the geologists on earth. Scott identified Hadley Rille and the Apennines, which curved around the landing site from north to west to south. The 15,000-foot Mount Hadley reared up to the northwest, the 13,000-foot Hadley Delta to the south. St. George Crater was visible in the flank of Hadley Delta. After a half-hour of this description, Scott re-entered the cabin, and the cabin was repressurized. The astronauts removed their space suits, had dinner and then slept. Scott and Irwin awoke at about 5 a.m. EDT July 31. After tending to technical, housekeeping and communication chores, they donned their space suits, depressurized the cabin and opened the hatch at 9:15 a.m.

Scott led the way down *Falcon*'s ladder and set foot on the moon at 9:26 a.m. Irwin stepped off the ladder at 9:34. They described the the soil they walked on as "like soft powdered snow."

The astronauts gathered several "contingency" samples of lunar surface material in case they should have to make an emergency departure. Then, in view of a TV camera attached to the LM, they lowered the 10-foot-long lunar rover from its cradle in *Falcon*'s descent stage. Scott climbed aboard the rover at 10:09, drove the car around the LM and reported that the vehicle's front-wheel steering apparatus did not operate but that he could direct the car adequately with the rear-wheel steering apparatus.

The astronauts set off in the rover at 11:20 a.m. for a 2-hour, 5-mile drive at speeds of up to 8 mph. In 25 minutes they reached the rim of Elbow Crater at the foot of Hadley Delta. At this first stop they picked up samples of surface material. The next stop was made partly up the slope of Hadley Delta near the rim of 2-mile-wide St. George Crater, where they picked up more surface material and took core samples to depths of about 3 feet. (They picked up a total of about 51 pounds of lunar material July 31.) The bottom of Hadley Rille was clearly visible to the astronauts and to their rover's TV camera, and it was reported to be surprisingly flat.

On returning to the LM, the astronauts deployed their $26 million ALSEP (Apollo lunar surface experiment package), a conglomeration of instruments designed to radio a variety of scientific data back to earth. The devices, set up in an area about 300 feet from the LM, included a seismometer, a laser reflector, instruments to detect atmosphere and charged particles, a magnetometer and a spectrometer. The astronauts had been told to curtail their EVA (extra-vehicular activity) by a half-hour the first day because Scott's metabolic rate and oxygen expenditure were reported to be higher than anticipated. They therefore returned to the LM early and closed the hatch at 3:49 p.m. after an EVA of 6 hours 34 minutes.

The Aug. 1 EVA lasted for a record 7 hours 13 minutes. It began with the opening of the hatch at 7:47 a.m. EDT and ended with its closing at 3 p.m. The astronauts drove their rover some 6 miles during a 4-hour excursion and found a rock that geologists at the Manned Spacecraft Center in Houston said might be part of the original lunar crust. The find was made about 600 feet up a slope of Hadley Delta near a crater named Spur. By the time the day's EVA was completed the astronauts had collected a record total of about 103 pounds of lunar material.

The mission's 3d and final EVA, which took place Aug. 2, began at 4:52 a.m. EDT and lasted for 4 hours 50 minutes until 9:42. It increased the total duration of the mission's EVAs on the moon to a record 18 hours 37 minutes. During the Aug. 2 EVA the astronauts drove the rover nearly $6\frac{1}{2}$ miles and brought the vehicle's total for its 3 excursions to nearly $17\frac{1}{2}$ miles. The astronauts Aug. 2 also took back at least 72 pounds of surface material; they thus raised the total for the mission to more than 226 pounds. The Aug. 2 collection included a 9-to-10-foot core sample that the astronauts were able to drill and remove only with great difficulty.

The goal of the Aug. 2 trip on the rover was the rim of Hadley Rille, where the astronauts observed in greater detail a phenomenon they had noted previously—the horizontal layers on layers of rock that formed the sides of the deep rille. Geologists in Houston said the layering indicated that the dry lunar "seas" had been formed not by a single eruption of lava but by many volcanic incidents.

Scott Aug. 2 performed before the TV camera a scientific demonstration similar to those performed nearly 400 years ago by astronomer-physicist Galileo Galilei from the Leaning Tower of Pisa. Holding a feather in one hand and a hammer in the other, he dropped them simultaneously to show that both would reach the moon's surface at the same time despite their difference in weight.

One of the astronauts' final tasks before re-entering *Falcon* was to drive the rover about 300 feet from the LM and position the TV camera so that it could photograph the departure of the astronauts from the moon the next day.

Falcon's ascent stage, with Scott and Irwin aboard, left the moon at 1:11 p.m. EDT Aug. 2 after a record stay of 2 days, 18 hours, 5 minutes. In blasting off, it used the 4-legged descent stage as a launching pad.

The astronauts piloted the LM up into lunar orbit and brought it to rendezvous with the CSM, which had been waiting in orbit with Worden aboard. The vehicles docked at 3:10 p.m. After ascertaining that *Falcon* and *Endeavour* were tightly linked together, Scott and Irwin brought their lunar samples, film and other mementos through the tunnel into *Endeavour*.

Because of a suspicion that there was a hatch leak, the jettisoning of *Falcon* was delayed for 49 minutes. The hatches, however, were found to be tightly sealed. With all 3 astronauts aboard *Endeavour*, the CSM and LM were separated at 9:04 p.m., and *Falcon* was sent crashing down to the moon 2 hours later. (The LM's impact on the lunar surface was recorded by seismometers left on the moon by *Apollos 12, 14* and *15*.)

The astronauts spent the next 2 days in lunar orbit aboard *Endeavour* performing additional experiments, taking photos and making observations to add to their scientific data on the moon and space. Worden, who had been conducting such experiments during his crewmates' stay on the moon, continued to have chief responsibility for this phase of the mission.

A $78\frac{1}{2}$-pound scientific "subsatellite" was ejected from *Endeavour* at 4:13 p.m. EDT Aug. 4 into a lunar orbit with an initial apocynthion of 86 miles and pericynthion of 63 miles. It was expected to stay in orbit for about a year. Its mission was to radio to earth data on gravitational and magnetic fields and on high-energy particles.

Earlier Aug. 4 the TV camera aboard the lunar rover on the surface of the moon was activated by radio signal from ground controllers at the Manned Spacecraft Center in Houston. The TV camera transmitted color telecasts for about 12 minutes. It then stopped and could not be restarted. Engineers speculated that the halt was due to overheating.

Endeavour was brought out of lunar orbit by a 2-minute 21-second blast of its main engine beginning at 5:22 p.m. EDT Aug. 5 while the CSM was behind the moon. The ignition increased the spacecraft's speed to 5,780 mph. and headed it towards the earth in a trajectory so accurate that at first no mid-course correction seemed necessary.

Worden made a 16-minute "space walk" starting at 11:40 a.m. EDT Aug. 5 while his spaceship was 196,000 miles from earth—a record distance for such an activity. The purpose of the EVA was to inspect instruments and to retrieve film cassettes from the instrument bay in the service module, which was to be jettisoned before re-entry. With Irwin tending a 24-foot tether attached to Worden, the latter made 3 trips between the cabin's hatch and the instrument bay. The cabin had been depressurized before the EVA, and all 3 astronauts wore space suits during the exercise.

The slight movements made during Worden's EVA apparently caused a slight deflection in the spacecraft's direction, and a very small course adjustment was decided on to correct this. The course correction was made at 1:30 p.m. EDT Aug. 7 as *Endeavour* approached the earth. The service module was jettisoned nearly 3 hours later.

With its heat shield facing in the direction of flight, *Endeavour*'s command module smashed into the upper fringes of the atmosphere at 4:32 p.m. over the mid-Pacific at an altitude of about 400,000 feet. The spaceship's speed was slowed by friction with the air. This friction heated up the air around *Endeavour* and the spaceship's surface to a temperature of about $5,000°$ F. and enveloped the spaceship in a blanket of ionized particles that temporarily halted radio communication between the capsule and the earth.

2 small parachutes were deployed at first to stabilize the spaceship, and the 3 83-foot-wide main parachutes popped out at an altitude of about 10,000 feet. Although one of the 3 main chutes later deflated, the remaining 2 were sufficient to lower the spacecraft safely to the water.

Endeavour splashed down in the Pacific at 4:46 p.m. EDT Aug. 7 some 333 miles north of Hawaii about $5\frac{1}{2}$ miles from its aiming point and about 7 miles from the recovery carrier *Okinawa*. Within 40 minutes the astronauts were brought aboard the *Okinawa* by helicopter, and initial medical checks showed that they had endured their 12-day space mission with no unexpected ill effects. The command module was hoisted aboard the carrier later.

The *Apollo 15* astronauts were the first moon explorers not required to undergo quarantine after their return from the moon. Experience with previous moon voyagers had convinced space medical officials that there were no dangerous lunar microorganisms that could be brought back from the moon.

Dr. Charles A. Berry reported in Houston Aug. 13 that *Apollo 15*'s astronauts had suffered heart irregularities and dizziness on the moon and had recovered unusually slowly from the effects of weightlessness after returning to earth. The minor spells of irregular heart rhythm were experienced by Scott and Irwin, who actually walked on the moon, and Irwin also had dizzy spells while in space. Scott, Irwin and Worden, who had remained in lunar orbit in the command ship, all had difficulty in recovering after coming back to earth, the space physician said. Berry speculated that the heart irregularities were caused by fatigue, and he held that they were probably unimportant.

2 Soviet Probes Orbit Moon

2 unmanned Soviet space vehicles, *Luna 18* and *Luna 19*, were launched toward the moon by Soviet space technicians Sept. 2 and Sept. 28. Both probes were carried up atop multi-stage rockets, which put them into parking orbit around the earth before sending them into lunar trajectory. In neither case was the launching announced in advance nor was the mission's objective made public.

Luna 18 reached the vicinity of the moon Sept. 7 and went into an orbit about 60 miles above the lunar surface. The orbit, inclined at an angle of 35° to the moon's equator, had a period of one hour 59 minutes. After 54 revolutions around the moon, *Luna 18* crashed into the lunar surface at 10:48 a.m. Moscow time Sept. 11 in what Tass described as an "unfavorable" landing in a mountainous area near the moon's dry Sea of Fertility. "The program of research with the aid of the *Luna 18* automatic probe has been completely carried out," Tass said. It reported that while in orbit, *Luna 18* had "performed numerous maneuvers" for "testing automatic near-moon navigation methods and securing a landing on the lunar surface."

Luna 19 went into lunar orbit Oct. 3 at an altitude of about 84 miles above the moon's surface. The inclination was given as 40° 35' and the period as 2 hours one minute 45 seconds. The announcement of *Luna 19*'s launching had hinted that a lunar landing was not planned.

Moon Car Stops Working

The USSR's 8-wheeled *Lunokhod 1* vehicle stopped working Oct. 4, 10 months and 17 days after it had been landed on the moon by the Soviet lunar probe *Luna 17*. The remote-controlled car's moving parts ceased operation after 11 "lunar days" when the fuel of its nuclear heater was exhausted.

Dr. Boris Nepoklonov, scientific chief of the earth-based team that controlled *Lunokhod 1*'s movements on the moon, said in an interview distributed by Tass Nov. 2 that the vehicle had explored more than 95,000 square yards of the moon's arid Sea of Rains and had produced data making it "possible to compile an adequate scheme on the terrain covering an area of more than half a million square yards." "We have over 500 lunar panoramas" transmitted to earth from *Lunokhod 1*'s cameras, Nepoklonov said, "over 20,000 television pictures, 25 chemical analyses of the ground and hundreds of probes of the physical-mechanical characteristics of the upper layer of the Sea of Rains." He noted that "geologists and selenologists are convinced that the Sea of Rains is one of the oldest formations of the lunar sphere that extends from the equator to the moon's north pole."

Although the vehicle's moving parts were inoperable, Nepoklonov said, the French-made laser reflector mounted on *Lunokhod 1* would still be used as a fixed astronomical beacon on the moon.

Lunar Findings

Dr. Gary Latham of Columbia University reported in Houston Aug. 19 that seismometers left on the lunar surface by *Apollos 12, 14* and *15* had pin-pointed a 6-mile-wide region 400 miles below the Ocean of Storms as the source of 80% of the quakes on the moon.

Col. David Scott, *Apollo 15*'s commander, reported at the National Press Club in Washington Aug. 23 that the $8\frac{1}{2}$-foot-deep core sample he had brought back from the moon had been found to have 57 separate layers of soil built up over 2.4 billion years. The rock he had brought back and described as possibly "genesis rock" was reported by scientists at the State University of New York at Stony Brook Sept. 17 to be about 4.15 billion years old—give or take 200 million years. It was therefore perhaps 450 million years younger than the moon, whose age was estimated at 4.6 billion years. The estimates were prepared from experiments designed and/or conducted by Drs. Oliver A. Schaeffer, Liaquat Husain and John F. Sutter.

Dr. Marcus Langseth reported at the Manned Spacecraft Center in Houston Sept. 1 that temperature probes drilled into the lunar surface during the *Apollo 15* mission had showed that the interior of the moon was hot—and that the heat increased about 1° F. with each foot of depth. *Apollo 15* instruments during orbit had detected a 600-mile-wide radioactive "hot spot" at the eastern edge of the Sea of Storms.

The apparent discovery of short-lived water vapor geysers on the moon was reported Oct. 15 by 2 Rice University physicists, Drs. John W. Freeman Jr. and H. Kent Hills, on the basis of data from suprathermal ion detectors set up on the moon by the crews of *Apollo 12* and *Apollo 14*. The vapor jets apparently vented from cracks in the lunar surface at the sites of both lunar probes Mar. 7 after a series of small moonquakes. The venting continued for about 14 hours.

OTHER MANNED FLIGHT DEVELOPMENTS

Cosmonauts Dock With Space Station

The USSR achieved 2 dockings of manned spacecraft and unmanned space stations during 1971, but one of the missions ended in death for 3 cosmonauts.

In the first mission, 3 Soviet cosmonauts blasted into orbit aboard *Soyuz 10* Apr. 23. They docked their craft Apr. 24 with a previously launched unmanned "orbital scientific station," the *Salyut* (*Salute*). The linked vehicles flew together in orbit for $5\frac{1}{2}$ hours and then separated. *Soyuz 10*, with its 3 cosmonauts aboard, was brought out of orbit and landed safely in Kazakhstan Apr. 25 while the unmanned *Salyut* continued on in orbit.

The mission had started at about 4:30 a.m. local time Apr. 19 with the launching of *Salyut* atop a multi-staged carrier rocket at the Baikonur cosmodrome in Kazakhstan. *Salyut* achieved an orbit with an apogee of 138 miles, perigee of 124 miles, period of 88.5 minutes and 51.6° angle of inclination. Communist Party Gen. Sec. Leonid Brezhnev, speaking in Sofia Apr. 21 at the 10th Congress of the Bulgarian Communist Party, said that *Salyut* marked "not only a new stride in the exploration and conquest of outer space but also an important stage in a further advance in this field...."

Soyuz 10, with its 3 cosmonauts aboard, was launched from the Baikonur cosmodrome at 2:54 a.m. Moscow time Apr. 23. 9 minutes later it was in an orbit with an apogee of 153 miles and perigee of 129 miles. *Soyuz 10*'s crew consisted of Red Air Force Col. Vladimir A. Shatalov, 43, the spaceship commander, who had already made 2 previous space flights; civilian flight engineer Aleksei S. Yeliseyev, 36, who had flown in space twice previously; and civilian test engineer Nikolai N. Rukavishnikov, 39, who was making his first space flight. Shatalov told the ground controllers that they had sent them into an orbit that was "a bit too high," and he later brought *Soyuz 10* down to a slightly lower altitude.

The docking of *Soyuz 10* and *Salyut*, which took place at 4:47 a.m.

Moscow time Apr. 24, was accomplished in 2 stages, the first automatic, the 2d manual. "During the first stage," Tass reported, "the ship [*Soyuz 10*] was automatically brought to a distance of 180 meters [195 yards] from the station [*Salyut*]. Further approach and berthing was carried out by the crew." "After fulfilling the planned experiments," Tass said, "the crew separated *Soyuz 10* from the station and moved away from it."

In an amplifying statement, Tass reported: "Scientific-technical experiments and work on docking and undocking were carried out. . . . The principles of the automatic station were tested, and new docking systems were checked out together with a radar set."

Soyuz 10 was brought out of orbit Apr. 25, and it landed safely by parachute at 2:40 p.m. local time some 75 miles northwest of Karaganda, Kazakhstan. All 3 cosmonauts aboard were reported to be in good health after their 2 days in space.

Although the USSR had given no advance word of the mission, it had been reported in Moscow as early as Apr. 14 that, according to "scientific sources," the Soviets were expected to send up at least 2 manned spaceships within the next few days as part of a program for putting together an orbiting laboratory. It was known beforehand that Soviet newsmen had already left for Baikonur to report the launching.

The new, never-named "chief designer of spaceships" (director of Soviet space operations) had been quoted in the Moscow economic daily *Sotsialisticheskaya Industriya* Mar. 14 as saying that the USSR was getting ready to send men into orbit for a long mission in the program for a permanent orbital laboratory. (Theodore Shabad reported in the *N.Y. Times* Mar. 15 that, "from indirect evidence," the chief designer was "believed" to be Mikhail K. Yangel, 50, an engineer and member of the Soviet Academy of Sciences. Yangel was reported Oct. 25 to have died after a heart attack.)

Despite the various hints and news "leaks," most basic information about the mission was kept secret. Each launching was made public only after it took place. The docking was disclosed after it had been completed and the spaceships had separated. The decision to bring the cosmonauts down after what Western observers considered a surprisingly short time was not revealed until the announcement that they had landed safely.

One of the first hints about the size of the orbital laboratory was given by Yelisev after *Soyuz 10*'s landing, when he told newsmen that docking with *Salyut* "was a little like a train entering a railroad terminal; that's how we felt as our rather big Soyuz eased up to the station." Another hint came from cosmonaut-engineer Konstantin P. Feoktistov.

Tass quoted Feoktistov as saying that it was more difficult to dock a relatively small Soyuz spaceship with a huge vehicle such as *Salyut* than to dock 2 equally sized space vehicles.

3 Cosmonauts Die as Flight Ends

After setting a 24-day endurance record for space flight, 3 Soviet cosmonauts died early June 30 on their return to earth from the orbiting laboratory *Salyut*. The deaths were announced on Moscow radio at about 8:15 a.m. Moscow time, less than 6 hours after a recovery team had opened the hatch of the spaceship *Soyuz 11* and had found the cosmonauts "in their seats without any signs of life."

Soviet authorities admitted initially that they were mystified by the deaths of the spacemen, and a special commission was appointed late June 30 to investigate the reasons for the fatal end of the mission. The inquiry commission reported July 12 that the "sudden death" of the cosmonauts had been caused by a "rapid" air pressure drop due to a "loss of the ship's sealing" 30 minutes before landing. Soviet journalist Victor Louis, frequently used by the Soviet government to "leak" information to the Western press, had reported July 2 that, "as a result of "human error and mechanical failure," "the spacecraft's hatchway opened slightly—enough to suck the cosmonauts' air supply into space." Because of a design fault in *Soyuz 11*, Louis said, the cosmonauts had "failed to seal the hatch of their spacecraft properly."

Until the discovery of the cosmonauts' deaths, the record-breaking *Salyut-Soyuz 11* mission had appeared to be an outstanding success. The return to earth from space had also seemed to be going as scheduled.

Tass reported that Lt. Col. Georgi T. Dobrovolsky, 43, and his 2 civilian crew members—flight engineer Vladislav N. Volkov, 35, and test engineer Viktor I. Patsayev, 37—had "completed the flight program in full" June 29 and had been "directed to make a landing." They "transferred the materials of scientific research and the logs to the transportation spaceship—*Soyuz 11*—for return to earth," Tass said, seated themselves in *Soyuz 11*, "checked the onboard systems" and separated *Soyuz 11* from *Salyut* at 9:28 p.m. Moscow time June 29.

After *Soyuz 11* was in separate flight, the cosmonauts reported to their ground controllers that "the unlinking operations [had] passed without a hitch and all the systems were functioning normally."

The spaceship, with the cosmonauts aboard, was brought out of orbit by the firing of the retro-rocket at 1:35 a.m. June 30, and Tass reported that this braking engine "functioned throughout the estimated time." "At the end of the operation of the braking engine," Tass continued, "communication with the crew ceased." Apparently the ground controllers never re-established contact with the *Soyuz 11* crew. The loss of communication at this point was considered unusual by American observers, since communication with returning U.S. astronauts usually continues 15–20 minutes after the end of the retro-rocket phase and does not end until re-entry into the atmosphere creates an ionized envelope of air around the spaceship and cuts off communication for up to about 12 minutes.

Tass reported that, "according to the [*Soyuz 11* re-entry] program, after [the] aerodynamic braking in the atmosphere, the parachute system was put into action, and before landing, the soft-landing engines were fired. The flight . . . ended in a smooth landing in the pre-set area" near Karaganda, Kazakhstan.

A helicopter-borne recovery team landed "simultaneously with the ship [*Soyuz 11*]," Tass said. Members of this team opened the hatch and found the 3 cosmonauts lifeless in their seats. Soviet press sources said the dead men appeared to be sleeping peacefully, their faces were calm and there were no signs of frantic struggle for life.

Messages of sympathy were sent to the USSR June 30 by Pres. Nixon, U.N. Secy. Gen. U Thant, Pope Paul VI, other U.S. and world leaders and the astronaut corps of the U.S.

Soyuz 11, with its 3 cosmonauts aboard, had been launched from the Baikonur cosmodrome atop a multistage booster rocket at 7:55 a.m. Moscow time June 6 while *Salyut* was ending its 779th revolution around the earth. Soviet authorities had made no advance announcement of the mission and had disclosed it only after they had determined that *Soyuz 11* was safely in orbit. The spaceship achieved an initial orbit whose parameters were not announced. At 1:50 p.m. the cosmonauts "corrected" the orbit. The apogee was then given as 217 kilometers (135 miles), the perigee as 185 kilometers (115 miles), the inclination as 51.6° and the period as 88.3 minutes.

The docking of *Soyuz 11* with *Salyut* June 7 was accomplished in 2 phases: An automatic phase began at 7:26 a.m. Moscow time after the spaceship had been brought to within 4 miles of the orbiting laboratory. This automatic maneuver brought the vehicles to within a few hundred feet of each other. Then, at 7:50 a.m., Dobrovolsky began the manned phase, in which he linked the 2 spacecraft nose-to-nose. At link-up, Tass reported, the *Salyut-Soyuz 11* assembly was in an orbit with an apogee of 155 miles, perigee of $131\frac{1}{2}$ miles and period of 88.2 minutes.

When the vehicles were linked, they were connected by a "tunnel" used by the cosmonauts to pass from one vehicle to the other. After the pressure in *Salyut* had been adjusted to equal that in *Soyuz 11*,

Patsayev crawled through the tunnel into *Salyut* at 10:45 a.m. He was followed by Volkov. Dobrovolsky remained for a short time aboard *Soyuz 11* to handle communications, and then he too entered *Salyut*.

Soviet scientists and engineers who broadcast commentaries on the event provided some hitherto unrevealed information about *Salyut*. Cosmonaut-engineer Konstantin Feoktistov said in a televised news conference June 7 that the linked spaceships weighed 25 metric tons (55,000 pounds) and were 20 meters (65 feet) long. Since *Soyuz 11* was known to weigh about 15,000 pounds, the weight of Salyut was presumed to be about 40,000 pounds. Feoktistov reported that *Salyut*'s pressurized module, which included the crew quarters, had a habitable volume of 3,500 cubic feet and a maximum diameter of four meters (13 feet). *Salyut* also had an unpressurized service module with a propulsion system.

2 increases in the orbit's altitude were made June 8 and 9 after slight deteriorations in the orbit. The first increase raised the apogee to 165 miles and the perigee to 148 miles. The 2d increased the apogee to 175 miles and the perigee to 161 miles.

Tass reported that the crew's principal tasks were to check out the space lab, test its navigation instruments and study the earth's geography and atmosphere and the effects of prolonged space flight on men. In a telecast June 9, the cosmonauts demonstrated "penguin suits" they wore to force them to exert their muscles and thus prevent the weakness that had overtaken previous space fliers after prolonged periods of weightlessness in space. During their unprecedented stay aboard the space laboratory, they reported consistently that they were in excellent health. They exercised vigorously and at length with specially designed equipment to counteract the debilitating effect of long weightlessness. Much of their experimental work was designed to show the effects of long space flights on men and to provide data on how men could live and work for long periods in the environment of space. Patsayev marked his 38th birthday aboard *Salyut* June 19 with a celebration in which his crewmates toasted him with tubes of prune paste.

In one experiment expected to be of early practical value, the cosmonauts June 15 took spectral photos of Caspian coastal areas for use in farming, land improvement, geodesy and cartography. The data was coordinated with photos obtained simultaneously by Soviet planes.

The cosmonauts June 24 exceeded the previous space endurance record of 17 days, 16 hours, 58 minutes and 50 seconds, which had been set by Soviet cosmonauts Andrian Nikolayev and Vitali Sevastyanov in *Soyuz 9* June 1-19, 1970. (Nikolayev and Sevastyanov had needed a month to recover after nearly 18 days of weightlessness.)

Shuttle's Military Role

Col. John G. Albert disclosed at a space conference in Phoenix, Ariz. Mar. 15 that the U.S. Defense Department planned ultimately to use manned reusable rocket planes—the space shuttles currently under development—as the sole boosters "for all military space operations." Albert, director of space development for the Air Force, reported that "the Defense Department is putting its faith in the shuttle, and as a result we are not developing any other space rocket beyond the present Titan-3." He said plans called for unmanned tests of the shuttle beginning in 1976, manned tests beginning in 1977 and delivery of fully operational vehicles beginning in 1979.

Cosmic Ray Threat

U.S. findings released Apr. 8 indicated that cosmic rays could cause serious damage to astronauts on long space missions unless additional and more effective shielding were provided. The warning was based on a study of "cosmic ray tracks" found in plastic helmets worn by Apollo astronauts on trips to the moon. A report on the research, headed by Drs. Robert L. Fleischer and P. Buford Price at the General Electric Research & Development Center in Schenectady, was published Apr. 9 in the journal *Science*.

The researchers agreed that unless astronauts were given more effective shielding, a 2-year mission to Mars could result in the destruction by cosmic rays of .12% of the cells of the cerebral cortex, .05% of the cells of the retina and more than 1.5% of the larger cells of the nervous system. These cells, according to the researchers, were "irreplaceable." Damage during Apollo-style short trips to the moon, however, was considered too slight to worry about.

(Fleischer and Price were named Apr. 8 as among the winners of the Atomic Energy Commission's Ernest O. Lawrence Award for their research in this cosmic-ray field.)

INTERPLANETARY PROBES

U.S. Satellite in Mars Orbit

The unmanned probe *Mariner 9* went into orbit around Mars at 7:33 p.m. EST Nov. 13 after travelling some 247 million miles since its launching from Cape Kennedy, Fla. May 30. It was the first man-made object to orbit another planet.

The $65 million spacecraft had started taking photos of Mars at a distance of 535,000 miles as it approached the planet Nov. 10. These photos and those taken at the beginning of the orbital phase of the mission showed the planet to be largely obscured by a tremendous yellow dust storm that had been raging for a month and a half.

Mariner 9's nearly $5\frac{1}{2}$-month journey had followed the calculated course so precisely that the probe's controllers canceled a course correction scheduled for Oct. 26. The spacecraft was only 60 kilometers ($37\frac{1}{4}$ miles) off its aiming point after a flight of 1/4 billion miles as it reached the vicinity of Mars Nov. 13.

As *Mariner 9* passed over Mars' southern hemisphere at an altitude of 940 miles, the probe's 17-pound rocket engine was fired for 15 minutes at a thrust of 300 pounds. This ignition reduced the spacecraft's speed from 11,000 mph. to 8,300 mph. As a result, *Mariner 9*, captured by the planet's gravity, went into an initial orbit in which it dipped to a low point of 868 miles above Mars' surface and rose to a high point of 11,135 miles. The orbit crossed the planet's equator at an angle of 64°, and each revolution took 12 hours 34 minutes.

In addition to its 2 cameras—one with a wide-angle lens and the other with a narrow-angle telephoto lens—*Mariner 9* carried an infra-red radiometer to measure Mars' surface temperature, an infra-red spectrometer to provide data on the composition of the planet's surface and to measure surface and atmospheric temperatures, and an ultra-violet spectrometer to study atmospheric composition, structure and pressures. Radio signals, which took 6 minutes 43 seconds to travel the 75 million miles from *Mariner 9* to the earth, indicated that all instruments were operating properly. The data were being analyzed by NASA's Jet Propulsion Laboratory (JPL) in Pasadena, Calif., which had built *Mariner 9* and was directing the mission.

Mariner 9 was the 2d of 2 virtually identical 2,200-pound probes that U.S. space officials had hoped to put into Martian orbit in November. The first of the 2, *Mariner 8*, launched from Cape Kennedy May 8, fell into the Atlantic because of a malfunction of the 2d stage of the 2-stage Atlas-Centaur booster rocket. After several delays due to technical problems, *Mariner 9* was launched at 6:23 p.m. EDT May 30.

Dr. Ellie D. Miner of JPL reported Nov. 15 that *Mariner 9*'s infra-red radiation sensor had detected an unsuspected 15-square-mile "hot spot" on the Martian surface at Latitude 10° S., Longitude 120° W. The "hot spot" averaged 12° F. above the normal -55° to -60° F. temperature of the surrounding areas. JPL scientist Jack Lorell reported Nov. 19 that analysis of *Mariner 9*'s orbit and other data had showed Mars' gravitational field to be very irregular. The analysis had indicated that Mars bulged at the equator, especially at Tharsis, an equatorial area at 110° W. Longitude. The planet's flattened poles had been noted previously. (The diameter of Mars was reported to be 24 miles longer across the equator than through the poles.)

Mariner 9's orbit had been "trimmed" at 9:45 p.m. EST Nov. 15 by a rocket blast that reduced the orbit's high point to 10,585 miles, the low point to 863 miles and the period to 11 hours 59 minutes. This orbit would take the probe between Mars' 2 small moons and keep *Mariner 9* circling the planet, it was estimated, for at least 17 years.

Mariner 9 began photographing Mars' moons Nov. 26. Deimos, the smaller of the moons, was photographed first, from a distance of 5,450 miles. Dr. Carl Sagan, director of Cornell University's Laboratory for Planetary Studies, described Deimos from the photo as shaped like a potato. The larger moon, Phobos, was photographed Nov. 29 and 30 from distances of 9,000 and 3,400 miles. It was described by a JPL spokesman Dec. 1 as "just a big chunk of rock, sort of oblong." The irregular, crater-marked Phobos, about twice the size of Deimos, was reported to be about 16 miles long and 13 wide. Its biggest crater was about 4 miles wide.

U.S. officials in Washington Oct. 20 had announced a U.S.-Soviet agreement for each country to telegraph immediately to the other all findings "of special interest" made by their current Mars probes—the U.S.' *Mariner 9* and the USSR's *Mars 2* and *Mars 3*, which were scheduled to reach the vicinity of the planet a week after *Mariner 9*.

In observance of this data-exchange agreement, Dan Schneiderman of the JPL Nov. 16 sent to Dr. I. A. Zhulin in Moscow a Telex reporting that Mars' "surface features are almost entirely obscured" by what was believed to be "a dust storm which covers the entire planet." Although the USSR had not confirmed that it planned to land its probes on Mars, Western observers had reported indications that the Russians did plan a landing, and U.S. officials had expressed fear that the dust storm might endanger such a project.

Soviets Land Capsules on Mars

The 2 Soviet unmanned Mars probes joined *Mariner 9* in orbit around Mars Nov. 27 and Dec. 2. Both probes deposited capsules on the Martian surface and continued in orbit around the planet.

The Soviet probes—*Mars 2* and *Mars 3*—had been launched May 19 and 28, respectively, on their 290-million-mile curving paths to Mars.

(The U.S.' *Mariner 9*, launched May 30, had a flatter trajectory covering 247 million miles.) The virtually identical Soviet probes, each weighing 4,650 kilograms (10,230 pounds), were sent up without advance notice. *Mars 2* went up at 7:23 p.m. Moscow time May 19, *Mars 3* at 6:36 p.m. May 28.

Before entering its Martian orbit Nov. 27, *Mars 2* sent to the planet's surface a capsule carrying "a pennant showing the USSR's coat of arms." This capsule, the first manmade object to reach the surface of Mars and the 2d to land on any planet other than the earth (a Soviet capsule had touched down on Venus in Dec. 1970 and had burned), apparently crash-landed.

Mars 3 sent down a TV-equipped capsule that soft-landed on Mars Dec. 2 and transmitted TV pictures for 20 seconds before contact was lost.

Mars 2's delivery of the pennant to Mars Nov. 27 was first disclosed by Tass Nov. 30. Tass reported that the probe had carried out 3 course corrections June 17, Nov. 20 and Nov. 27 as it hurtled towards Mars and that it approached to within 776 miles of the planet's surface Nov. 27 before going into an orbit with a low point of 858 miles from the surface, a high point of 15,534 miles, a period of 18 hours and a 48° angle of inclination to Mars' equator.

The soft-landing of the *Mars 3* capsule Dec. 2 was first revealed by Tass Dec. 7. The capsule landed at Latitude 45° S., Longitude 158° W. between the regions of Electris and Phaetonis. It came down in a hurricane that drove dust particles at speeds of up to 300 mph. "Video signals received from [the capsule on] the surface of Mars were of short duration and then suddenly ceased," Tass disclosed. It was not until Dec. 19 that the USSR revealed that the TV transmission had lasted only 20 seconds.

The Dec. 7 account disclosed that the capsule's signals were relayed to the orbiting *Mars 3*, which stored them and later transmitted them to earth during radio communications sessions held Dec. 2–Dec. 5. (Radio signals took about 9 minutes to travel the 100 million miles of space that then separated the earth and Mars.) Tass reported that the low point of *Mars 3*'s orbit was about 800 miles and the period about 11 days.

The Dec. 19 report revealed that the descent of the *Mars 3* capsule took "only a little more than 3 minutes." A pilot parachute first slowed the capsule to the speed of sound, the main parachute then took over and the soft-landing engine was finally ignited at an altitude of 65 to 95 feet.

The USSR notified U.S. Jet Propulsion Laboratory officials in

Pasadena, Calif. Nov. 30 of the orbiting of *Mars 2* and of the delivery of the pennant on the planet's surface. The Russians made their report by means of the Teletype system set up under the Oct. 20 U.S.-Soviet agreement.

EARTH SATELLITES

International Communications Satellites

The first 2 satellites in a new series of Intelsat-4 commercial communications spacecraft were orbited during 1971. The first was sent up Jan. 25 by means of an Atlas-Centaur rocket assembly launched from Cape Kennedy by NASA at 7:36 p.m. The 3,094-pound unmanned spacecraft, capable of carrying as many as 9,000 simultaneous transatlantic phone conversations, went into a preliminary orbit with an apogee of 22,700 miles and perigee of 165 miles. A radio signal from Andover, Me. Jan. 27 started it moving toward its synchronous orbit over the equator at an altitude of 22,300 miles above the Atlantic.

NASA launched the $13\frac{1}{2}$ million satellite on contract with Comsat (Communications Satellite Corp.), the U.S. member of Intelsat (International Telecommunications Satellite Consortium). Intelsat had agreed to pay NASA $16 million for the launching.

The 2d Intelsat-4, a 3,100-pound payload, was launched by NASA from Cape Kennedy Dec. 19 on contract with Comsat. The $29\frac{1}{2}$ million 17-foot satellite went up at 8:10 p.m. in the nose of an Atlas-Centaur rocket. Its mission was to relay phone and TV communications between the U.S. and Canada in the Western hemisphere and points in Europe and North Africa. It was to replace a smaller Intelsat-3 satellite in synchronous orbit 22,300 miles above the Atlantic at the equator.

Intelsat membership had reached 81 with the admission of Ghana Nov. 15. Members of Intelsat were Algeria, Argentina, Australia, Austria, Belgium, Brazil, Britain, Cameroon, Canada, Ceylon, Chile, China, Colombia, Denmark, Dominican Republic, Ecuador, Egypt, Ethiopia, France, Gabon, Ghana, Greece, Guatemala, India, Indonesia, Iran, Iraq, Ireland, Israel, Italy, Ivory Coast, Jamaica, Japan, Jordan, Kenya, Kuwait, Lebanon, Libya, Liechtenstein, Luxembourg, Malagasy Republic, Malaysia, Mauritania, Mexico, Monaco, Morocco, the Netherlands, New Zealand, Nicaragua, Nigeria, Norway, Pakistan, Panama, Peru, Philippines, Portugal, Saudi Arabia, Senegal, Singapore, South Africa, South Korea, South Vietnam, Spain, Sudan, Sweden, Switzerland, Syria, Tanzania, Thailand, Trinidad and Tobago, Tunisia, Turkey, Uganda, U.S., Vatican City, Venezuela, West Germany, Yemen, Yugoslavia, Zaire and Zambia.

(In a letter dated Sept. 1 and made public Nov. 1, U.S. State Undersecy. U. Alexis Johnson had informed Chairman Theo Lefevre of the European Space Conference that the U.S. would make its rockets available for West European projects—on a purchase basis—without the user nations being required to participate in post-Apollo U.S. space programs. European leaders had expressed suspicion that U.S. commercial pressure would keep the U.S. government from providing rockets for communications projects competing with the U.S.-supported Intelsat. Johnson said that should U.S. launch assistance be requested for international public telecommunications satellites on which Intelsat had not made "a favorable recommendation . . . we expect that we would provide launch assistance for those systems which we had supported within Intelsat" provided that the requesting agency "has met its relevant obligations. . . ." In cases where there was no "favorable Intelsat recommendation and the U.S. had not supported the proposed system," Johnson continued, the U.S. decision would take into account "the degree to which the proposed systems would be modified" to meet Intelsat objections.)

Canadian Communications Min. Eric Kierans had announced Apr. 20 that Canada and the U.S. had signed a $27 million agreement to launch a powerful, experimental communications satellite in 1974. The experimental model of the satellite would be designed and built in Canada and would be launched by the U.S. National Aeronautics & Space Administration. The project was an attempt to develop operations satellites that could transmit to simpler, cheaper ground stations.

An international space communications organization named Intersputnik was created in Moscow Nov. 15 under an agreement signed by the USSR and 8 other Communist countries: Bulgaria, Hungary, East Germany, Cuba, Mongolia, Poland, Rumania and Czechoslovakia. The agreement asserted that "all states of the world," Communist or otherwise, could join. Intersputnik was to "coordinate its activities with the International Telecommunications Union as well as with other organizations whose activities are related to the use of communications satellites."

China's 2d Satellite

Communist China's 2d earth satellite was sent into orbit Mar. 3 by a booster rocket launched from the Shuang-cheng-tze space center in

Kansu Province at 7:15 a.m. EST. The launching was reported the same day by the North American Air Defense Command.

The first Chinese announcement of the launching was made 13 days later, when Hsinhua, the Chinese press agency, announced the shot Mar. 16 and gave these details: Weight of the satellite—486 pounds; apogee (high point) of orbit—1,826 kilometers (1,132 miles), orbit's perigee (low point)—266 kilometers (165 miles); period—106 minutes. Hsinhua reported that the orbiting satellite "successfully sent back scientific data on various experiments . . . and is now continuing the planned work of scientific experiment."

Chinese plans to launch the satellite had been reported as early as Feb. 23 by George C. Wilson in the *Washington Post* on the basis of "feverish activity" observed at the Gobi Desert space center, 30 miles south of the Chinese city of Haerhpin and some 300 miles south of the Soviet border. This was the launching site of China's first satellite, which went into orbit Apr. 24, 1970. The site was kept under systematic photographic observation by U.S. reconnaissance satellites.

In announcing Norad's detection of the launching, the Pentagon disclosed Mar. 3 that a U.S. Navy pilot had apparently seen the upper stage of the carrier rocket as it boosted the satellite into orbit. The "bright plume . . . at a high altitude" was observed by the pilot as he returned to the carrier *Kitty Hawk* after a mission over Southeast Asia.

(Defense Secy. Melvin R. Laird had reported Mar. 9 that Communist China might have tested an intercontinental ballistic missile in a short-range flight in 1970.)

Japan Scores 2 Successes

Japan sent its 2d and 3d satellites into space during 1971. The first of the 2, a 138.6-pound payload dubbed *Tansei (Light Blue)* for Tokyo University's school color, was sent into orbit Feb. 16 by means of a 4-stage Mu rocket launched by Tokyo University scientists from the Uchinoura space center. Because the pacifistic space team was reluctant to develop militarily useful guidance and control equipment, a "gravity-turn" device was used to guide the satellite into orbit.

A 44-ton Mu-48 rocket fired by a Tokyo University team from Uchinoura Sept. 28 sent a 145-pound satellite called *Shinsei (New Star)* into orbit. The payload, Japan's first operating scientific observation satellite, was designed to develop data on cosmic rays and solar electric waves. As with the previous Japanese satellites, the *Shinsei* launching used the "gravity-turn" technique rather than more reliable guidance and control devices. It was reported, however, that the Tokyo University space scientists, faced with complaints that the gravity-turn system was too unreliable for serious scientific satellite work, had started research on conventional guidance and control. The Japanese government's separate rocketry program had always included such research.

U.S. Launchings

A 635-pound scientific satellite, *Explorer 43*, was sent into orbit Mar. 13 by means of a 3-stage Delta rocket launched from Cape Kennedy. The $8 million satellite, also known as IMP 8 (for interplanetary monitoring platforms), was equipped to collect data on cosmic rays, the solar wind, electric and magnetic fields and radio astronomy. Its orbit had an apogee of about 121,000 miles and perigee of some 145 miles.

Oso 7 (for orbiting solar observatory), a 1,400-pound satellite, was sent into a faulty orbit Sept. 29 by means of a 2-stage Delta rocket launched from Cape Kennedy at 5:45 a.m. Instead of going into a circular 345-mile-high orbit, *Oso 7* achieved an eccentric orbit with an apogee of 357 miles and perigee of 206 miles because of a malfunction in the carrier rocket's 2d stage. The malfunction caused the satellite to wobble initially and resulted in a $7\frac{1}{2}$-hour delay in the process of "locking on" the sun so that its solar batteries could be charged. The wobbling and sun-orientation faults were corrected by means of radio commands sent from Cape Kennedy and the Goddard Space Flight Center in Greenbelt, Md. The satellite was equipped to provide data on the sun's corona, solar flares and other solar energy phenomena.

The *Oso 7* carrier rocket also orbited a 45-pound satellite to be used in training exercises by Apollo spaceship tracking stations.

A scientific satellite was sent into polar orbit Oct. 17 by means of a Thor-Agena rocket launched from Vandenberg Air Force Base, Calif. The satellite's mission was to collect data on high energy particles, solar energy, other space phenomena and satellite communications. It carried a 32-by-6-foot solar power cell that went into space rolled up in a cylinder but was unrolled after the satellite was in orbit.

A $10\frac{1}{2}$ million improved weather satellite was destroyed 97 minutes after launching Oct. 21 from Vandenberg Base. A Delta carrier rocket had brought the 685-pound Tiros into a preliminary 200-mile-high polar orbit as scheduled. But when the rocket was restarted an hour later to carry the satellite into a higher orbit, it was pointed down instead of up. The satellite, therefore, plunged back into the atmosphere and was burned up over Devon Island, Canada.

Soviet Launchings

3 Soviet weather satellites were launched Apr. 17, July 16 and Dec. 29. The first and 3d were in the Meteor series, whose mission was

to help in weather forecasting and in measuring the heat reflected from the earth. The Apr. 17 satellite went into a near-circular orbit about 300 miles high. The orbit of the Dec. 29 Meteor had an apogee of 905 kilometers (562 miles), perigee of 880 kilometers (847 miles), inclination of 81.2° and period of 102.7 minutes.

The USSR launched 64 Cosmos satellites during 1971, beginning Jan. 12 with *Cosmos 390* and ending Dec. 16 with *Cosmos 454*. 8 of the satellites—*Cosmos 444* through *451*—were sent up atop a single carrier rocket Oct. 13 and put into orbits with apogees of 962 miles and perigees of 874 miles. Richard D. Lyons reported in the *N.Y. Times* Sept. 6 that 4 Cosmos satellites—*Cosmos 379* sent up Nov. 24, 1970, *Cosmos 382* Dec. 2, 1970, *Cosmos 398* Feb. 26, 1971 and *Cosmos 434* Aug. 12—had apparently been launched to test components for a manned landing on the moon. *Cosmos 434* was said to have had a variable-thrust restartable engine with 6 times the thrust of the engine of a standard Soyuz manned spacecraft.

The 18th and 19th Soviet Molniya-1 communications satellites were launched July 28 and Dec. 20. The satellite launched July 28 achieved an orbit with an apogee of 39,300 kilometers (24,420 miles) in the Northern hemisphere, perigee of 470 kilometers (292 miles) in the Southern hemisphere, period of 11 hours 45 minutes and 65.4° angle of inclination.

The first satellite in a new Soviet communications satellite series, Molniya-2, was launched Nov. 24 into an orbit with an apogee of 24,435 miles in the Northern hemisphere and perigee of 285 miles in the Southern hemisphere.

(The *Washington Post* reported Feb. 5 that the USSR had launched 29 reconnaisance (spy) satellites and the U.S. 9 during 1970. The Soviet total compared with 29 sent up in 1968 and 26 in 1969. The U.S. had launched only 2 in 1968 and 6 in 1969. There were a total of 57 satellites launched by the USSR and 16 by the U.S. during 1970 for specialized military missions.)

Other Satellites

France's 7th scientific satellite, a 211-pound payload, was sent into orbit Apr. 15 by means of a Diamant-B rocket launched from the French Guiana space center at Kourou. The mission of the D-2-A satellite, dubbed *Tournesol (Sunflower)* Apr. 17, was to produce data on the distribution of hydrogen in the solar field.

France's 213-pound D-2A Polaire, launched from Kourou space center Dec. 5, failed to achieve orbit. Officials attributed the failure to malfunction of the Diamant-B carrier rocket's 2d stage. The satellite's mission was to study the distribution of hydrogen in the atmosphere.

An Italian-designed and-built satellite was sent into equatorial orbit Apr. 24 by means of a U.S. 4-stage Scout rocket launched by a U.S.-Italian team from the San Marco space platform, which stood on stilts in the Bay of Formosa 3 miles off the coast of Kenya. The 360-pound scientific satellite, designed to collect data on the upper atmosphere, was the 3d launched from the San Marco site and the 4th in a cooperative program conducted by the U.S. NASA and the Aerospace Research Center of Rome University.

A 114-pound U.S. satellite was sent into orbit Nov. 15 by means of a rocket launched by an Italian team from the San Marco space platform. The satellite's mission was to study the earth's inner magnetosphere.

A 145-pound British instrument package dubbed *Prospero* was sent into polar orbit Oct. 28 by means of a Black Arrow rocket launched by a British team at the Woomera space center in South Australia. Britain thus became the 6th nation to launch its own satellite with its own carrier rocket. Its 4 previous satellites had been launched by the U.S. Further use or development of the Black Arrow had been canceled by British Aerospace Min. Frederick Corfield.

An attempt to put an 800-pound West European test satellite into synchronous orbit failed Nov. 5 when explosions in the Europa-2 launching rocket's British-built Blue Streak first stage brought the rocket and satellite down into the Atlantic shortly after its launching from France's Kourou space center. The 110-ton rocket, developed by the 7-nation European Launcher Development Organization (Eldo), from which Britain applied to withdraw Dec. 16, had 4 stages built, respectively, by Britain, France, West Germany and Italy. The failure was a particular blow to France and West Germany, the Eldo nations most committed to the effort to develop a West European space launching system independent of the U.S. system.

OTHER DEVELOPMENTS

U.S. & Soviet Scientists Cooperate

U.S. and Soviet scientific delegations agreed in Moscow Jan. 21 to exchange about 3 grams of lunar surface material. The agreement was signed by Acting NASA Administrator George M. Low and Prof. Mstislav V. Keldysh, president of the Soviet Academy of Sciences. Low had arrived in Moscow Jan. 16 with a 6-man U.S. delegation to discuss

U.S.-Soviet space science cooperation following successful U.S.-Soviet negotiations on space rescue and on standardizing docking devices. The U.S. delegation left Moscow Jan. 21 after agreeing to the formation of international working groups on synchronizing research.

In another demonstration of U.S.-Soviet space cooperation, Dr. Aleksandr P. Vinogradov, vice president of the Soviet Academy of Sciences, attended the 2d annual Lunar Science Conference, held by NASA in Houston, Tex. Jan. 11-14. The U.S.-Soviet lunar-soil exchange was indorsed by Vinogradov at the Houston meeting Jan. 14. Vinogradov, reporting on the analysis of the USSR's lunar samples, said the age of the material brought back by *Luna 16* had been estimated at 4.45 to 4.65 billion years. This sample had been dug up from a depth of 13 inches, which was as deep as *Luna 16*'s robot equipment could dig before hitting hard rock.

Treaty Drafted

A draft of an international convention covering payment of compensation for damages caused by objects launched into space was completed in Geneva during 1971 after a 7-year effort. Under the proposed convention, a nation would be "absolutely liable" to pay any damage, including personal injury, caused on the earth's surface or to aircraft in flight by an object it sends into space. Either of the nations concerned could ask for the creation of a 3-man claims commission to rule on the merits and amount of the compensation claims. The commission's ruling would be binding only if both parties agreed beforehand. The Soviet Union had ended a long impasse by reluctantly agreeing to use of an international claims commission to fix the cost of damages.

NASA Budget Declines

Pres. Nixon's fiscal 1972 budget request for space research and technology, submitted to Congress Jan. 29, totaled $3,300,635,000. It included several new projects that would involve increased spending in later years. The major such item was a space shuttle; $55 million to $60 million was requested for the development of its 550,000-pound thrust engine and $100 million for developing its frame.

The Nerva nuclear rocket program of NASA and the AEC was cut back from the $86 million level of fiscal 1971 to $15 million for each agency. The projected final moon landing flight was set back 5 months to Dec. 1972 as another economy move. The request for planetary exploration was $180 million, a $145 million increase; the Viking project to send 2 unmanned spacecraft to land on Mars (by 1975) and $30 million requested for work to begin the development of a series of spacecraft to study the outer planets.

Before giving approval to the space program for fiscal 1972, Congress increased NASA's total authorization by about $50 million. Attempts by Rep. Bella S. Abzug (D., N.Y.) to eliminate funding for the space shuttle was rejected by the House June 3 and 30. A similar attempt in the Senate June 29 by Sen. Walter F. Mondale (D., Minn.) was defeated 64-22.

The approved NASA authorization, totaling $3,354,950,000, was cleared by Congress by voice votes of the House July 27 and the Senate July 28. Pres. Nixon signed the bill Aug. 6. The authorization bill carried restrictions (a) to bar, unless the Administration made an exception, use of the funds to universities prohibiting armed forces recruitment and (b) to prohibit institutions of higher education from using NASA funds for individuals contributing to campus disruptions.

Soviet Space Expert Defects

The Soviet scientist Anatoly Fedoseyev, said to be a major figure in Soviet space research, was given permission to remain in Great Britain June 20 after defecting at the Paris Air Show May 26. Home and Foreign Office spokesmen gave no indication June 20 of how Fedoseyev had entered Britain, his whereabouts or his plans. Unconfirmed reports said Fedoseyev had been flown from Paris to NATO headquarters in Brussels and then to Washington before being granted asylum in Britain.

Mikhail N. Smirnovsky, Soviet ambassador to Britain, requested June 21 that a member of his staff be allowed to speak with Fedoseyev. Smirnovsky's request was made to Sir Denis Greenhill, permanent undersecretary at the Foreign Office, who noted that Fedoseyev was free at any time to contact the Soviet embassy.

The Soviet Union expelled 2 British diplomats June 22 and accused them of complicity in Fedoseyev's defection. Those expelled were Martin Nicholson, first secretary, and Patrick Jackson, 2d secretary, of the British embassy in Moscow. Britain, which maintained that Fedoseyev had acted on his own initiative, retaliated June 22 by ousting Lev Sherstnev, first secretary, and Valery Chousovitin, 3d secretary, of the Soviet embassy in London.

INDEX

A

A-1 (French earth satellite)–35–6
ABERNATHY, Rev. Ralph David–181–2
ABRAHAMSON, Maj. James A.–121
ABZUG, Rep. Bella S. (D., N.Y.)–307
ACADEMGORODOK (Academic Town), USSR–232
ADAMS, Maj. Michael J.–15, 153
ADEY, Dr. W. Ross–230
AEROBEE-150A (U.S. sounding rocket)–37
AEROJET-General Corp.–124–5
AERONAUTICAL & Space Sciences Committee (U.S. Senate)–89–92, 250
AERONAUTICS & Astronautics, American Institute of–273
AGENA (U.S. rocket)–11, 40–2, 45–6, Agena-D–47–9, 50
AGNEW, Spiro T.–179, 181, 190, 222–4, 246
AGRICULTURE, (U.S.) Department of–195
AIR FORCE, U.S.–7, 14–6, 25–8, 65, 67–72, 76, 108, 114, 116, 118, 121, 124–5, 151, 153–6, 181, 221, 226–30, 232, 234–6, 241, 263–4, 267, 269, 296
AKADEMIK Shirshov (Soviet research ship)–251
ALBANIA–105
ALBERT, Col. John G.–296
ALDRIN Jr., Maj. (later Col.) Edwin Eugene–1, 50–2, 169, 179–96
ALEKSEYEV, Valentin–63
ALGERIA–300
ALLEN, Joseph P.–121
ALLENBY Jr., Dr. Richard J.–150
ALLEY, Dr. Carroll O.–198
ALOUETTE (Canadian scientific satellites): Alouette 1–238. Alouette 2–37
AMERICAN Institute of Aeronautics & Astronautics–273
AMERICAN Telephone Co.–123
ANDERS, Maj. William A.–137–42
ANIK (Canadian communications satellite)–270–1
AOSO (advanced orbiting solar observatory, U.S.)–29–30
APOLLO, Project (U.S. manned moon-landing program)–1, 16, 25, 29, 39, 53, 62, 70–2, 85–93, 99, 118, 131–42, 149–50, 158, 169–204, 212–4, 219–23, 229, 234, 243–50, 260–3, 275–91, 296, 303, 306. Cost estimates–195. Fatal fire & investigation–86–93.
APOLLO Applications Program–219, 221, 233
APOLLO Lunar Exploration Office–150
APOLLO missions: Apollo 6–133. Apollo 7–133–6. Apollo 8–137–42, 231, 234, 244. Apollo 9–170–4, 232. Apollo 10–170, 174–9. Apollo 11–176–96, 198, 212–4, 222–3, 229, 234. Apollo 12–198–204, 246, 263, 284, 290–1. Apollo 13–243–50. Apollo 14–276–82, 290–1. Apollo 15–282–90
APPROPRIATIONS Committee (U.S. Senate)–53, 266
APPROPRIATIONS Committee (U.S. House of Representatives)–124
ARGENTINA–300
ARIEL (British satellite series)–120
ARIZONA, University of–56
ARMSTRONG, Neil Alden–1, 40–1, 153, 169, 179–96, 201
ARMY, U.S.–154, 228
ASSOCIATED Press (AP)–155
ASTRONAUTICAL Congresses, International–14, 16–7, 22, 74, 76–7, 106–7, 136–7, 274
ASTRONAUTICS, International Academy of–274
ATDA (augmented target docking adapter, U.S.)–42–3
ATLAS (U.S. rocket)–40, 41, 42–3, 45, 47, 50, 155, 156. Atlas-D–26, 72. Atlas-Agena–25, 30, 71, 99, 104, 114, 153, 155, 227, 264. Atlas-Agena-B–68. Atlas-Agena-D–62, 72, 95, 98, 115, 117. Atlas-Centaur–26, 30, 56–7, 59, 96, 99–100, 148, 155–6, 230, 264, 300
ATOMIC Energy, Joint Congressional Committee on (U.S.)–127
ATOMIC Energy Commission, U.S.–195, 219, 306
ATS (applications technology satellite series)–71–2, 115, 117, 155–6, 230–1, 241. ATS 1–71–2. ATS 2–115. ATS 3–117. ATS 4–155–6. ATS 5–230–1. ATS 6–241.
ATWOOD, J. Leland–90–1

309

AUERBACH, Stuart–265
AUGER, Gen. Pierre–109
AUSTRALIA–105, 112, 118-20, 197, 262, 300
AUSTRIA–300
AVIATION & Cosmonautics (Soviet publication)–32
AVIATION Week & Space Technology (U.S. magazine)–219
AZUR (West German satellite)–241

B

BAIKONOUR cosmodrome (Tyuratam, Kazakhstan), USSR–22, 54, 64, 78, 93-4, 142, 146, 159, 205, 207, 212, 214, 218-9, 236, 251, 254, 266, 291, 294
BAIRD, Dr. Keith–249
BAKER, Dr. Robert M. L.–151
BAMBI (U.S. Air Force satellite killer program)–267
BARABASHOV, Nikolai P.–55
BARON, Thomas R.–91
BARON, Mrs. Thomas R. (Marlene)–91
BARTH, Dr. Charles A.–217-8
BAUDOUIN, King (Belgium)–179
BEAN, Cmndr. (later Capt.) Alan LaVern–198-204
BEECH Aircraft Corp.–250
BELAI, V. E.–76
BELGIUM–75, 109, 164, 300
BELL Telephone Co. of Canada–163
BELOKONEV, Aleksei–93
BENDIX Corp.–235
BENJAMIN Franklin (U.S. submarine)–11
BENN, Anthony Wedgwood–164
BEREGOVOI, Col. (later Maj. Gen.) Georgi Timofeyevich–142-4, 231
BERGEN, William B.–92
BERRY, Dr. Charles Alden–5, 13, 140, 181, 249, 288-9
BERTRAND, Jean-Jacques–238
BIOSATELLITE (space biology study satellite series, U.S.)–72, 152, 229-30. Biosatellite 1–72, 118. Biosatellite 3–229-30
BIRMINGHAM, University of (Britain)–37
BISPLINGHOFF, Dr. Raymond L.–150

BLACK Arrow (British rocket)–270, 305
BLAGONRAVOV, Anatoly A.–55, 76
BMEWS (ballistic missile early warning system, U.S.)–126
BOBKO, Maj. Karol J.–233
BOEING Co.–92, 101, 235
BOGGS, Rep. Hale (D., La.)–223
BOLENDER, Brig. Gen. Carroll–93
BOLKOW company (West Germany)–129
BONNY (pigtail monkey, U.S.)–229-30
BOONE, Adm. W. F.–150
BORBA (Belgrade newspaper)–107
BORISOV, T.–87
BORMAN, Edwin–231
BORMAN, Lt. Col. (later Col.) Frank A.–9-14, 88-90, 137-42, 190, 213-4, 231-2, 252
BORMAN, Mrs. Frank (Susan)–231
BORMAN, Frederick–231
BOWDOIN College–122
BRANDT, Willy–246
BRAZIL–300
BRETT, Dr. Robin–196
BREZHNEV, Leonid–291
BUDGET Bureau, U.S.–53
BULGARIA–78, 105, 165, 237, 266, 268, 301
BULGARIAN Communist Party–291

C

CADITI (Communist Chinese sounding rocket)–125
CALIFORNIA, University of–198
CAMEROON–300
CAMPO de Arenosillo, Spanish launch site–77
CANADA–35-7, 105, 127-8, 162-3, 237-8, 262, 270-1, 300-1
CANADIAN National Railway–163
CANADIAN Pacific Railway–163
CANARY Bird (Intelsat satellite)–108
CARABELA-4 (Spanish rocket)–77
CARDINAL, Jean-Guy–238
CARNEGIE Endowment for International Peace–240
CARPENTER, M. Scott–122
CARPENTIER, Dr. William R.–193
CBS TV (Columbia Broadcasting System)–195

INDEX

CENTAURE (French rocket)—52, 83
CERNAN, Lt. Cmndr. (later Cmndr.) Eugene Andrew—41, 174-9
CEYLON—300
CHAFFEE, Lt. Cmndr. Roger B.—86, 190
CHAPMAN, Dr. Dean R.—197
CHAPMAN, Dr. John H.—127
CHAPMAN, Philip K.—121
CHILE—300
CHINA (Communist), People's Republic of—105, 125, 163, 264-6, 301-2
CHINA (Nationalist), Republic of—300
CHOUSOVTIN, Valery—307
CHRISTIAN Science Monitor (U.S. newspaper)—36
CHRYSLER Corp.—235
CHURCHILL, Canada—128
CLARK, Evart—79, 159
CLARK University—222
CNES (France's National Center for Space Studies)—36-7, 82
CNET (France's National Telecommunications Center)—37
COAST & Geodetic Survey, U.S.—153
COHN, Victor—233
COLLINS, Maj. (later Lt. Col.) Michael—45-7, 180-96
COLOMBIA—300
COLUMBIA (Apollo 11 command & service modules)—180, 185-6, 191
COLUMBIA Broadcasting System (CBS)—195
COMMERCE, U.S. Department of—66, 115, 153, 195
COMMUNICATIONS Satellite Corp. (Comsat)—107-8, 123, 166, 239-40, 272, 300
CONGRESS, U.S.—15, 29-30, 52-3, 88-92, 123-5, 127, 149, 151, 163, 179, 181, 190-1, 219-20, 225, 250, 260-1, 266, 306-7
CONRAD Jr., Lt. Cmndr. (later Capt.) Charles (Pete)—4-8, 47-9, 198-204
COOPER Jr., Lt. Col. Leroy Gordon—4-8, 10, 282
CORFIELD, Frederick—305
CORTRIGHT, Edgar M.—250
COSMIC ray dangers—296
COSMIC Research (Soviet publication)—76

COSMOS (Soviet earth satellite series)—24, 32-4, 73-4, 78-81, 109-14, 126-7, 158-62, 235-6, 266-7, 304
COTTRELL, R. F.—125
CREWS, Lt. Col. Albert H.—233
CREWS, Maj. Robert M.—15
CRIPPEN, Lt. Cmndr. Robert L.—233
CROSSFIELD, A. Scott—153
CUBA—78, 105, 165, 301
CUNNINGHAM, R(onnie) Walter—134-6
CZECHOSLOVAKIA—78, 105, 165, 198, 237, 266, 268, 301

D

DAI Chi Chinci (Japanese freighter)—237
DANA, William H.—152
DAY, LeRoy E.—221
DEATHS (astronauts and cosmonauts)—1, 85, 86-94, 275, 292-5
DEFENSE, U.S. Department of—9, 15, 116, 127, 150, 153-4, 156, 195, 219, 228, 296
DEFENSE Research Board (Canada)—37
De GAULLE, Charles—36, 78
DEIMOS (Mars' moon)—298
DELTA (U.S. rocket)—28, 66, 99, 108-9, 117-8, 154, 156-7, 227, 229, 239-41, 271-2, 303. Delta-DS-V3E—153. Delta-N—230
DEAN, Sir Patrick—105
DEMORET, Robert B.—274
DENMARK—105, 109, 300
DIADEME (French satellites)—119
DIAL (West German satellite)—269
DIAMANT (French rockets)—35-6, 81, 119. Diamant-B—128, 269, 304-5
DIAPASON (French satellite)—81-2
DISCHER, John—221
DOBROVOLSKY, Lt. Col. Georgi T.—293-5
DOBRYNIN, Anatoly F.—105
DOCKINGS (in space)—39-41, 45-50, 109-10, 158-9, 170-3, 177, 182, 192, 199, 202, 204-7, 246, 272-3, 275, 277, 283, 287, 291-2, 294-5
DODGE (Department of Defense gravity experiment, U.S. satellite)—116
DOLGOV, Piotr Ivanovitch—93
DOMINICAN Republic—300
DOUGLAS Aircraft Co.—14

DRAGON (French sounding rocket for radiation studies)–36
DRURY, Charles M.–162
DRYDEN, Dr. Hugh L.–92
DuBRIDGE, Dr. Lee A.–224
DYAL, Dr. Palmer–204
DYNASOAR (U.S. Air Force program)–15

E

EAGLE (Apollo 11 lunar module)–180, 184-5, 191
EARLY Bird (U.S. communications satellite)–44, 78, 107
ECUADOR–300
EGGERS, Dr. Alfred J.–151
EGYPT–300
EISELE, Maj. (later Lt. Col.) Donn Fulton–134-6
EINSTEIN, Albert–190
EISENHOWER, Dwight D.–190
ENGLAND, Anthony W.–121
ENVIRONMENTAL Satellite Center, National–111
ENGLE, Maj. Joe H.–153
EPSTEIN, Julius–93
ERNO (West German company)–163
ESRO (satellite series of European Space Research Organization)–108-9, 164. Esro 1–164. Esro 2–108-9, 164
ESSA (weather satellite series of U.S.' Environmental Science Serivces Administration)–66, 71, 115-7, 156-7, 228. Essa 1–66, 71. Essa 2–66, 71. Essa 4–115. Essa 5–115-6. Essa 6–117. Essa 7–156. Essa 8–157. Essa 9–228
ESSEX (U.S. aircraft carrier)–136
ETHIOPIA–300
EUROPA (European Launcher Development Organization rockets)–74-5, 109, 129, 165, 242, 305. Europa-1–74-5, 109, 165, 242. Europa-2–305
EUROPEAN Launcher Development Organization (ELDO)–74-5, 109, 119, 129, 163-5, 242, 305
EUROPEAN Space Conference–301
EUROPEAN Space Research Organization (ESRO)–83, 108-9, 163-5
EUROPEAN Space Technology Center (ESTEC)–165

EWING, Dr. Maurice E.–203
EXPLORER (U.S. satellite series)–27-8, 99, 116, 153-5, 229, 264, 270, 303. Explorer 1–264. Explorer 29 (Geos 1)–27. Explorer 30–27-8. Explorer 31–37. Explorer 32–67-8. Explorer 33–63-4. Explorer 34–116. Explorer 35–99, 101. Explorer 36 (Geos 2)–153. Explorer 37–153-4, 155. Explorer 39 (Air Density Explorer)–155. Explorer 40 (Injun Explorer)–155. Explorer 41 (IMP series)–229. Explorer 43 (IMP 8)–303

F

F2 (U.S. rocket)–154
FABIOLA, Queen (Belgium)–179
FAGET, Dr. Maxime–88
FAIRLEY, Peter–219
FAITH 7 (Project Mercury space capsule)–5
FEDERAL Communications Commission, U.S. (FCC)–108
FEDOSEYEV, Anatoly–307
FEOKTISTOV, Dr. Konstantin P.–231-2, 292-3, 295
FILIPCHENKO, Lt. Col. Anatoly Vasilievich–207-10
FINCH, Robert H.–223
FINLAND–105
FINLEY, Lt. John L.–16
FINNEY, John W.–265
FLEISCHER, Dr. Robert L.–296
FOBS (fractional orbital bombardment system, USSR)–125-7, 159-60, 236, 264, 266-7
FORD Foundation–108
FORD Motor Co.–69
FOSTER Jr., John S.–127, 266
FRANCE–35-7, 75, 81-3, 105, 109, 112, 118-20, 162-3, 237-9, 257-8, 265, 268-90, 300, 304-5
FREEDOM 7 (Project Mercury space capsule)–276
FREEMAN Jr., Dr. John W.–290
FRIA (French scientific satellite)–37
FRUTKIN, Arnold W.–273
FRYE, Penny–91
FRYE, Robin–91
FULLERTON, Maj. Charles G.–233
FULTON, Rep. James G. (R., Pa.)–125

INDEX

G

GABON–300
GAGARIN, Col. Yuri Alexeyevich–94, 166-7, 190
GALLUP Poll–225
GARRIOTT, Owen K.–16
GAST, Dr. Paul–196-7
GATLAND, Kenneth–267
GAZENKO, Oleg G.–77
GEER, E. Barton–88
GEMINI, Project (U.S. manned space flight project)–1, 3-14, 39-52, 86, 89, 109, 121, 134, 137, 195, 232, 252. Missions: Gemini 3–3-4, 47, 86; Gemini 4–4, 86; Gemini 5–4-9; Gemini 6–9-14; Gemini 7–9-14, 50, 137, 252; Gemini 8–39-41; Gemini 9–41-5; Gemini 10–47; Gemini 11–47-9; Gemini 12–49-52, 137
GENERAL Electric Co.–69, 76, 152, 229, 296
GEODETIC Satellite Program, U.S. National–153
GEOLOGIC Survey, U.S.–97
GEOPHYSICAL Research, Journal of–101
GEOS (geodetic earth orbiting satellites, U.S.)–27, 153. Geos 1 (Explorer 29)–27. Geos 2 (Explorer 36)–153
GERMANY, East–237, 268, 301
GERMANY, West–35, 38, 75, 78, 82, 109-10, 119, 129, 163, 164, 236, 241, 266, 269, 301, 305
GHANA–300
GIBSON, Edward K.–16
GILRUTH, Dr. Robert F.–92, 273
GIVENS Jr., Maj. Edward G.–87
GLOD, G. D.–76
GOLD, Dr. Thomas–197
GOLDBERG, Arthur J.–105
GORBATKO, Lt. Col. Victor Vasilievich–207-10
GORDON Jr., Lt. Cmndr. (later Capt.) Richard Francis–47-9, 199-204
GOVERNMENT Operations Committee (U.S. House of Representatives)–15
'GRAND TOUR' of planets–261-2
GRATZEV, Aleksis (or Grachev, Aleksei)–93
GRAVELINE, Dr. Duane E.–16
GREAT Britain–36, 75, 77, 83, 105, 109, 118-9, 163-4, 269-70, 300, 305, 307
GREECE–300
GREENHILL, Sir Denis–307
GREENWOOD, Dr. William R.–197
GRIGOROV, Nikolai L.–31-2
GRISSOM, Maj. (later Lt. Col.) Virgil Ivan–4, 86-7, 190
GROSE, Peter–236
GRUMMAN Aircraft Engineering Corp.–132, 157, 195, 235
GUADALCANAL (U.S. helicopter carrier)–45-6, 174
GUAM–191
GUAM (U.S. aircraft carrier)–49
GUATEMALA–300
GURNEY, Rep. Edward J. (R., Fla.)–89

H

HAISE Jr., Fred Wallace–244-50
HALL, Dr. Robert L.–151
HALLEY'S Comet–226
HAMMAGUIR, Algeria (French launch site)–35-6, 52, 81-2, 119, 128, 269
HARDER, Dr. James A.–151
HARRIS (Louis) poll–223
HARTSFIELD, Maj. Henry W.–233
HAWKER Siddeley Dynamics (British company)–163
HECHLER, Rep. Ken (D., Va.)–91
HEIS Island (Soviet launch site)–112, 119
HENIZE, Karl G.–121
HENTY, Norman H. D.–75
HEOS-A (highly eccentric orbit satellite, ESRO)–109. Heos 1–164-5
HERR, Dr. Kenneth C.–216
HERRES, Lt. Col. Robert T.–121
HESS, Dr. Wilmot N.–234
HILLS, Dr. H. Kent–290
HIRASAKI, John K.–193
HODGE, John–9
HOLIFIELD, Rep. Chet (D., Calif.)–15
HOLMBURG, Mervin–91
HOLMQUEST, Donald L.–121
HORD, Dr. Charles–218
HORNET (U.S. aircraft carrier)–193, 203-4
HOROWITZ, Dr. Norman H.–215, 218
HOULDING, John–270-1

HSINHUA (Communist Chinese news agency)–105, 265-6, 302
HUGHES Aircraft Co.–72, 107, 228, 270-1
HUMPHREY, Hubert H.–87, 122, 179
HUMPHREYS Jr., Maj. Gen. James W.–122
HUNGARY–78, 105, 165, 237, 266, 301
HUSAIN, Dr. Liaquat–290
HYNEK, Dr. J. Allen–151

I

ICBMs (intercontinental ballistic missiles)–35, 126, 265, 266
IMP (interplanetary monitoring platform, U.S.) series–116, 229, 303
INDEPENDENT Television News (British)–219
INDIA–112, 118-9, 241, 300
INDONESIA–35, 37, 300
INTELSAT (International Telecommunications Satellite Consortium)–107-8, 123, 165-6, 238-40, 271-2, 300-1. Intelsat 1 (communications satellite)–107. Intelsat 2 (Lani Bird 2)–107-8. Intelsat 3–239, 300. Intelsat 3A–166-7. Intelsat 4–300
INTERCOSMOS (USSR and East European nations' unmanned satellites)–236-7, 268. Intercosmos 1–236-7, Intercosmos 2–237. Intercosmos 3–268. Intercosmos 4–268
INTERIOR, U.S. Department of–195
INTERNATIONAL Academy of Astronautics–274
INTERNATIONAL Astronautical Congresses–14, 16-7, 22, 74, 76-7, 106-7, 136-7, 274
INTERNATIONAL Astronautical Federation–76-7, 106-7, 136-7
INTERNATIONAL Astronomical Union–141
INTERNATIONAL Committee for Space Research–36
INTERNATIONAL Quiet Sun Year (IQSY)–27-8
INTERNATIONAL Telecommunications Satellite Consortium–See INTELSAT
INTERNATIONAL Telecommunications Union–301
INTERNATIONAL Telephone & Telegraph Corp.–123
INTERSPUTNIK (Communist nations' communications network)–165, 240, 301
IRAN–300
IRAQ–300
IRELAND–300
IRWIN, Lt. Col. James Benson–283-9
ISRAEL–300
ISIS 1 (Canadian scientific satellite)–238
ITALY–36, 75, 109, 118-9, 163, 270, 300, 305
ITOKAWA, Dr. Hideo–37, 82
IVANCHENKO, Vladimir–65
IVORY Coast–300
IWO Jima (U.S. aircraft carrier)–249
IZVESTIA (Soviet newspaper)–128, 147, 255, 259, 265-6

J

JACKSON, Sen. Henry M. (D., Wash.)–127
JACKSON, Patrick–307
JAMAICA–300
JAMES, Jack N.–19
JAPAN–82-3, 105, 112, 118-9, 214, 237, 241-2, 262, 265, 268-9, 300, 302-3
JET Propulsion Laboratory (JPL) (of California Institute of Technology)–17, 19, 30, 59, 76, 90, 98, 100, 104, 198, 215-6, 297, 299
JODRELL Bank radio-telescope observatory (Manchester, England)–22-3, 53-6, 61, 63-5, 103-4, 142, 145, 147-8, 211, 213-4
JOHNS Hopkins University–37, 123, 153
JOHNSON, Daniel–163-4
JOHNSON, Lyndon B.–8, 10, 13-5, 19, 30, 54, 56, 72-3, 87, 94, 105-6, 122-4, 138, 150-1, 181, 190
JOHNSON, Mrs. Lyndon B.–181
JOHNSON, U. Alexis–301
JOHNSTON, Richard S.–76
JOINT Congressional Committee on Atomic Energy, U.S.–127
JORDAN–300
JOURNAL of Geophysical Research (U.S. journal)–101

INDEX

JOURNEYMAN (Argo D-8, U.S. rocket) –24
JUDY-Dart (U.S. meteorological rocket) –77
JUPITER (planet)–152, 155, 212, 225-6, 262

K

KACHUR (or Kascheur), Ivan–93
KAMININ, Lt. Gen. (later Col. Gen.) Nikolai Petrovich–94-5, 128
KAMINSKI, Dr. Heinz–110, 146, 208
KAPPA-8 (Indonesian rocket)–37
KAPRYAN, Walter J.–232-3
KARNOW, Stanley–265
KASPUTIN Yar (Soviet launch site)–79
KELDYSH, Dr. Mstislav Vsevolodovich– 14, 16-7, 65, 74, 95, 103, 110, 210, 213-4, 253, 273, 305
KENNEDY, Sen. Edward M. (D., Mass.) –222-3
KENNEDY, John F.–179, 190, 222
KENYA–300
KERALA (Indian launch site)–120
KERWIN, Lt. Cmndr. (later Cmndr.) Joseph P.–16, 249
KETTERING Grammar School (England)–79, 110
KHARKOV University observatory–55
KHRUNOV, Lt. Col. Yevgeni Vasilyevich–205-7
KIERANS, Eric–271, 301
KING, Dr. Elbert–234
KIRUNA (Sweden launch site)–83
KITTY Hawk (U.S. aircraft carrier)–302
KLIORE, A. J.–18, 217
KNIGHT, Maj. William J.–153
KOMAROV, Col. Vladimir Mikhailovich –93-5, 110, 142, 158, 190, 207
KOMMERER, William–197
KOMSOMOLSKAYA Pravda (Soviet newspaper)–32, 210
KONSTANTINOV, Aleksei F.–209
KOREA, South–300
KOSYGIN, Aleksei N.–165
KOUROU (French Guiana launch site)– 36, 269, 304
KRAFT Jr., Christopher Columbus–6, 11
KRASNAYA Zvezda (Red Star) (Soviet Defense Ministry newspaper)–9, 79, 128
KUBASOV, Valeri Nikolayevich–207-10
KUIPER, Dr. Gerard P.–56
KUWAIT–300

L

LAIRD, Melvin R.–302
LAKE Champlain (U.S. aircraft carrier)– 5, 8
LAMBDA (Japanese rocket)–119. Lambda-4S1–82. Lambda-4S2–83. Lambda-4S4–241. Lambda-4S5–268
LANGHAM, Dr. Wright H.–122
LANGSETH, Dr. Marcus–290
LATHAM, Dr. Gary–290
LAWRENCE Jr., Maj. Robert H.–87, 121, 290
LAWYER, Capt. Richard E.–15
LEBANON–300
LEBEDINSKY, Aleksandr–60
LEE Jr., Chester J.–250
LEE, William A.–93
LEFEVRE, Theo–164, 301
LEIGHTON, Dr. Robert B.–17, 19-20, 217-8
LEIKIN, Grigory–22
LENINGRAD Military Research Committee (USSR)–32
LENOIR, William B.–121
LEONARD F. Mason (U.S. destroyer)– 41
LEONOV, Lt. Col. Aleksei Arkhipovich –4, 17, 214
LIBERTY Bell (Project Mercury space capsule)–86
LIBYA–300
LICK Observatory, Mount Hamilton, Calif.–198
LIECHTENSTEIN–300
LIFE (magazine)–194-5
LIND, Dr. Don L.–233-4
LLEWELLYN, John A.–121
LLOYD'S of London–272
LOADS (low altitude density satellite, U.S.)–155
LONDON Daily Express (newspaper)– 55
LOPUKHIN, A. S.–166
LORELL, Jack–297-8
LOUIS, Victor–293

LOVELL, Sir Bernard—22-3, 54, 64-5, 104, 142, 145, 207, 213-5
LOVELL Jr., Cmndr. (later Capt.) James Arthur—9-14, 50-2, 137-42, 244-50, 252, 274
LOW, George M.—92, 232, 305
LUNA (Soviet moon probe series)—22-3, 53-6, 59-64, 129, 147-8, 180-1, 212-5, 253-8, 273, 289, 306. Luna 3 —22. Luna 7—22-3. Luna 8—23. Luna 9—53-6. Luna 10—60-1. Luna 11—61, 63. Luna 12—64. Luna 13—59-60. Luna 14—147-8. Luna 15—180-1, 212-5. Luna 16—253-6, 273, 306. Luna 17—256-7, 289. Luna 18—289. Luna 19—289
LUNAR Orbiter (U.S. moon probe series)—61-3, 95-9, 101-2, 195, 198. Lunar Orbiter 1—61-3, 101. Lunar Orbiter 2—96, 101. Lunar Orbiter 3—95-6, 101. Lunar Orbiter 4—97-8, 101. Lunar Orbiter 5—99
LUNAR Rover 1 (U.S. moon car)—282, 285-7
LUNDIN, Bruce T.—233
LUNOKHOD 1 (Soviet robot moon car) —253, 256-8, 289-90
LUST, Reimar—82
LUXEMBOURG—300
LYONS, Richard D.—195, 236, 266, 304

M

MacDONALD, Dr. Gordon J. F.—152, 226
MacINTOSH, Dr. R. M.—162
MACKASEY, Bryce—271
MACLEAY, Capt. Lachlan—16
MacLEOD, Alexander—36
MALAGASY Republic—300
MANNED Space Flight: Soviet—1, 3, 4, 85, 93-5, 142-4, 204-10, 250-3, 291-5. U.S.—1, 3-14, 39-52, 131, 133-42, 169-204, 243-50, 275-89
MANSFIELD, Sen. Mike (D., Mont.)—223
MAO Tse-tung—265
MARINER (U.S. unmanned space probes)—3, 17-21, 30, 65, 104, 215-8, 262, 296-8. Mariner 1—104. Mariner 2—24, 30, 104. Mariner 4—17-21, 30, 65. Mariner 5—104. Mariner 6—215-8. Mariner 7—215-8. Mariner 8—297. Mariner 9—296-8
MARS (planet)—2, 3, 16-21, 30, 150, 152, 181, 212, 215-8, 223-5, 260-2, 275, 296-300, 307
MARS (Soviet Mars probes)—298-300. Mars 2—298-300. Mars 3—298-300
MARTIN Marietta Corp.—14, 91-3
MASSACHUSETTS Institute of Technology (MIT)—116, 150
MATHEWS, Charles W.—41, 220-1
MATTINGLY 2d, Lt. Cmndr. Thomas K. —245
MALAYSIA—300
MALI—300
MARSHALL, Justice Thurgood—179
MATRA company (France)—163
MAURITANIA—300
MAX Planck Institute for Extraterrestrial Research—82
McCARTHY Jr., Dr. John F.—90
McDIVITT, Maj. (later Col.) James Alton—4, 170-4, 232
McDONALD, Dr. James E.—151
McDONALD Observatory, Mount Locke, Tex.—198
McDONNELL Aircraft Corp.—14
McKAY, John B.—153
McNAMARA, Robert S.—125-6
MERCURY (planet)—152, 197, 262
MERCURY, Project (U.S. manned space flight project)—1, 5, 14, 86, 89, 134, 195, 232, 276
METEOR (Soviet weather satellite series)—236, 251, 267-8, 303-4
METEORITES—166
MEXICO—300
MICHAEL Jr., William H.—63
MICHAILOV, Aleksandr—22
MICHAILOV, Gennady—93
MICHEL, Dr. Frank Curtis—16, 234
MIDDLETON, Rear Adm. Roderick O.—92
MILKY Way—155
MILLER, Rep. George P. (D., Calif.)—225, 265
MINASHIN, Vladimir—239
MINER, Dr. Ellie D.—297
MINUTEMAN (U.S. missile)—7, 11, 126, 127

INDEX

MIRAGE-4 (French atomic bombers)—35

MIRV (multiple, independently targeted re-entry vehicles)—127

MITCHELL, Cmndr. Edgar Dean—276-82

MOBS (multiple-orbit bombardment system)-159

MOE (an elk in Yellowstone National Park)—228

MOL (manned orbiting laboratory project, U.S. Air Force)—14, 16, 121, 233, 226

MOLNIYA-1 (Lightning Soviet communications satellite series)—31, 36, 77-9, 110-1, 129, 157-8, 236, 267, 304

MOLNIYA-2 (Soviet communication satellite)—304

MONDALE, Sen. Walter F. (D., Minn.)—307

MONGOLIA—78, 165, 301

MONACO—300

MOON—1, 3, 16-7, 21-2, 39, 53-64, 72-4, 76, 85, 90-1, 95-102, 105-6, 131-42, 144-9, 169-204, 243-50, 253-8, 260-3, 275-91, 305-6. First manned landing—169, 179-96. First manned orbit—137-42. Rock analysis —196-8, 204, 255, 290. Treaty proposal—72-3

MORGENTHALER, George W.—274

MOROCCO—300

MOSCOW University—31, 76

MRBM (medium range ballistic missile)—265

MU (Japanese rocket)—119, 302. Mu-1—82. Mu-48—302. Mu4SI—269

MUELLER, Dr. George E.—87, 220-1, 225, 232, 234

MURRAY, Bruce C.—20

MUSGRAVE, Franklin Story—121

N

NAPOLITANO, Dr. Luigi C.—76

NARDIN, Gen. Pierre—35

NATIONAL Academy of Sciences—16, 151-2, 226

NATIONAL Aeronautics & Space Act of 1958—190

NATIONAL Aeronautics & Space Administration (NASA)—10-2, 14, 16-7, 19, 24-31, 37-8, 41, 43, 52-3, 56-8, 62-3, 66-72, 87-93, 95, 98-9, 101, 108-9, 114-8, 121-2, 124-5, 132-3, 141, 149-57, 164, 166, 178, 181, 186, 193-5, 196-7, 215, 219-23, 226-35, 238-41, 243-4, 249-50, 260-5, 271, 278, 300-1, 305-7. Budget—52-3, 149-50, 219, 222-3, 226, 243, 260-1, 306-7. Employment—260-1. Spending on space—219

NATIONAL Aeronautics & Space Council—20-1, 87

NATIONAL Environmental Satellite Center—111

NATIONAL Investigations Committee on Aerial Phenomena (NICAP)—235

NATIONAL Research Council—122

NATIONAL Science Foundation—195

NATO (North Atlantic Treaty Organization)—307

NAVAL Research Laboratory (U.S.)—28, 154

NAUGEBAUER, Dr. Gerry—217

NAVY, U.S.—8, 44, 46, 133, 153, 193, 302

NEPOKLONOV, Dr. Boris—289-90

NEPTUNE (planet)—152, 225-6, 262

NERVA (U.S. nuclear-powered rocket engine)—150, 219, 306

NESS, Dr. Norman F.—101

NETHERLANDS—75, 109, 300

NEUBECK, Capt. F. Gregory—16

NEW Orleans (U.S. aircraft carrier)—282

NEW YORK, State University of (Stony Brook)—290

N.Y. TIMES (newspaper)—24, 53, 79, 86, 149, 155, 159, 195, 236, 265, 266, 292, 304

NEW Zealand—300

NICARAGUA—300

NICHOLSON, Martin—307

NICKS, Oran—19

NIGER—105

NIGERIA—300

NIKE-Apache (U.S. rocket)—28

NIKOLAYEV, Col. (later Maj. Gen.) Andrian Grigoryevich—250-3, 273-4, 295

NIMBUS (U.S. weather satellite series)—67, 228, 263. Nimbus 1—67. Nimbus

2-67, 228. Nimbus 3-263. Nimbus 4-263. Nimbus-B-154
NIXON, Richard M.-178-9, 181, 185, 189, 193-4, 199, 224-5, 249, 260-2, 280, 294, 306-7. Space goals-261-2
NIXON, Mrs. Richard M.-199
NIXON, Stuart-235
NIXON, Tricia-199
NORTH American Air Defense Command (NORAD)-112, 302
NORTH American Aviation, Inc.-88-92
NORTH American Rockwell Corp.-250
NOVOSIBIRSK, USSR-231-2
NOVOSTY (Soviet press service)-179
NORWAY-300

O

OAO (orbiting astronomical laboratory, U.S.) series-66-7, 156-7. OAO-A-66-7. OAO 2 (Stargazer)-156-7. OAO 3-157. OAO 4-157
O'BRYANT, William T.-150
OFFICE of Department of Defense & Interagency Affairs (NASA)-150
OGO (orbiting geophysical laboratory, U.S.) series-26-7, 68, 117, 153, 228-9. Ogo 1-27. Ogo 2-26-7. Ogo 3-68. Ogo 4-117. Ogo 5-153. Ogo 6-228-9
OGONYOK (Soviet magazine)-94
O'HAIR, Mrs. Madalyn Murray-234
OHSUMI (Rising Sun Satellite 1, Japanese)-268-9
OKINAWA (U.S. aircraft carrier)-288
O'LEARY, Brian T.-121
ORBA (British satellite)-270
OSCAR 4 (orbiting satellite carrying amateur radio)-29
OSO (orbiting solar observatory, U.S.) series-26, 115, 227, 230, 303. Oso 3 -115. Oso 5-227. Oso 6-230. Oso 7-303
OTH (over-the-horizon radar, U.S.)-126
O'TOOLE, Thomas-258
OV (orbiting vehicle) series-67, 72, 117-8. OV 1-9-72. OV 1-10-72. OV3-1-67. OV3-6-117-8
OVERMYER, Maj. Robert F.-233

P

PAC (package attitude control) system-230

PACIFIC 3 (Intelsat satellite)-108, 239-40
PACKARD, David R.-226
PAGE, Dr. Thornton-235
PAGEOS 1 (passive geodetic earth-orbiting satellite, U.S.)-69
PAIGE, Dr. Hillard W.-76
PAKISTAN-300
PAINE, Dr. Thomas O.-178, 185, 221-2, 224-5, 250, 260-2
PAMEUNGPEUK (Indonesia launch site) -37
PANAMA-300
PARIN, Vasily V.-74, 210
PARKER, Robert A.-122
PATSAYEV, Viktor I.-293-5
PAUL VI, Pope-138, 244, 294
PEANUTS (cartoon strip)-174
PEGASUS (U.S. 'winged' satellite)-25-6, 173
PEKING Aeronautical Institute (China) -125
PERM, USSR-4
PERRY, Geoffrey-110
PERU-300
PETERSEN, Capt. Forrest S.-153
PETERSON, Maj. Donald H.-121, 233
PETRONE, Rocco A.-232-3
PETROV, Boris Nikolayevich-129, 210, 255-6, 273
PETROVICH, Georgi V.-32
PHILCO-Ford Corp.-115, 269
PHILCO Western Development Laboratories-69
PHILIPPINES-197, 300
PHILLIPS, Maj. Gen. (later Lt. Gen.) Samuel Cochran-90, 134, 221, 232
'PHILLIPS report'-90, 91
PHOBOS (Mars' moon)-298
PIC du Midi d'Ossau Observatory, France-258
PICKERING, Dr. William H.-19-20, 76, 225
PIMENTAL, Dr. George C.-215-6
PIONEER (U.S. interplanetary probes)-28, 70, 118, 156, 226-7. Pioneer 6-28. Pioneer 7-70. Pioneer 8-118. Pioneer 9-156. Pioneer 10-226-7
PLANETARY Explorer program (proposed unmanned flights to planets)-219

INDEX 319

PLESETSK (USSR launch site)–79
PLUTO (planet)–225-6, 262
PODGORNY, Nikolai V.–14, 54, 138, 231-2
POLAIRE (French satellite)–304-5
POLAND–78, 165, 237, 268, 301
POLARIS (U.S. missile)–11, 126
POLAROID CORP.–92
PONOMAREV, Leonid–8-9
PONTECORVO, Bruno–128
POOR People's Campaign–182
POPE Paul VI–138, 244, 294
PORTUGAL–300
POSEIDON (U.S. missile)–127
POWER Corp. of Canada–162
PRAVDA (Soviet newspaper)–31, 61, 63, 74, 94, 111, 129, 144, 146-7, 159, 209-10, 212, 255
PRAVETSKY, V. N.–74
PRESS, Dr. Frank–203
PRICE, Dr. P. Buford–296
PRINCETON (U.S. aircraft carrier)–178
PROJECT 647 (U.S. missile-warning satellite)–264
PROSPERO (British satellite)–305
PROTON (Soviet unmanned satellite series)–31-2, 48, 69, 78, 158. Proton 1 –31, 69. Proton 2–32. Proton 3–48, 78. Proton 4–158
PUERTO Rico–191

Q

QUEBEC (province), Canada–162-3, 237-8

R

RADIO Corp. of America (RCA)–123, 228
RADIO-Quebec–162-3
RADVS (radar altimeter and doppler velocity sensor, U.S.)–58
RANGER (U.S. moon probe series)–56, 195
RCA Ltd., Canada–270-1
RED Banner (Communist Chinese Red Guard publication)–125
RED Star (Krasnaya Zvezda) (Soviet Defense Ministry newspaper)–9, 79, 128
REDSTONE (U.S. rocket)–120
REGENCY Corp.–232
REUTERS (news agency)–94, 214, 218
RENDEVOUS Evaluation Pod (REP, U.S. unmanned instrument package)– 6-7, 9
RESEARCH & Technology Advisory Council (NASA)–150
ROBERTS, Judge Jack–234
ROBERTS, Dr. Walter O.–235
ROHINI RH-75 (Indian sounding rocket) –120
ROME, University of–270, 305
ROMNEY, George W.–223
ROOSA, Maj. Stuart Allen–276-82
ROSTOW, Eugene V.–123
ROYAL Aircraft Establishment (British) –267
RUBIS (French rocket)–82
RUKAVISHNIKOV, Nikolai N.–291-2
RUMANIA–78, 165, 237, 266, 301
RUMSFIELD, Rep. Donald (R., Ill.)– 89-90
RUSHWORTH, Col. Robert A.–153
RUSK, Dean–75, 105
RYAN, Rep. William F. (D., N.Y.)–91

S

SAAB company, Sweden–163
SAGAN, Dr. Carl–151
SAINT (U.S. Air Force satellite inspection system)–267
SALYUT (Salute, Soviet orbital scientific station)–291-5
SAMOA, American–191
SANDPIPER (U.S. missile)–151
SAN Marco (Italian satellite)–120
SARABHAI, Vikram A.–106
SATELLITES–See satellite names
SATURN (planet)–152, 225-6, 262
SATURN-1 (U.S. rocket–16, 25, 70-1, 86. Saturn-1B–53, 132, 134, 221
SATURN-4B (S4B) (U.S. rocket)–69-71
SATURN-5 (U.S. rocket)–30, 53, 69, 92, 124, 133, 138, 149, 170, 174-5, 182, 195, 199, 219-21, 245-6, 276-7, 283
SAUDI Arabia–300
SECAM (French color TV system)–36
SCHAEFFER, Dr. Oliver A.–204, 290
SCHERER, Lee R.–150
SCHIRRA Jr., Capt. Walter Marty–9-14, 134-6, 232
SCHMITT, Harrison–16, 141, 233
SCHNEIDERMAN, Dan–298

SCHRIEVER, Gen. Bernard A.–15
SCHULTZ, Charles M.–174
SCHWEICKART, Russell Louis (Rusty) –170-4
SCIENCE (U.S. journal)–197, 296
SCIENCE, American Association for the Advancement of–235
SCIENCE & Astronautics Committee (U.S. House of Representatives)–52-3, 88, 89-92, 149, 151
SCIENCE Council of Canada–127
SCOTT, Maj. (later Col.) David Randolph–40-1, 170-4, 290, 293-9
SCOUT (U.S. rocket)–26-7, 37, 67, 70-1, 108, 120, 154-5, 164, 241, 270, 305
SEAMANS Jr., Dr. Robert C.–90, 92-3, 109, 151, 224, 234-5
SECOR (sequential collation of range satellite, U.S. Army)–26, 154, 228, 263
SEDOV, Leonid I.–76-7, 106-7, 136-7
SEE, Elliott M.–9
SENEGAL–300
SENTINEL (U.S. anti-ballistic missile defense system)–126
SERYOGIN, Col. Vladimir Sergeyevich– 167
SEVASTYANOV, Vitali Ivanovich– 250-3, 273, 295
SHABAD, Theodore–292
SHAPIRO, Henry–208
SHARP, Mitchell–237-8
SHARP, Dr. Robert–218
SHATALOV, Col. Vladimir Aleksandrovich–205-10, 291-2
SHEA, Dr. Joseph F.–92
SHEPARD Jr., Capt. Alan Bartlett–232, 276-82
SHERBROOKE, University of–162
SHERSTNEV, Lev–307
SHIBORIN, Serenty (or Terentiy)–93
SHINKLE, Maj. Gen. John G.–92
SHINSEI (New Star, Japanese satellite)– 302-3
SHOEMAKER, Dr. Eugene M.–97, 198
SHONIN, Lt. Col. Georgi Stepanovich– 207-10
SHUANG-cheng-tze (Chinese space center)–301-2
SIERRA Leone–105

SIGMA 7 (U.S. space capsule)–14
SILVERSTEIN, Dr. Abe–233
SIMFEROPOL, USSR–231
SIMMONS, Dr. Gene–233
SINGAPORE–300
SINKIANG (Chinese space research center)–125
SISAKYAN, N. M.–74
SJOBERG, Sigurd Arnold–249
SKURIDIN, Dr. G. A.–32-3
SKYLARK (British rocket)–83
SKYLAB, Project (U.S. manned orbiting workshop proposal)–260, 262
SKYNET (British military communications satellite)–241, 269
SL-3 (U.S. rocket engine)–124
SMART, Gen. Jacob E.–150-1
SMIRNOVSKY, Mikhail N.–307
SMITH, Dr. Bradford A.–217
SMITH, William D.–92-3
SMITHSONIAN Astrophysical Observatory–157
SMITHSONIAN Institution–196
SOCIÉTÉ Européene pour l'Étude et l'Intégration des Systèmes Spatiaux (Setis)–129
SOLANDT, O. M.–271
SOLRAD (solar radiation spacecraft, U.S.)–154
SOTSIALISTICHESKAYA Industriya (Soviet economics newspaper)–292
SOUTH Africa–300
SOVIET Academy of Medicine–74
SOVIET Academy of Sciences–14, 17, 33, 61, 65, 74, 76, 95, 103, 110, 166, 210, 212, 253-4, 258, 292
SOVIET-American Relations Institute (USSR)–231
SOVIET Commission for Exploration & Use of Outer Space–55
SOVIET Committee for the Study of the Physical Conditions of the Moon–55
SOVIET State Commission for Space Exploration–94
SOYUZ (Union, Soviet manned flight series)–93-5, 110, 142-4, 158, 204-10, 236, 250-3, 273, 291-5. Soyuz 1– 93-5, 110, 158. Soyuz 2–142, 143. Soyuz 3–142-4. Soyuz 4–204-7. Soyuz 5–204-7. Soyuz 6–207-10. Soyuz 7–207-10. Soyuz 8–207-10.

INDEX 321

Soyuz 9–250-3, 273. Soyuz 10–291-3. Soyuz 11–293-5
SPACE: From Sputnik to Gemini (book)–1, 4
'SPACE bus' proposal–127
'SPACE shuttle' proposal–102, 220-1, 225, 260-2, 296, 306
'SPACE station' proposal–220-1, 225, 231-2, 253, 255, 261-2
SPACE Task Group (NASA)–224-5, 232, 261-2
SPACE vehicles–See vehicle names
'SPACE walks'–3-4, 40, 42-8, 50-1, 76, 86, 172, 206, 288
SPAIN–77, 109, 300
SPOOK Bird (U.S. spy satellite)–155
SPUTNIK 1 (Soviet satellite)–1, 105, 129
STAFFORD, Maj. (later Col.) Thomas Patten–9-14, 41-5, 174-9
STANFORD University–93
STARGAZER (U.S. astronomical satellite)–264
STARBIRD, Lt. Gen. Alfred D.–126
STATE, U.S. Department of–237-8, 301
STOLYEROV, Gen. Anatoly–129
STORMS Jr., Harrison A.–92
STRANG, Col. Charles F.–88
SUDAN–300
SUN–1, 51-2, 65, 70, 97, 155, 197. Eclipse–51-2, 97
SUPREME Court, U.S.–234
SURVEYOR (U.S. moon probe series)–26, 56-9, 62, 96-102, 148-9, 195, 197, 200-1, 263. Surveyor 1–56-9. Surveyor 2–59. Surveyor 3–96-7, 200-2, 263. Surveyor 4–98-9. Surveyor 5–99-101. Surveyor 6–100-1. Surveyor 7–148-9, 197
SUTTER, Dr. John F.–290
SWEDEN–109, 163, 300
SWIGERT Jr., John Leonard–245-50
SWITZERLAND–109, 300
SYMPHONIE (French-West German communications satellite)–119
SYNCOM (U.S. communications satellite)–78
SYRIA–300

T

TACOMSAT (tactical communications satellite, U.S.)–227-8
TAID (thrust-augmented improved Delta rocket, U.S.)–27, 63, 70-2, 107-8, 115-6, 238
TAMAN (Soviet oceanographic ship)–258
'TANG Fang Hung' ('The East Is Red,' Communist Chinese song)–265
TANSEI (Light Blue, Japanese satellite)–302-3
TANZANIA–300
TASS (Soviet news service)–8-9, 21-3, 31, 34, 54, 56, 60-1, 63, 74, 78, 80, 93, 102-4, 110-2, 128, 137, 146-8, 157-9, 205-6, 208, 209-15, 218, 237, 252, 254-5, 257-9, 266-8, 289-290, 293-4, 299
TAYLOR, Capt. James M.–16
TAYLOR, Dr. S. Ross–204
TD-1 (Esro satellite)–163
TD-2 (Esro satellite)–163
TEAGUE, Rep. Olin E. (D., Tex.)–88
TELESAT Corp. (Canadian)–271
TERESHKOVA, Valentina Nikolayeva–251
TETR (test & training satellite, U.S.)–156, 227
TEXAS, University of–189, 198
THAILAND–301
THANT, U–138, 294
THIOKOL Chemical Corp.–124
THOMPSON, Dr. Floyd L.–87, 89
THOMPSON, Milton O.–153
THORNTON, William E.–122
THOR-Able-Star (U.S. rocket)–26
THOR-Agena (U.S. rocket)–114, 153, 227, 303. Thor-Agena-B–37, 67. Thor-Agena-D–26, 69, 71, 117, 153
THOR-Delta (U.S. rocket)–26, 66, 67, 115, 164-6
THORAD-Agena-D (U.S. rocket)–154, 228
TIMOFEYEV, Dr. Boris Vasilyevich–166
TIROS (TV infra-red observation satellite, U.S.)–24-5, 66, 303
TITAN (U.S. rockets): Titan B/Agena-D–227. Titan-2–5, 8, 12, 40-1, 43, 45, 47, 50, 93. Titan-3–28-9, 154, 296. Titan 3-Agena-D–114. Titan-3B-Agena–71, 153, 263. Titan-3C–14, 27, 32, 68-9, 114-6, 227-8, 263-4

TITOV, Lt. Col. Gherman S.–74, 137, 231
TOKYO University–83, 119, 241, 268, 302
TOLUMBKO, Col. Gen. Vladimir F.–15, 35, 128–9
TOS (Tiros operational satellite)–66
TOURNESOL (Sunflower) (French satellite)–304
TRANSIT (U.S. Navy satellite navigation system)–118, 122–3. Transit 4B–118
TREATIES, Space–85, 102, 105–6, 163, 214, 237, 306
TRINIDAD & Tobago–301
TRUD (Soviet trade union newspaper)–87, 129
TRULY, Lt. Cmndr. Richard H.–16, 233
TRW, Inc.–166
TUNISIA–301
TURKEVICH, Dr. Anthony–101
TURKEY–301
20TH Century Fund–240

U

UCHINORA (Japanese launch site)–82–3, 119, 241, 268–9, 302
UFOs (unidentified flying objects)–129, 151, 234–5
UGANDA–301
UGOLYOK (Blackie) (Soviet dog)–73–4
ULTIMAS Noticias (Mexico City newspaper)–137
UNION of Soviet Socialist Republics (USSR)–See specific project (*e.g.*, VOSHKOD) or person
UNITED Aircraft Co.–151
UNITED Arab Republic–129
UNITED Nations–72–3, 102, 105–6, 120, 165, 191, 214, 237. Space treaties–72–3, 105–6, 214, 237
UNITED Press International (news service)–208
UNITED States–See specific project (*e.g.*, APOLLO), agency (*e.g.*, NATIONAL Aeronautics & Space Administration or person
URANUS (planet)–152, 225–6, 262
UREY, Harold C.–215

V

Van ALLEN, Dr. James A.–226
Van DOLAH, Dr. Robert W.–88

VASSILYEV, P. V.–76
VATICAN City–301
VECHERNY Leningrad (Soviet publication)–166
VELA (U.S. nuclear test detection satellites)–116, 228, 263
VENERA (Venus) (Soviet Venus probe series)–23–4, 64–5, 102–4, 211–2, 258–60. Venera 1–23, 102. Venera 2 23–4, 64, 102. Venera 3–23–4, 64–5, 102–3. Venera 4–102–4, 212. Venera 5–211–2. Venera 6–211–2. Venera 7 –258–60
VENEZUELA–301
VENUS (planet)–1, 23–4, 28, 30, 64–5, 85, 97, 102–4, 152, 211–2, 258–60, 262, 299
VENUS (Soviet Venus probe series)–See VENERA
VESTA (French rocket)–119
VETEROK (Breezie, Soviet dog)–73–4
VIET Nam, North–264
VIET Nam, South–115–6, 154, 300
VIKING, Project (proposed U.S. unmanned landings on Mars)–306–7
VINOGRADOV, Dr. Aleksandr P.–306
VIRGIN Islands–191
VOLKOV, Vladislav Nikolayevich–207–10, 293–5
VOLYNOV, Lt. Col. Boris Valentinovich–205–7
Von BRAUN, Dr. Wernher–16
VOSHKOD (Soviet manned spaceship series)–2–4, 34–5, 94–5. Voshkod 1–94, Voshkod 2–4
VOSTOK (Soviet manned space flight series)–34–5, 251
VOYAGER (U.S. interplanetary spacecraft)–30

W

WALKER, Joseph A.–153
WARSAW Radio–94–5
WASHINGTON Post (newspaper)–159–60, 221, 233, 236, 258, 262–3, 265–7, 302, 304
WASP (U.S. aircraft carrier)–9, 13, 42, 44–5, 52
WEATHER Bureau, U.S.–25
WEBB, James Edwin–13–4, 52–3, 89–93, 95, 124, 149–50
WESTERN Union International–123

WESTINGHOUSE Electric Corp.–157
WHITE, Justice Byron–179
WHITE 2d, Maj. (later Lt. Col.) Edward Higgins–4, 86–7, 190
WHITE Jr., George C.–88
WHITE, Col. Robert M.–153
WILFORD, John Noble–149
WILLIAMS, Maj. Clifton C.–87
WILLIAMS, Hosea–182
WILLIAMS, John J.–88
WILSON, George C.–159–60, 236, 266–7, 302
WISCONSIN, University of–157
WOOMERA (Australian launch site)–74–5, 109, 120, 165, 242, 270, 305
WORDEN, Maj. Alfred Merrill–283–9
WORLD Meterological Organization (WMO)–106
WORLD Weather Watch (WWW)–106
WRESAT 1 (weapons research establishment satellite, Australia–120–1
WYDLER, Rep. John W. (R., N.Y.)–89

X

X-15 (U.S. experimental rocket plane)–152–3
X-RAY Explorer (U.S. satellite)–270

Y

YANG Cheng-wu–125
YANGEL, Mikhail K.–292
YEGEROV, Dr. Boris B.–74
YELISEYEV, Dr. Aleksei Stanislavovich–205–10, 291–2
YEMEN–301
YORKTOWN (U.S. aircraft carrier)–141
YOUNG, Lt. Cmndr. (later Cmndr.) John Watts–4, 45–7, 174–9
YUGOSLAVIA–301

Z

ZAIRE–301
ZAMBIA–301
ZA Rubezhom (Life Abroad, Soviet magazine)–15
ZHULIN, Dr. I. A.–298
ZIEGLER, Ronald L.–224
ZOND (Soviet space probe series)–21–2, 137, 144–7, 218, 258. Zond 3–212. Zond 4–144–5, 147. Zond 5–145, 147. Zond 6–146–7. Zond 7–218. Zond 8–258
ZOVODOVSKY, Vassilyevitch–93
ZUCKERT, Eugene M.–15
ZVEZDNY Gorodok (Stellar Town, or Star City), USSR–128, 231, 252, 273

DATE DUE

MAY 1 '72			
NOV 1 1985			
OCT 2 7 '90			
APR 2 6 '95			
OCT 2 6 1999			
DEC 1 1 2002			

GAYLORD — PRINTED IN U.S.A.